Solid mechanics research for QNDE

Solid mechanics research for quantitative non-destructive evaluation

Proceedings of the ONR Symposium on Solid Mechanics Research for QNDE, Northwestern University, Evanston, IL, September 18–20, 1985

Edited by

J.D. Achenbach

The Technological Institute
Northwestern University
Evanston, IL, USA

Y. Rajapakse

Mechanics Division
Office of Naval Research
Arlington, VA, USA

1987 **MARTINUS NIJHOFF PUBLISHERS**
a member of the KLUWER ACADEMIC PUBLISHERS GROUP
DORDRECHT / BOSTON / LANCASTER

Distributors

for the United States and Canada: Kluwer Academic Publishers, P.O. Box 358, Accord Station, Hingham, MA 02018-0358, USA
for the UK and Ireland: Kluwer Academic Publishers, MTP Press Limited, Falcon House, Queen Square, Lancaster LA1 1RN, UK
for all other countries: Kluwer Academic Publishers Group, Distribution Center, P.O. Box 322, 3300 AH Dordrecht, The Netherlands

Library of Congress Cataloging in Publication Data

ISBN 90-247-3428-2

PRINTED IN THE NETHERLANDS

PREFACE

Within the Solid Mechanics Program at the Office of Naval Research (ONR), our primary mission is to establish a basic research program which addresses the fundamental issues in solid mechanics where a clear scientific understanding is lacking. Our approach involves first identifying the various scales at which material and structural response and failure occur. Within each level of behavior we address the basic mechanical phenomena for which a clear physical description is not available. ONR's program emphasizes experimental research to identify and quantify the interacting behavior and response mechanisms. Theoretical and computational approaches are developed to explain the details of the physical processes and to establish the technology necessary to control the thermomechanical behavior of materials and structures.

Within the Department of Defense, it is a natural evolution that all new systems must generally operate in more demanding environments than the systems they replace. Thus, structural designers are pushed towards lighter weight, precision structures utilizing new materials. In such an environment, structural design margins simultaneously shrink and become more critical. Such trends make it essential that a well founded scientific base for the nondestructive detection and assessment of subcritical flaws in structural materials and structures exist. Within the ONR Solid Mechanics Program we are interested in both the identification of flaws and assessment of their degree of criticality. Because of our strong emphasis on fundamental research we see strong synergism between our efforts in Quantitative Nondestructive Evaluation (QNDE) and many of the other areas in our program. As an example, the fundamentals of elastic wave transmission, reflection, and dispersion in composite materials is of critical importance in both our structural acoustics and QNDE programs. Similarly, flaw criticality assessment in QNDE is integrally coupled to our fracture mechanics and constitutive/damage law programs.

During the last several years, constitutive and damage law development together with micromechanics of material deformation has seen significant growth within our program. As we plan for the future directions of our program, it is apparent that an expanded effort in establishing a science base for QNDE is essential if we are to effectively take advantage of our increased ability to describe the failure of advanced engineering materials.

Alan S. Kushner
Program Manager
Solid Mechanics

INTRODUCTORY COMMENTS

J.D. Achenbach and Y. Rajapakse

Non-destructive evaluation (NDE) procedures play an important role in materials processing, as well as in subsequent material testing, product design, analysis of service-life expectancy, manufacturing, and quality control of manufactured products. They are also essential to on-line monitoring of the integrity of structural elements and complex systems. Rational accept and reject criteria should be based on NDE tests. Critical safety, efficiency and operational features of large-scale structures depend on adequate NDE capabilities.

The severe environment in which Naval structures may operate, together with the utilization of advanced materials, create a clear need for reliable methods of non-destructive evaluation. Flaws and defects are introduced into materials during processing and into structural systems during service. The presence of defects could severely hamper the structural integrity and effective performance of Naval systems. Improved methods for the detection and quantitative characterization of defects are, therefore, of great interest to the Navy.

The objective of non-destructive evaluation frequency go beyond the detection of inhomogeneities. The capability to detect is a first and basic requirement. However, with the development of more sophisticated equipment and faster and more advanced signal processing methods, it has become feasible to detect very small inhomogeneities. Since some of these inhomogeneities may be harmless, it has become essential that test results make it possible to discriminate, for example, between cracks, pores and inclusions. Furthermore the effects of an inhomogeneity on the overall strength of a component depends not only on its mere presence and its general nature, but also on its location, size, shape and orientation, and this information must also be obtained from the non-destructive test. This need for detailed information has given rise to a more rigorous and fundamental approach which is called quantitative non-destructive evaluation (QNDE).

Research in QNDE is typically concerned with quantitative relationships for the interaction of penetrating radiation with relevant material inhomogeneities (flaws, welds, cladding, grain structure, porosity, etc.). These relationships are validated in model experiments, and they are then used to interpret measured data to establish the influence of potentially dangerous inhomogeneities on material and structural performance. Considerable interest lies in identifying and sizing microscopic or macroscopic flaws which would ultimately lead to failure. Other measurements of direct interest relate to residual stresses and material properties such as fracture toughness, preferably by direct methods.

Quantitative NDE spans such disciplines as mechanics of solids, materials science, electrical engineering, applied physics, applied mathematics and computer science. The ultimate aim is the prevention of mechanical failure. This aim is achieved by combining knowledge of the material state obtained by QNDE methods with considerations of the failure modes and information on the service conditions, to develop procedures to assess the safety of structural components, and to make a performance prediction. Hence, the component provided by solid mechanics plays a singularly important role. A proper understanding and exploitation of the interplay between detection and characterization methods with considerations of defect geometry, stress distribution, constitutive behavior and fracture mechanics is of the greatest importance.

The propagation of mechanical disturbances in solids and the conditions for failure of solid bodies are very active research topics in the general area of solid mechanics. They are also important components of the research program of the Mechanics Division of the Office of Naval Research. One of the aims of the Symposium was, therefore, to explore the contributions of solid mechanics research to QNDE, particularly to ultrasonic techniques. From a broader point of view, ultrasonic techniques have many advantages. They are relatively simple to apply, and mechanical waves can penetrate a material to substantial depth. Moreover, ultrasonic measurements of mechanical properties and defects are generally very directly related to useful life and eventual failure of a component.

Much work remains to be done before the methods of quantitative nondestructive evaluation will become fully implemented. At the present time the field is one of intensive research activity. Deficiencies still exist in many areas, including our ability to reliably interpret NDE measurements for the purpose of extracting quantitative information on defects and on the related failure characteristics of structural components and systems. Significant progress has, however, been achieved, and there is no doubt that important further advances are forthcoming. The Symposium presented an opportunity to take stock of the current state of the art in QNDE and to survey methods and techniques that are now being investigated, particularly with regard to ultrasonic techniques.

The presentations and backgrounds of the eighty-six participants of the Symposium reflected the interdisciplinary nature of work in QNDE. The organizers would like to thank the Chairmen of the Sessions: L.E. Hargrove, J.A. Simmons, D.E.

Eitzen, O. Buck, D.E. Yuhas, W.A. Ellingson, Y. Weitsman, L.B. Welsh, A.V. Clark, Jr. and S.K. Datta. The extended discussions during separately scheduled discussion sections were a special feature of the Symposium. The contributions of the leaders of these discussions: R.E. Green, M. Hamstad, D.O. Thompson, G.D. Sendeckyj, R.D. Weglein, J.E. Gubernatis, K. Salama and G.C. Johnson, are gratefully acknowledged.

TABLE OF CONTENTS

9. CONTRIBUTED PAPERS

DOD INTERESTS IN QNDE

NAVAL SHIP SYSTEM REQUIREMENTS IN QUANTITATIVE NONDESTRUCTIVE EVALUATION

Robert DeNale
David Taylor Naval Ship Research & Development Center
Ship Materials Engineering Department
Metals and Welding Division
Annapolis, Maryland

Abstract

This paper presents an overview of the importance of quantitative nondestructive evaluation in ship systems of the U.S. Navy. The paper describes current quantitative nondestructive evaluation issues, research and development thrust areas and highlights selected programs which incorporate the elements of quantitative nondestructive evaluation. Finally, the paper identifies research and development needs to enhance quantitative nondestructive evaluation for ship systems.

1. Background

Quantitative nondestructive evaluation (QNDE) impacts both the construction and maintenance of ship structures and systems in the U.S. Navy. The impetus to fabricate and repair ships and submarines with increased reliability, increased performance capabilities, and increased availability to the fleet while decreasing overall life cycle costs requires that the nondestructive testing methods be both quantitative and accurate.

Advances in nondestructive evaluation will reduce costs associated with fabrication by allowing timely correction of defects during the actual fabrication operation as well as reducing costs associated with repairing discontinuities that are benign.

2. Issues

The nondestructive evaluation techniques used during construction and overhaul of naval vessels are radiography, magnetic particle inspection, penetrant inspection, ultrasonics, eddy current inspection, and visual inspection. The methodology of these inspection techniques with respect to construction and overhaul of ships is governed by MIL-STD-271 Nondestructive Testing Requirements for Metals. Based on use of this document the following are considered issues to be reconciled for inspection to be prudent and cost effective.

Radiography of weldments is currently time consuming, a possible safety hazard because of the use of radioisotopes, and dramatically impacts scheduling.

Magnetic particle inspection of welds can result in non-relevant indications which require further evaluation or unnecessary rework, thus increasing costs.

Welding processes for higher strength steels will be predominantly automated, thereby changing the most likely encountered discontinuity from volumetric (porosity) to planar. Additionally, higher strength materials will have reduced flaw tolerances and smaller size discontinuities will become critical.

Finally, inspection of thick section composite materials by currently permitted methods would be difficult at best and ignores the potential of new technology for enhanced inspection.

3. Needs

Table 1 summarizes the NDE requirements of naval ship systems. In the areas of defect detection and sizing the ability to locate and size must have have an accuracy of 1/32 to 1/16-inch. This includes depth and length determinations of both surface breaking and buried discontinuities. In the area of material characterization a measurement of a change in the notch toughness of HY steels of 5 ft-lb as measured by the Charpy V-notch specimen is required. A quantitative, nondestructive measure of the degradation of nylon rope used in a marine environment is required. Quantitative correlations between various defects and mechanical properties of thick section composite materials is necessary. The ability to detect and quantify the integrity of thermally sprayed coatings or elastomer/ceramic/metallic interfaces are also near term needs.

Table 1: Ship systems requirements in nondestructive evaluation

THRUST AREA	CURRENT NEED
Defect Sizing	Crack depth measurement in steam casings Discontinuity sizing in condensers and heat exchangers Sizing of discontinuities in welds Sizing of surface discontinuities in structural welds Crack sizing in shaft material
Enhanced Detection	Detection of planar discontinuities in welds Detection of surface discontinuities in welds and pipe Detection of planar discontinuities in bar stock – receipt inspection Inspection of thin welds by UT
Material Characterization	Nylon rope, HY-80 Castings
Acceptance Criteria	Ultrasonic criteria Radiographic criteria Surface criteria – depth
Composites	Detection, charactrization and sizing of discontinuities Defect – property correlation
Bond Integrity	Elastomer – metallic interfaces Sprayed coatings

4. Thrust Areas

Based on these Navy needs, a series of focused thrust areas were developed. The thrust areas are defect sizing, enchanced detection, material characterization, acceptance criteria, composites, and bond integrity. The details of each thrust area are discussed.

Defect sizing incorporates five current needs which include the general topics of welds and machinery components. Specifically, the needs are that of: crack depth measurement in steam casing; discontinuity sizing in condensers and heat exchangers; volumetric and surface discontinuity sizing in structural welds; and fatigue crack sizing of shaft material.

The thrust area of enhanced detection incorporates four current needs, three of which deal with weld inspection. These current needs include: detection of planar discontinuities in welds; detection of surface discontinuities in hull structural welds and pipe welds; detection of planar discontinuities in bar stock; and inspection of structural welds in thin sections by ultrasonics.

The third thrust area is broadly titled material characterization. In this thrust area a simple question is asked – is this component/material acceptable for use within the specified boundary conditions? There are currently two needs identified within this category; a nondestructive testing technique to determine the degree of degradation in nylon rope and a nondestructive testing technique to ascertain the notch toughness of HY-80 castings.

The thrust area of acceptance criteria encompasses the establishment of acceptance criteria based on a fracture mechanics approach to design. The current needs in this thrust area include: creating an ultrasonic acceptance criteria based on depth, proximity, and character of the discontinuity; modifying the current radiographic acceptance criteria to embrace fracture mechanics instead of workmanship; and developing a surface inspection criteria that includes depth measurements.

The final two thrust areas, composites and bond integrity, are generic in nature. This is due to composite materials being in the conceptualization stage for ship structures and systems. The nondestructive evaluation technology for composites and bonds are generic since there is no failure mechanism, threat, or loading situation yet identified for Navy composite systems. The current needs in the area of composites are detection, charactrization, and sizing of discontinuities and defect-property correlation. Bond integrity has two needs – detecting unbonded areas between interfaces and inspection of thermally sprayed coatings on machinery components for unbonded regions.

5. Highlighted Programs

The David Taylor Naval Ship Research and Development Center is developing the alternating current potential drop (ACPD) technique for depth measurement of surface breaking linear discontinuities. This technique relies on the principle that, due to the skin effect, alternating current is carried in a thin layer at the surface of the material under test. The depth of this current-carrying layer is controlled by the frequency of the alternating current. Therefore, for a given frequency, the potential drop between any two points in this AC field is proportional to the bulk resistivity of the material and the path length between the two contacts. The presence of a surface breaking flaw within this AC field represents an increase in the resistance to the AC flow by causing an increase in the surface path length. This phenomenon allows potential drop measurements in the vicinity of the flaw to be directly related to the flaw depth.

Figure 1 depicts a surface breaking crack. After an AC electric field is established in the metal surface, the ratio of the voltage drop measured on sound metal beside the crack, V_1, to the voltage drop over the crack, V_2, can be directly related to the depth of the crack.

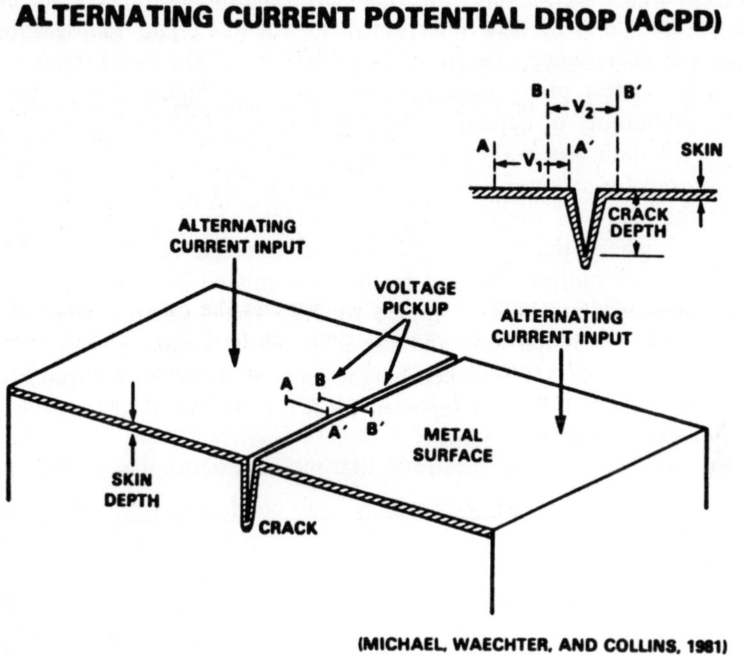

Figure 1: Schematic representation of the ACPD technique

The objective of the ACPD defect sizing program is to develop an instrument for field use which can reliably measure depths of surface breaking linear defects in boiler drums and headers. The materials of interest include carbon steel plate, cast HY-80, and 2 1/4 Cr - 1 Mo castings and forgings. The goal of the program is to achieve an accuracy of the measured depth to within 10 percent. Results indicate that the instrument can detect potential differences due to crack depths as shallow as 1/64-in.; however, crack surface lengths must exceed 1/2-in. for accurate measurements with the current probe design.

In the area of enhanced detection a task group consisting of representatives from the construction yards, Navy shipyards, and Navy laboratories are pursuing a program to institute ultrasonic inspection as an alternative to radiography for volumetric inspection of hull structural welds. The task group members include Electric Boat Division, Newport News Shipbuilding, Charleston Naval Shipyard, Mare Island Naval Shipyard, Naval Research Laboratory, Naval Coastal Systems Center, Naval Surface Weapons Center, and David Taylor Naval Ship Research and Development Center.

Plates with purposely introduced flaws were fabricated, inspected using six radiographic techniques (x-rays, Ir 192, and Co 60 with M or AA film) in accordance with MIL-STD-271E and ultrasonically inspected in accordance with NAVSEA-0900-LP-006-3010 and interpreted to Class 1. The radiographs were interpreted to the acceptance criteria of NAVSHIPS 0900-LP-003-9000 Class II. The results of these test plate inspections are shown in Figure 2 as a bar graph comparing the probability of detection and rejection as a function of the inspection technique. The UT data was evaluated to Class 1 and the RT data was evaluated to Class II as required for structural welds. The data indicates that UT detects more discontinuities than RT. Of the discontinuities detected by at least five inspectors, UT rejected 45% where the various RT method rejected between 21 and 33%. There were 26 discontinuities detected by RT and missed by UT and another 26 discontinuities detected by UT and missed by RT. The discontinuities missed by RT were greater than 0.5-inch long and varied in signal amplitude from just over 20% screen height to over 150% screen height (according to the UT reports). The discontinuities missed by UT were reported by RT to be 0.1 to 0.3-inch long and were generally classed as slag or porosity.

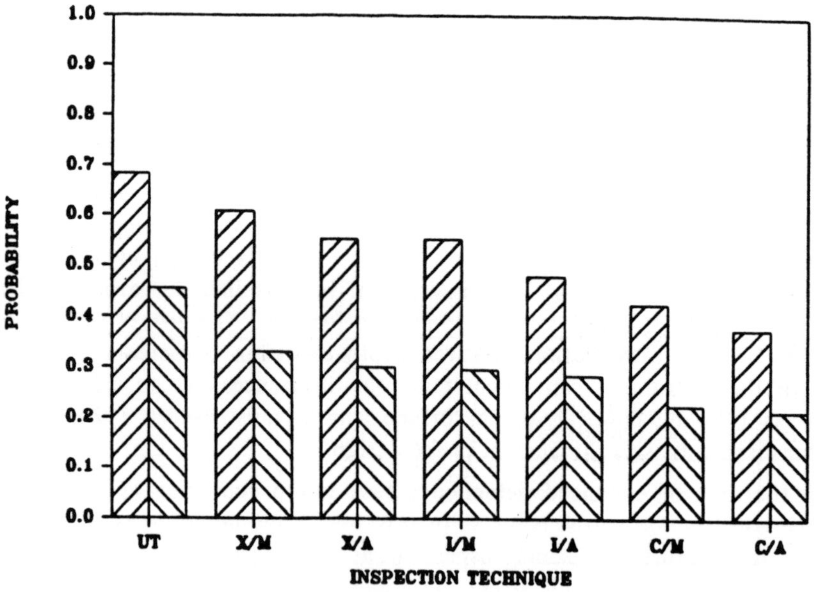

Figure 2: Probability of detection and rejection as a function of inspection technique (UT-Ultrasonics; X-X-ray; I-Iridium 192; C-Cobalt 60; M or AA film).

In the thrust area of material characterization the Johns Hopkins University Applied Physics Laboratory has studied the magnetic properties of stressed and unstressed nylon rope fibers as part of a program to develop a nondestructive test of rope strength. Irradiation (UV and x-ray) and mechanical fracture (stressing to failure, cutting, and machining) have been used to generate free radicals associated with broken bonds in the polymer chains. These paramagnetic defects have been examined using electron paramagnetic resonance spectroscopy at various temperatures and in different atmospheres. Some long lasting radicals have been identified using a broad magnetic field sweep. Figure 3 compares the electron paramagnetic resonance as a function of the magnetic flux for three cases of nylon rope. As indicated in the figure, a large change in the EPR signal occurs as the number of cut fibers increases. Additionally, the peak to peak measurement of the magnetic flux ($\Delta H_{pp} = 450G$) is broad. The intensity of this line was correlated with damage done to the rope by stressing to the breaking point; however, the characteristics of the resonant line led to the conclusion that the mechanism of stabilization of this free radical was due to spin trapping by inorganic impurities present in the rope. For this technique to consistently work the inorganic purities would be required to be present – a situation which cannot be relied upon. The EPR technique has been demonstrated to be consistent when a doping compound is introduced into the nylon rope. The effects of the doping compound on the mechanical properties of the rope have yet to be characterized.

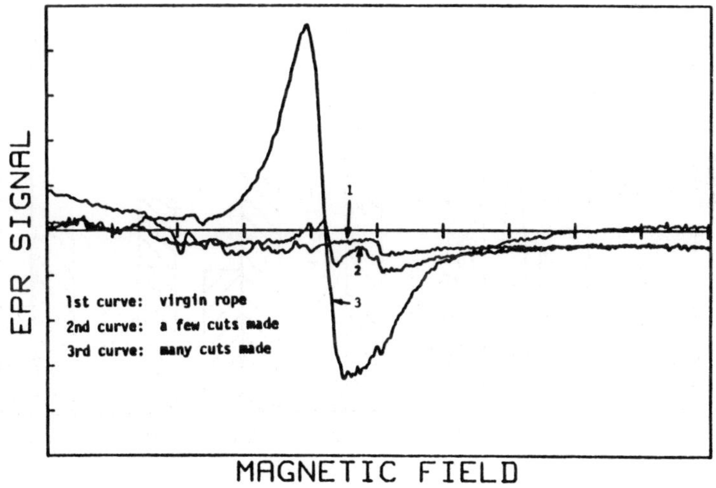

Figure 3: Electron spin resonance traces for virgin and degraded nylon rope

An on-going program at the Naval Research Laboratory is ultrasonically characterizing highly attenuative fiber reinforced composites by mapping the relative attenuation of the composite [1]. A transmission technique is used to map the relative attenuation since conventional pulse-echo techniques cannot easily be applied to characterize these materials due to lack of multiple echos. The attenuation was measured with specimens in the shapes of wedges or steps in order to eliminate errors due to energy losses at the interfaces. The mapping of attenuation provides a technique to nondestructively predict the periodicity (or lack of it) of the weave, the heterogeneities of the fibers, and to a lesser degree, of the matrix, the misalignment of the fibers and other irregularities. Figure 4 represents two cases of attenuation curves for the composite material in a "good" case, that is, when the fiber pattern and the concentration are regular and a corresponding curve in a typical "bad" case. The "good" case of Figure 4(a) depicts an exponentially decaying envelope of the maxima and minima of the signal which correspond to regions of lowest and highest fiber concentrations in a regular pattern. Conversley, the "bad" case of Figure 4(b) represents the irregularities of the fiber pattern as shown by the lack of periodicity in the maxima and minima.

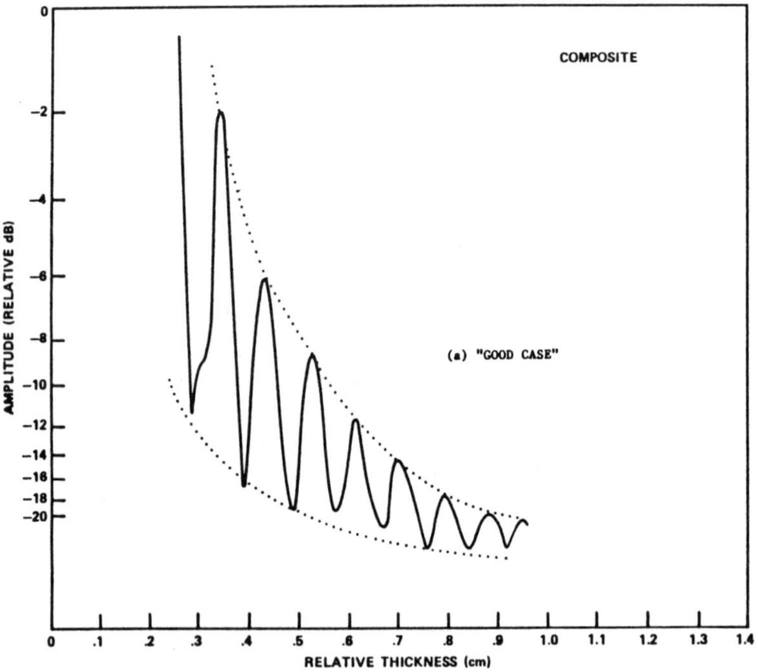

Figure 4(a): Attenuation curves for composite wedge specimens – "GOOD CASE"

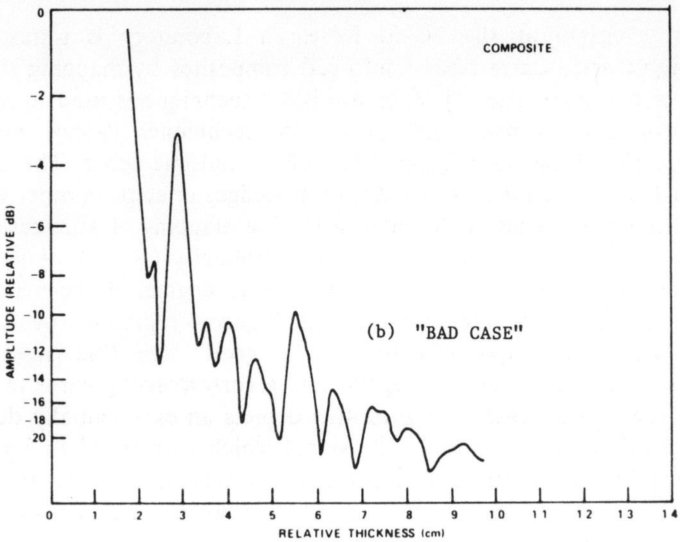

Figure 4(b): Attenuation curves for composite wedge specimens – "BAD CASE"

6. Summary

This paper has presented an overview of the importance of quantitative nondestructive evaluation in ship systems of the U.S. Navy by describing current quantitative nondestructive evaluation issues and detailing research and development thrust areas. The paper also highlighted selected programs which have incorporated the elements of quantitative nondestructive evaluation. Finally, the paper identified research and development needs to enhance quantitative nondestructive evaluation for ship systems.

Reference

[1] Batra, N.K., Chaskelis, H.H. and DeLsanto, P.P., "Ultrasonic Characterization of Highly Attenuative Fiber – Reinforced Composites", *Ultrasonic Symposium Proceedings*, 1985, pp. 975-978.

AIR FORCE REQUIREMENTS FOR NDE OF COMPOSITES

D.E. Chimenti
Materials Laboratory
Wright-Patterson Air Force Base OH 45433-6533

Abstract

This paper covers Air Force requirements and future needs for nondestructive evaluation in composite materials. Although composites have received only limited use in current airframe structures, vertical and horizontal stabilizers of fighter aircraft for example, aircraft on the drawing board or in testing employ graphite-fiber reinforced composites much more extensively. To anticipate the inspection requirements of these complex materials is not an easy task. Yet, many years of materials testing and the experience gained through limited service application certainly indicate the directions this work should take.

1. Discussion

In Figure 1 is a photo of the F-16 XL prototype aircraft being rolled out. Its family resemblance to the F-16 is apparent, but the wing design is rather different. In this advanced delta wing configuration the dark portions of the structure are fabricated from composites. The materials on the wing skin, vertical stabilizer, strake, and portions of the overwing fairing are graphite-BMI. At the wing root the layup is over 160 plies thick, trailing to 30 plies at the tip and on the strake. The substructure is bolted aluminum.

Figure 1: F-16 XL prototype aircraft

Another example is afforded by the new experimental X-29 aircraft, which has just begun flight testing. The wing cover, shown in Figure 2, is a single 8-meter-long piece of graphite-epoxy, aeroelastically tailored to control wingtip divergence. Because of the unusual aerodynamic forces on such a wing, its extreme stiffness-to-weight ratio is an essential feature of the design. Without composite materials the design is not practical. The tooling, accurate to 0.8 mm, is visible in the background.

The partly completed wing box in the assembly fixture is seen in Figure 3. Also visible is the aluminum and titanium substructure. Leading edges are titanium as are the root ribs, for added stiffness. The box is one piece tip to tip, with a 1.1 meter chord length at the wing root.

Assembled aircraft, Figure 4, shows the extent of composite materials in this prototype airplane. Tension pads cover a portion of the wing and were used for the initial proof testing, in spite of the fact that careful manufacturing NDE was also applied! The structures in front of the wing, called canards, serve as horizontal stabilizers. These have been fabricated from metallic materials.

Figure 2: Experimental X-29 aircraft

An example of current state-of-the-art industry inspection of composites at Martin-Marietta is given in Figure 5. Dual robots operating in tandem manipulate transducers with squirter attachments to perform through-transmission inspections of complicated part geometries. Results of inspections are analyzed by computer and displayed on a CRT or archived in hard copy.

Figure 6 shows the advanced composite wing of the AV-8B Harrier II in a computerized inspection station at McDonnell Douglas Corp. Careful initial NDE is necessary for quality assurance in this large complicated structure. A computer-reconstructed C-scan of damage caused by impact on a composite panel is seen in Figure 7. The bright area at the center of the indication is damage visible on the surface, while the dark zone surrounding it is a delamination near the impact surface. A larger region, elliptical in shape, shows the delaminated zone at the back surface.

In a metal component a crack grows in cyclic fatigue from a size that is barely detectable to one that can cause structural failure. The rate of growth may depend on the material, temperature, and stress environment. In composites, on the other hand, defect size tends not to increase with usage at a comparable rate, as

Figure 3: Wing box in assembly fixture

indicated by the data of Figure 8. Consequently, inspection methods and philosophy for composite aircraft may need to be modified.

Fatigue performance of defect-free and impact-damaged graphite-epoxy composite in a quasi-isotropic laminate under fully cyclic loading is given in Figure 9. The effect of a 3-Joule impact on the fatigue life is shown to be minimal. Incidentally, after 10**6 cycles the two curves do not cross.

The principal defects which occur in composites are delaminations, porosity, and inclusions. Delaminations can be introduced during manufacture or may result from the impact of a foreign object on the laminate. Porosity and inclusions are clearly manufacturing defects and generally cause a reduction in strength properties of the composite.

The major causes of delaminations are impact or improper cure of the layup. So long as they are small and occur in the subcritical portions of the structure, delaminations are considered not to be of significance. Exterior composite materials are designed to tolerate moderate impulsive loading from foreign object impact. From the inspection standpoint, this design principle means that an impact-induced delamination will not approach critical size if there is no visible surface damage.

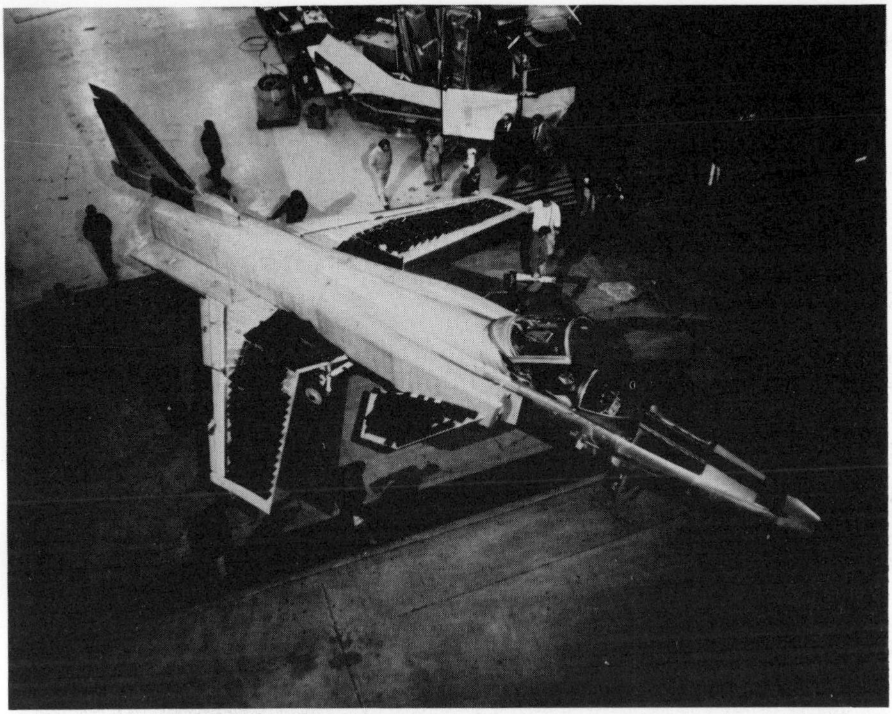

Figure 4: Composite materials in X-29

In practice, this allowable is occasionally exceeded if the structure is weakened by some other mechanism or if the mission profile of the aircraft changes. One type of structure is susceptible even to small delaminations. Since compressive loading strongly favors the growth of a delaminated area by buckling, structures in that category must be carefully inspected.

Porosity is associated with a variety of negative performance characteristics. It has been shown to degrade compressive strength and service life. It can be uniform in distribution or random. Porosity shows up mostly at interfaces and behaves like micro-delaminations. Physically, it is tubular in shape with its long axis oriented along the fibers. The average tube diameter is approximately one-half the ply thickness.

Inclusions, another manufacturing defect, are usually portions of the protective materials shipped with the pre-preg which are inadvertently left in the cured layup. These plastic films have a low sticking coefficient to the epoxy, by design, and therefore reduce the shear strength of the composite to very low values. They can be detected in pulse-echo C-scan. Other types of foreign matter will generally appear in either X-ray or standard ultrasonic inspections. Another class of material

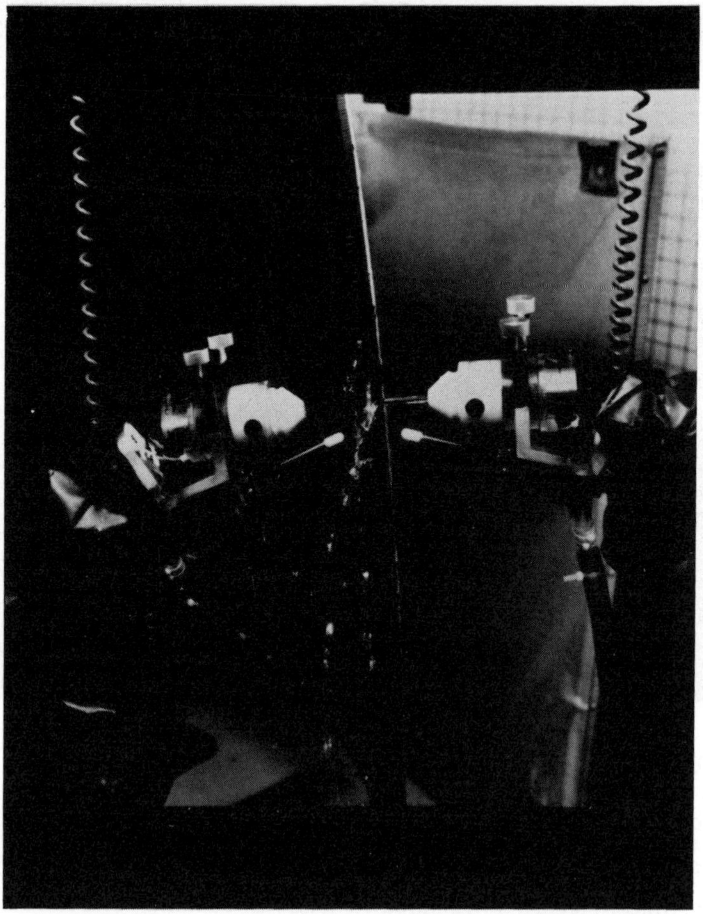

Figure 5: Dual robots

inclusion is moisture absorbed by the matrix. While some moisture retention by the matrix is desired to increase toughness, excessive amounts can lead to matrix strength degradation. Moisture also prohibits the use of high-temperature curing in the repair of damaged composites.

A summary of Air Force needs in NDE of composites is presented below:

Figure 6: Composite wing of AU-8B Harrier II

- Detect "token" bonds in co-cured structure
- Detect peel plies laminated into a part
- Detect voids, porosity in co-cured structure at bondline
- Quantify porosity simply and reliably at 1/2 to 5% level
- Measure elastic moduli point-to-point
- Verify layup stacking sequence before cure
- Scan large areas very rapidly for moderate-size flaws
- Measure moisture content and gradient

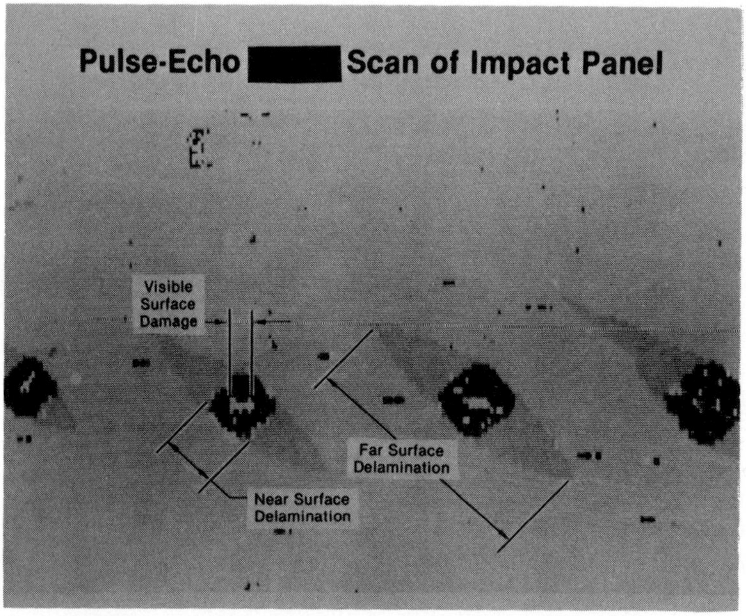

Figure 7: *C-scan*

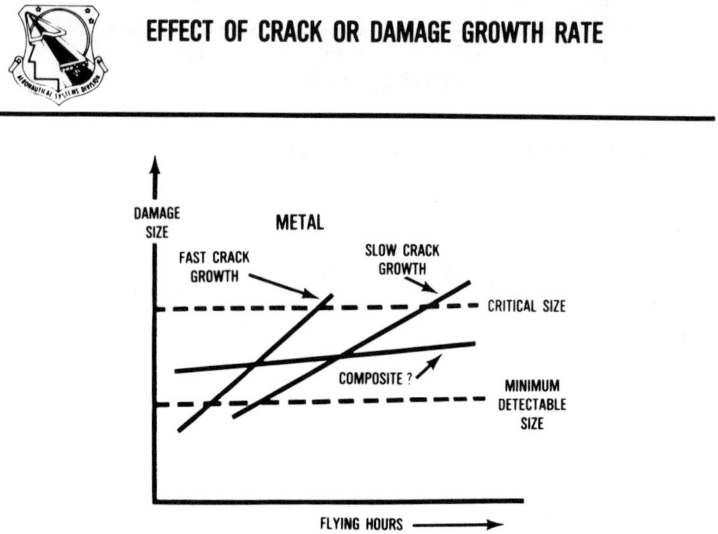

Figure 8: *Effect of crack on damage growth rate*

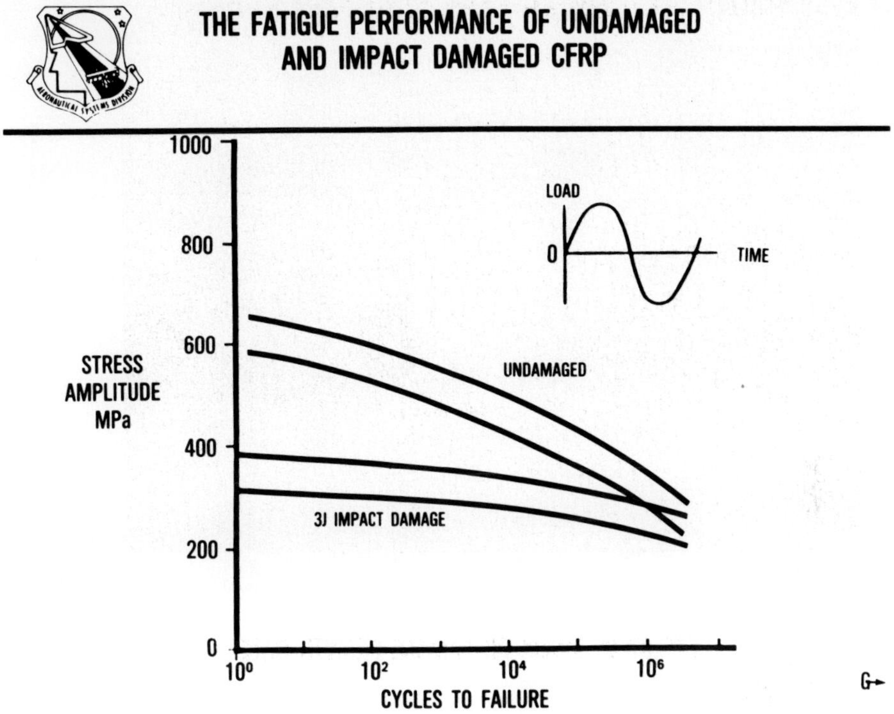

Figure 9: Fatigue performance of composite laminate

2. Conclusions

In conclusion, regardless of improvements in materials the Air Force will always need reliable, economical NDE at manufacture and simple, rugged NDI for in-service inspections.

ACOUSTIC EMISSION

ACOUSTIC EMISSION: A QUANTITATIVE NDE TECHNIQUE FOR THE STUDY OF FRACTURE

H.N.G Wadley
Institute for Materials Science and Engineering
National Bureau of Standards
Gaithersburg, MD 20899

Abstract

Acoustic emission, an NDE technique that shows promise for detecting and locating cracks in engineering structures, has been used as an experimental technique for the basic study of fracture. Examples of the use of acoustic emission for this latter purpose are reviewed, and future research opportunities identified.

1. Introduction

Acoustic emission refers to the transient elastic waves sometimes spontaneously generated by a body when subjected to stress. The frequency of these elastic waves extend into the MHz region and thus they can be detected with sensitive ultrasonic transducers attached to the bodies surface. Dislocation motion and crackgrowth are thought to be the origin of this naturally created ultrasound. Intuitively this arises because a stressed body at mechanical (static) equilibrium, returns to equilibrium after an abrupt deformation or fracture event, by means of an exchange of elastic information with the bodies boundaries and loading system. This elastic information exchange is of course achieved through the propagation of elastic waves–the same elastic waves monitored as acoustic emission [1].

The monitoring of acoustic emission offers a potential to deduce the location and observe the dynamics of microscopic deformation and fracture events provided the emission is properly measured. Location is possible since the velocity at which elastic waves propagate is usually well-known. Then, the simultaneous recording of signals at different positions on a body enables a set of time of flight differences to be deduced, and from this, source location by triangulation is achieved. The accuracy of this location depends upon the shortest wavelength components observed in the detected signals. For recording systems available today (10-100 MHz bandwidth) this can be of the order of 30-300μm, comparable to the dimensions of the source [2].

The dynamic source information contained within the signals is much more difficult to unfold due to the tensor nature of the problem. Conceptually, we can think of the signal as a time varying vector displacement caused by an abrupt localized change of stress (a second rank tensor). The magnitudes and time dependences of all the components of this stress change are the quantities we wish to determine. It is not possible to determine the six components (for an isotropic linear elastic medium) of stress from a single displacement waveform (though if all the components have the same temporal character, the temporal character alone may be approximately determined from a single signal and used as a "signature" of the event). Rather, it seems necessary to make six (and preferably more) displacement measurements at different positions around the source and to perform a simultaneous deconvolution to reveal the magnitudes of all six stress components. A further micromechanical modeling step would then enable the micromechanism responsible for the stress change to be quantitatively deduced.

Clearly then, acoustic emission has a great deal to offer the materials scientist attempting to probe the mechanisms by which materials respond to stress. This has been recognized for at least ten years, and the relative absence of widespread application of this tool indicates that difficulties have been encountered when attempts are made to realize this potential.

One problem is that not all the micromechanisms of interest are detectable [3]. This arises simply because the transducers used to measure the surface displacements associated with the elastic waves are not infinitely sensitive. Typically, the modern generation of piezoelectric transducers can only measure displacements greater than $10^{-13}m$ in the frequency range of ~20 kHz to ~2 MHz. More ideal broader bandwidth devices are available, but their sensitivities are even less. Narrow band (resonant) transducers with better sensitivities (down to $10^{-14}m$) exist, but they so distort the displacement signal that it is not presently feasible to use them for quantitative work unless they are very carefully calibrated, and then only provide information over a narrow range of frequencies. While advances in transducer sensitivity and bandwidth will no doubt eventually mitigate the problem, it turns out that a wealth of micromechanisms (from the propagation of single dislocations in perfect single crystals to cleavage microcracking of ferrite grains) are in fact detectable, (are in today's sensitivity-bandwidth envelope) and the more energetic of these may be measured with sufficient precision to be characterizable with today's technology.

The more serious problem is that of the characterization itself–a problem akin to the one in seismology of characterizing the faulting mechanism of an earthquake from remotely recorded seismograms. Current research at laboratories in the U.S.A., Great Britain, West Germany, France and Japan are attempting to apply the formulisms developed for earthquake characterization to the characterization of acoustic emission sources. The first successes of this are now emerging and it seems that if the experimentalist is prepared to properly perform acoustic emission measurements during conventional mechanical property testing, then indeed the dynamics of fracture micromechanisms are available to us through this new materials probe.

2. Phenomenological Observations

Since acoustic emission is, after all, no more than a naturally occurring phenomenon, it is reasonable to begin with a series of observations utilizing systematic microstructure control and intuitive reasoning to deduce the origin of the emission signals in terms of deformation and fracture micromechanisms. Perhaps the most controlled experiment would be that of the constant strain rate deformation of an f.c.c. crystal oriented for initially single slip. Numerous research groups have begun their studies at this point [4-7] and a typical result showing the acoustic emission power variation with strain is shown in Figure 1 [7]. In the experiment the sample was deformed at a slow strain rate ($10^{-4}\ s^{-1}$) and acoustic emission in the 0.1-1.0 MHz frequency range detected with a transducer whose sensitivity was $\sim 10^{-14}$ m.

It was observed that beginning during nominal elastic loading many individual signals were emitted and their summed power (intensity) quickly reached a maximum shortly after general yield, and that less emission is detected as the strain increases beyond this point, i.e. there is less emission per plastic strain increment. The explanation for this behavior lies in the microscopic nature of slip.

We know that the imposed strain rate, $\dot{\varepsilon}$ is satisfied by the formation and motion of dislocations i.e. $\dot{\varepsilon}=\rho b v$ where ρ is the mobile dislocation density, b the Burger's Vector and v the dislocation velocity.

Early during deformation, dislocations on slip planes at $\sim 45°$ to the tensile axis (maximum resolved shear stress planes) glide considerable distances at large velocity through an almost perfect lattice. The distance individual dislocations propagate in exceptionally pure material containing few grown-in defects can be as high as the slip plane diameter (a few mm's). However, toward the end of a mechanical test, a dislocation cell structure develops, and individual dislocations can only propagate distances of the order of the cell wall spacing, typically a few microns. Thus, the unimpeded coordinated motion of a few (low density) dislocations over large distances at high velocity is more probable to be the origin of a detectable signal than the motion of many dislocations moving in an uncoordinated

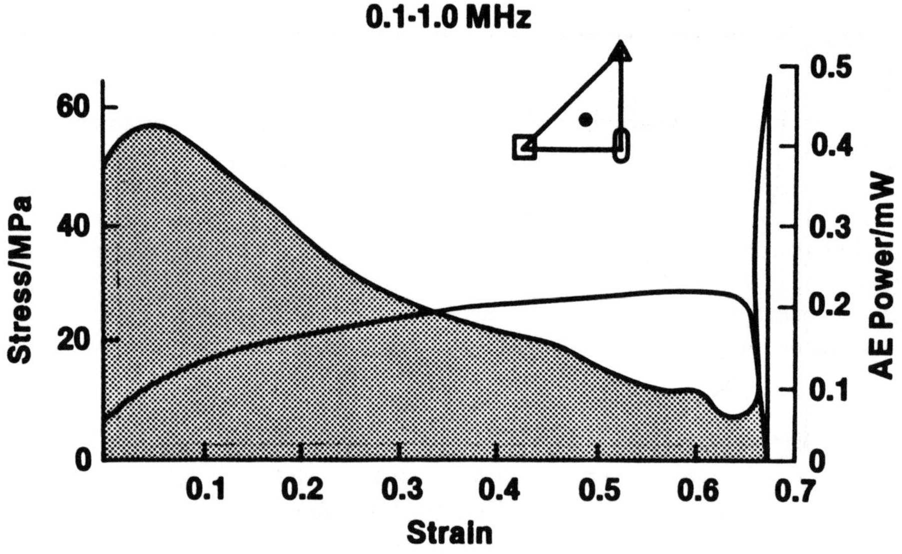

Figure 1: The acoustic emission (AE) intensity (power) detected during the tensile deformation of [149] single-crystal of aluminum decreases with increasing plastic strain during parabolic work hardening. The measurements were made over a frequency band from 0.1 to 1.0 MHz with an amplification of 98 dB. A smooth relationship is observed because many individually small signals are emitted within the response time of recording instrumentation, resulting in the observation of a mean emission intensity [7].

manner over short distances.

This hypothesis may be rather straightforwardly tested by systematically varying the spacing of obstacles that are impenetrable to dislocations and observing a standardized acoustic emission response. Grain boundaries are one possible barrier for this purpose, and the grain size can be varied over a wide scale by appropriate cold work/recrystallization treatments. The distribution of slip distances over some plastic strain range will now be a function of the grain size which, initially at least, controls the grain interior dislocation-dislocation interactions. When samples of aluminum polycrystals are tested under similar conditions to those of the single crystal, a similar dependence of the emission upon strain is observed to that of the single crystals reflecting again the development of a cell structure. However, the total emission detected depends sensitively upon grain size as shown in Figure 2 [7]. It can be seen that as the grain size decreases, then so does the amount of detectable acoustic emission, consistent with the view that the dislocation propagation distance has an important effect upon signal detectability.

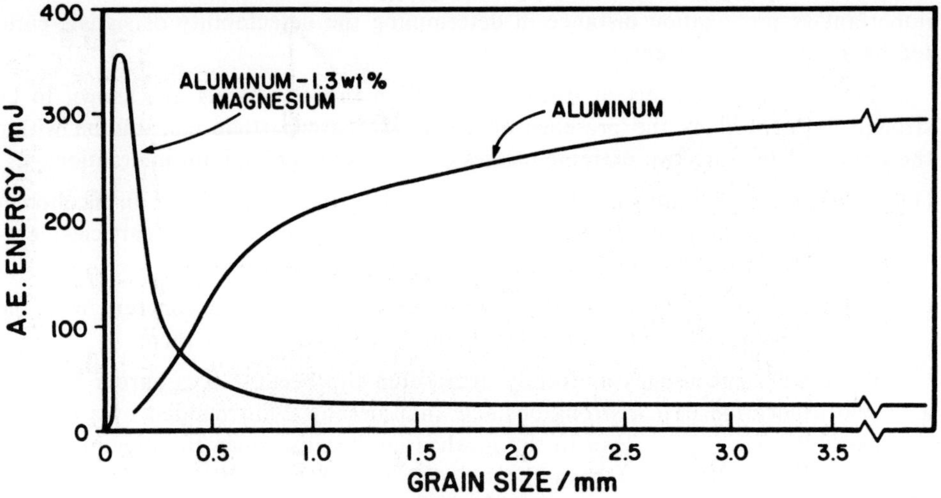

Figure 2: A comparison of the grain size dependence of acoustic emission (AE) from pure aluminum and an Al-Mg solid solution alloy shows that, for small grain sizes, solute atoms enhance the emission from dislocations while, for large grain sizes, the emission from dislocations is suppressed [7].

Interestingly, this observation is profoundly changed if substitutional impurities are introduced into the aluminum. In alloys of aluminum substitutional impurity segregation to dislocations occurs even when the bulk impurity concentration is below the solubility limit. During deformation this solute atmosphere either diffuses with the dislocation as it propagates or, if the stress is sufficiently high, the dislocation will escape, leaving behind the solute atmosphere. The breakaway phenomena manifests itself as a load drop during constant strain rate testing and is the origin of dynamic strain aging and yield point phenomena (upper and lower yield points, Luders band propagation) in aluminum alloys.

Figure 2 shows the effect of grain size upon the total acoustic emission of a (substitutional) aluminum - 1.3 wt.% magnesium alloy [7]. When the grain size was large (several mm's), almost no detectable acoustic emission was observed even though the total plasticity was comparable to that of the pure material. This material also did not exhibit any yield point or serrated yielding (dynamic strain aging) behavior. In contrast, higher yield strength samples with a smaller grain size of $\sim 80\ \mu m$ exhibited a maximum acoustic emission (much greater than that of equivalent grain size, pure Al) and a maximum yield point phenomenon and serrated yielding activity. Apparently, the origin of the emission is related to the break away of dislocations which depends upon grain size (probably due to a flow stress dependence upon grain size). In these alloys, the cooperative high velocity motion of groups of dislocations is the origin of the emission and emphasizes that dislocation velocity and the cooperative slip of numerous dislocations are just as important as propagation distance in determining the detectability of signals emitted by a deformation event.

The micromechanisms of dislocation motion in polycrystals are known to be strongly influenced by the presence of a second phase particle distribution within the grains. There are two extreme classes of dislocation-precipitate interaction:

(1) When the precipitates are large, strong, widely spaced and have an incoherent interface with the surrounding matrix, dislocations bow between particles, leaving behind a dislocation loop around the particle.

(2) When the precipitates are small, weak, closely spaced and coherent with the matrix, dislocations may penetrate the precipitates and shear them.

Process (1) encourages uniformly distributed slip because it is harder to force a second dislocation past a strong obstacle that already is surrounded by a dislocation loop. Process (2) can lead to slip localization because when the precipitates on a particular slip plane have been sheared once, subsequent shear on that plane is often easier resulting in a local strain instability. Studies in which the precipitate strength, size distribution, and coherency have been systematically varied in coarse grained aluminum alloys (where dislocation-impurity interactions are not active acoustic sources) indicate that process (1) leads to weak emission while process (2) leads to intense signals, Figure 3 [9,10].

From theoretical considerations of the surface displacement amplitude of acoustic emission signals it has been possible to deduce detectability criteria that a source must meet if the signal is to exceed background noise [11]. For the slip case, the criterion is of the form $nav \geqq k$ where a is the dislocation slip distance, v is the dislocation velocity, n is the number of cooperatively moving dislocations and k a constant that depends upon material elastic properties, and is proportional to the transducer sensitivity and source-receiver distance. For the aluminum single crystal tests $k \sim 0.035\ m^2s^{-1}$ for a transducer sensitivity of 10^{-14} m. Then, for $v \sim 350\ ms^{-1}$; na is $\geqq 100\ \mu m$ implying that individual dislocations, if they move fast enough, give rise to detectable events. These studies serve to show that high speed dislocation motion over distances of the order of 100 μm to 1 mm are the origin of detectable acoustic emissions during plastic deformation.

Following a similar procedure, detectability criteria for microfracture events have also been deduced [11]. For steel, the criterion is $\sigma a^2 v \geqq 5 \times 10^{14} rT$ for a signal of detectable strength to be emitted. Here, σ is the local stress, a the crack radius, v is the radical crack velocity, r the source-receiver distance and T the transducer sensitivity. Substituting realistic values gives

$$\sigma a^2 v \geqq 0.1 \text{ (watts)} \tag{1}$$

For cleavage microfracture of a ferritic steel, a is half a grain size (say

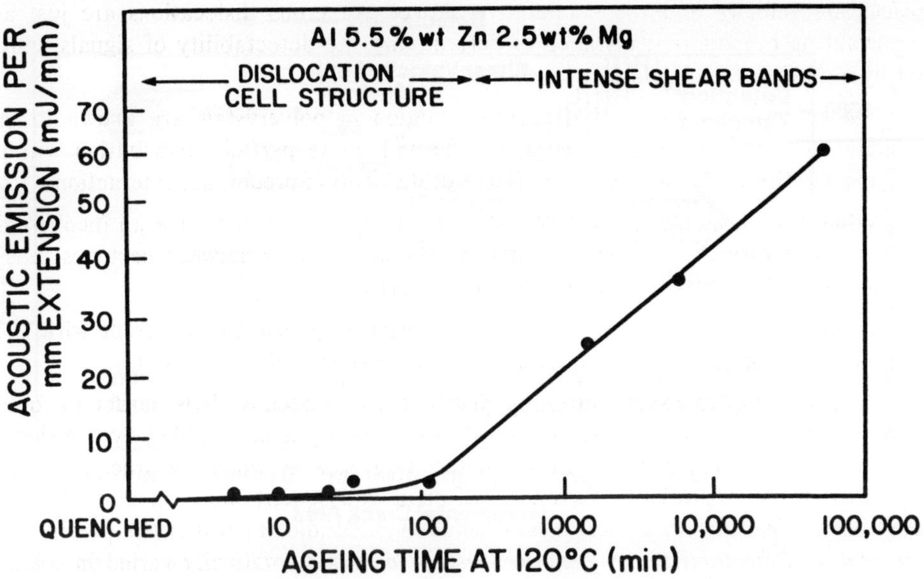

Figure 3: *Aging supersaturated aluminum alloys (in this case, Al-5.5 wt.% Zn-2.5 wt.% Mg) results in "shearable" precipitates which "focus" dislocation motion onto closely spaced slip planes (intense shear bands). The acoustic emission from this mechanism of dislocation motion is much more detectable than that leading to a cellular network in solid solution strengthened microstructures.*

$10\mu m$) $v \sim 10^3 ms^{-1}$ and $\sigma \sim 500$ MPa and thus $\sigma a^2 v = 50$ watts. Cleavage in steels (and other brittle materials) is clearly a source of readily detectable signals.

Other mechanisms of fracture have different a and v distributions and thus some will be detectable whilst others will not. A convenient way to summarize this information is to plot the ranges of a and v for each micromechanism on a map upon which is superimposed a detectability criterion [3]. In Figure 4 such a map is shown assuming $T \sim 10^{-12} m$. Only those events to the right of the detectability criterion in Figure 4 are detectable.

These studies serve to show that in contrast with stress-strain measurements or microhardness testing where an integrated static response to load is measured, acoustic emission signals eminate from individual microdeformation events, and these are extraordinarily sensitive to microstructure (compared to a stress-strain curve which may vary by a factor of 2 or 3). Furthermore, the passive nature of the technique (it does not perturb the source process) and its 100 percent volume coverage make it an ideal compliment to traditional materials testings tools. However, its full potential will only be realized if the dynamics of the source can be observed and for this a predictive theoretical formulation and a robust inverse modelling approach are required.

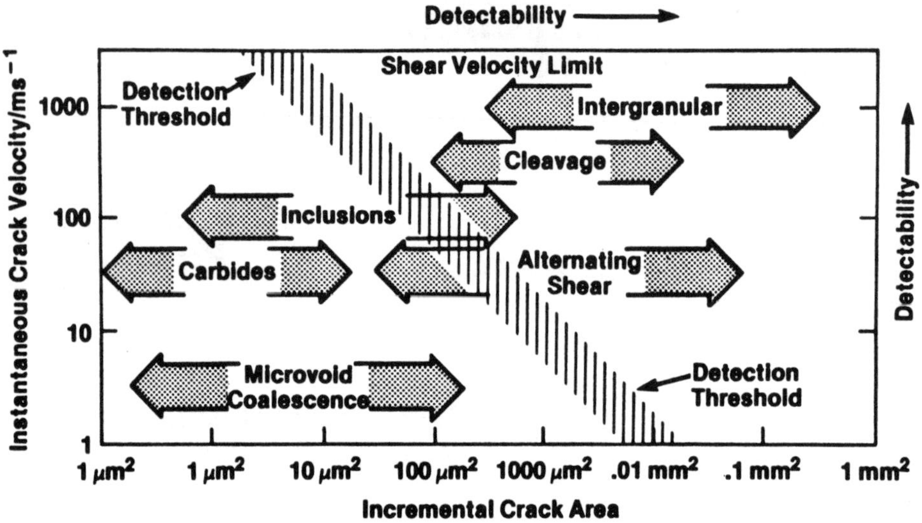

Figure 4: *Detectability of fracture events in steels. The detection threshold assumes a transducer sensitivity of 10^{-13} m, a distance between the source and the transducer of 0.1 m and a stress of 500 MN m^{-2} [3].*

3. Theoretical Formulation

The sequence of events giving rise to an acoustic emission signal are summarized in Figure 5. A casual sequence of processes occurs following the occurrence of the source event which can be likened to the creation of a dynamic force field at the source position. This is propagated as a mechanical disturbance through the body causing a surface displacement $\mathbf{u}(t)$ at the receiver location. The receiver detects the surface disturbance and produces an output voltage waveform that is subsequently electronically processed and digitally recorded. One goal of a theoretical formulation is the precise prediction of acoustic emission signals for prescribed sources, a second is the formulation of a robust approach to the inverse problem.

If the system in Figure 5 is a linear elastic one, then when viewed in the frequency domain, information is transmitted independently, frequency by frequency, from the source to the observed signal. Thus, the observed signal can be represented as a convolution between the source function, and the impulse responses of the body, transducer and electronic system (this latter assumed here a perfect delta function of unit strength for simplicity).

Then, for buried sources that can be modelled as dislocation combinations, the general formulation of Simmons and Clough [12] can be simplified and the displacement at $\mathbf{r}$ as a function of time t is given by:

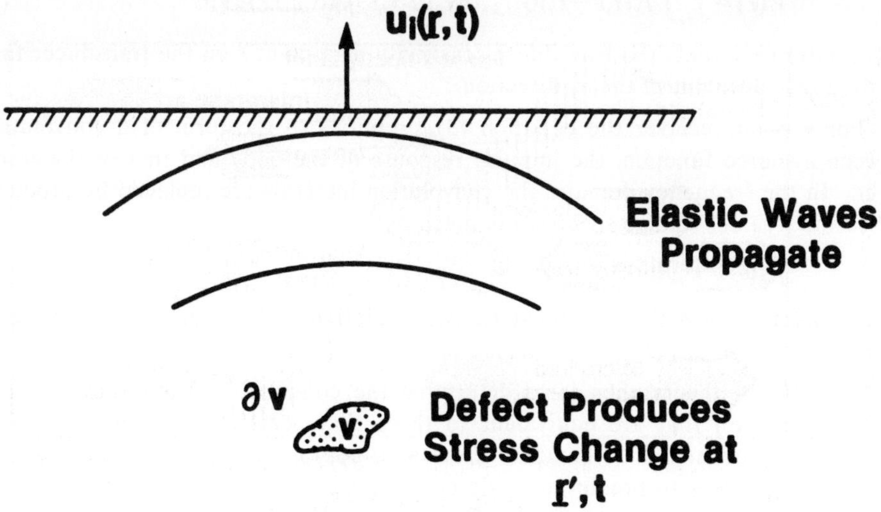

Figure 5: *Schematic illustration of the acoustic emission process.*

$$u_i(\mathbf{r},t)=\int d\mathbf{r}\int G_{ij,k'}(\mathbf{r},\mathbf{r}';t-t')\Delta\sigma_{jk}(\mathbf{r}',t')dt' \quad (2)$$

where $G_{ij}(\mathbf{r},\mathbf{r}',t)$ is the dynamic elastic Green's tensor representing displacement in the x_i-direction at $\mathbf{r}$ as a function of time due to a unit force impulse applied at $\mathbf{r}'$ and $t=0$ in the x_j direction. Thus, the Green's tensor is the solution to the wave equation for a unit force source. The corresponding solution for a dipole force of unit strength is obtained by spatial differentiation of G_{ij}, (denoted $_{,k'}$ above). $\Delta\sigma_{jk}$ corresponds to the stress change (dipole density) at $\mathbf{r}'$ and its volume integral for a point source $\overline{\Delta\sigma}_{jk}$, corresponds to the dipole (or moment) tensor of the source. Thus, for a point source equation (2) is a convolution between the dipole (or moment) tensor of the source and the dynamic elastic Green's function of the body.

This formulation provides the basis, in principle, for predicting surface displacement waveforms caused by homogeneous stress changes within a body. In practice, the formulation is computationally cumbersome for anything other than point sources in isotropic linear elastic bodies with either one or two flat parallel surfaces (infinite half space or infinite plate). The approach to sources that are spatially extended is to use a Taylor's expansion representation for the Green's tensor and to represent the source as a linear combination of moments (e.g., zeroth, first, second, etc.) [12].

The final step of the formulation is to introduce the effect of the transduction process upon the signal. For the case of "nondisturbing" transducers with a surface S_T the voltage as a function of time (t) due to a point dipole source $\overline{\Delta\sigma}_{ij}$ at position $\mathbf{r}_o{}'$ is:

$$V(t)=\int_{S_T}\int T_i(\mathbf{r},t-t')G_{ij,k'}(\mathbf{r},\mathbf{r}_o';t-t'')\overline{\Delta\sigma}_{jk}(t'')d\mathbf{r}\,dt'' \qquad (3)$$

where $T_i(\mathbf{r})$ represents the transducer sensitivity at a point $\mathbf{r}$ on the transducer face due to a displacement in the x_i-direction.

For a point receiver the equation (3) simplifies to the form of a convolution between a source function, the impulse response of the body and that of the transducer. In the frequency domain, the convolution integrals are replaced by products and so the (scalar) voltage spectrum is given by:

$$V(\omega)=\Gamma_{jk}(\omega)\overline{\Delta\sigma}_{jk}(\omega) \qquad (4)$$

where $\Gamma_{jk}(\omega)$ is now the combined transfer function of both the body and transducer.

The role of theory then is to determine the combinations of forces (dipoles, quadrupoles, etc.) that are equivalent to those that occur when various types of defect sources occur and to combine these with appropriate Green's functions and transducer responses to predict acoustic emission signals. An example of this for a simulated acoustic emission, a thermoelastic expansion source (due to pulsed laser absorption at the surface of an aluminum plate), is shown in Figure 6 [13]. Also shown for comparison is the experimentally measured waveform. The agreement is good for this elastically isotropic material. Discrepancies associated with grain scattering are, however, observed in more anisotropic materials such as stainless steel [13], and points to the need for research to deduce Green's tensor for realistic engineering materials.

4. Inverse Problem

From equation (4) it is clear that a deterministic approach to characterizing the acoustic emission is to evaluate $\Gamma_{jk}(\omega)$ and to divide $V(\omega)$ by Γ_{jk} on a frequency by frequency basis. This will reveal $\overline{\Delta\sigma}_{jk}(\omega)$ and the temporal behavior can be deduced by inverse transformation. While simple in concept, the approach must recognize several key issues.

Even for a point source in an isotropic linear elastic medium the dipole (or moment) tensor contains up to six independent components. However, a voltage is only a single parameter that depends upon an unknown combination of the source components. Ideally, six (or more) voltage waveforms should be recorded (with multiple transducers located around the source) and a simultaneous deconvolution performed to independently reveal the six unknowns.

Characterization schemes based upon the analysis of only a single transducer voltage are commonly incorporated in commercially available instruments. However, unless "a-priori" information exists to link the stress components to each other (i.e., constraining equations) so that there is only a single unknown quantity, it is difficult to understand what validity such characterization may have.

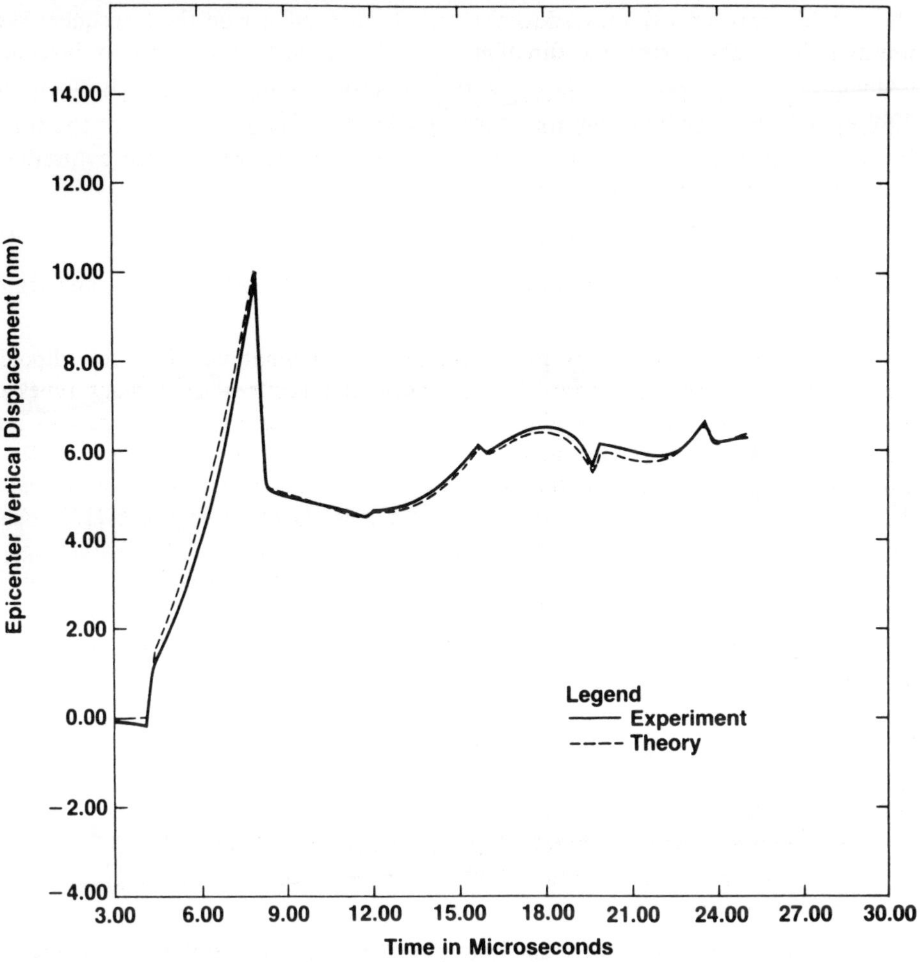

Figure 6: Comparison of experiment and theory for laser generated AE in 2024-T6 aluminum [13].

Schemes are evolving both for the full vector calibration of acoustic emission transducers [4-16] and the evaluation of Green's tensors for realistic bodies [17]. As a simple first step in applying the inverse modelling approach, Hsu, et al. [18], have determined the source function for a pseudo force step (the force unloaded when a glass capillary is broken by compression against a plate). Using a capacitance transducer responding only to out-of-plane displacement (u_3) the one-dimension problem to be solved was:

$$u_3(t)=\int_{o}^{t} G_{33}^{H}(t-\tau)\, F_3(\tau)d\tau \quad (5)$$

where $F_3(t)$ is the force time function to be determined. Using matrix inversion techniques they deduced the result shown in Figure 7 either from waveforms measured at epicenter or on the same plate surface as the source.

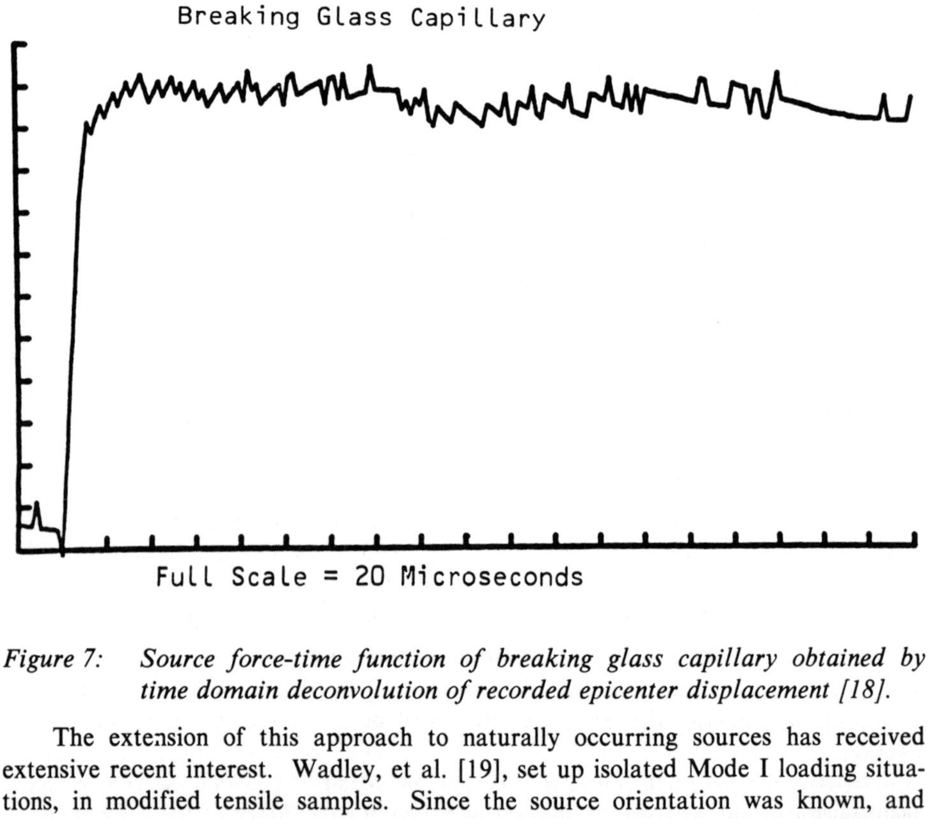

Figure 7: Source force-time function of breaking glass capillary obtained by time domain deconvolution of recorded epicenter displacement [18].

The extension of this approach to naturally occurring sources has received extensive recent interest. Wadley, et al. [19], set up isolated Mode I loading situations, in modified tensile samples. Since the source orientation was known, and elastic loading was assumed, constraining relations between source stress components could be derived so that only a single unknown (the crack diameter) was to be determined. Epicenter waveforms were then used to characterize the source in terms of the crack diameter and its rate of change (the radial crack speed). By using a very wide bandwidth (80 kHz-25 MHz), short duration events could be measured, and cracks propagating at significant fractions of the sound wave velocity resolved, Figure 8.

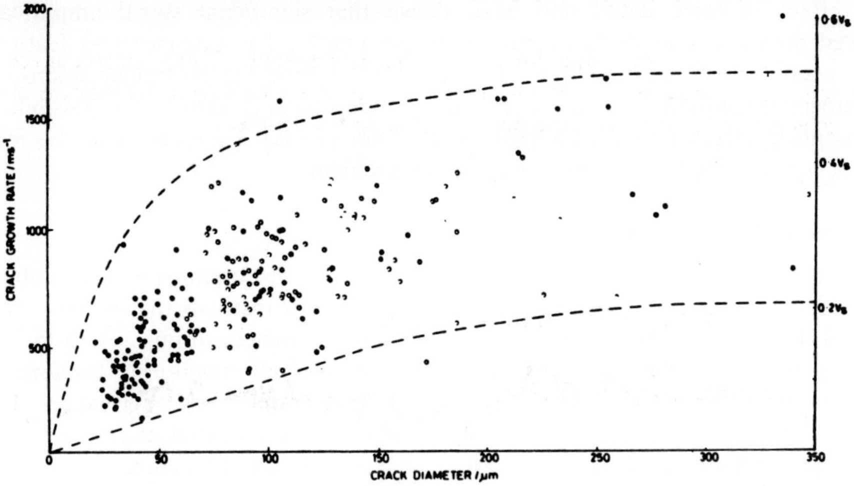

Figure 8: *Showing the crack diameter and growth rate for every emission recording during two fracture processes. •–Cleavage fracture of mild steel, o–intergranular fracture of electrolytic iron [19].*

Ohira and Pao [20] have recently reported experiments on compact tension samples (CTS) using a full six channel recording system. They have been able to detect signals from inclusion interface failure in A533B (several mm's) ahead of a pre-existing fatigue crack as they loaded the samples. They deduced the moment tensor for each microcrack and from this determined the mode (i.e., type I, II, or III) and plane of fracture.

This progress in the development of acoustic emission for such fracture characterization is encouraging. However, much still needs to be resolved before the technique can be considered routine. Perhaps of most immediate importance is the effect of a pre-existing crack upon the emitted acoustic emission signal.

A large pre-crack can be viewed either as perturbing the bodies Green's tensor or the mechanism of the source. If an uncracked body Greens tensor is used for deconvolution purposes the precrack effect will be present in the deduced source function. In experiments on CTS geometries Wadley and Scruby [21] indeed found that if the precrack presence was ignored during deconvolution, the deduced cleavage crack lengths were too large (by as much as a factor of ten). Physically, it was believed that the microscopic advance of the precrack (by a cleavage microfracture) caused a large precrack volume change (compared with that of the microcrack if isolated) and this in turn resulted in an overestimate of the microcrack crack length deduced assuming no precrack presence.

Achenbach, et al. [22], using a 2-d model have since theoretically investigated the effect in some detail and have shown that significant signal amplification indeed does occur in the presence of the pre-crack. The disconcerting feature at this point is that the amplification factor varies with the particular microcrack-macrocrack configuration and unless this is independently known, the possibility of introducing errors into the analysis exist. Very precise 3-d location of the microcrack position might however overcome the problem.

5. Future Research Needs

Research in basic measurement methodologies for acoustic emission has resulted in development of quantitative measurement and analysis methods that potentially can reveal the dynamics of deformation and fracture micromechanisms. With the technology available today, the possibility exists of passively monitoring fracture tests on CTS type geometries over a wide range of temperatures and environmental conditions. Coupled with traditional testing methods the potential exists to radically advance our understanding of the basic micromechanisms of fracture.

It is now possible to precisely locate the site of a microscope fracture event to a precision comparable to the dimensions of the source. The data to do this can be recorded sufficiently rapidly that this could be done at microfracture rates as high as a thousand per second (higher rates cause overlapping signals due to the setting up of sample reverberations). Thus, the critical sites where fractures first occur could be identified and correlated with microstructure features such as second phase or inclusion particles or abnormal impurity element concentrations. Additionally, it is feasible to observe, in three-dimensions, the evolution of a distributed damage state and its coalescence to form a macroscopic fracture, a competence that could be significant as we attempt to develop fracture criteria for composite materials.

With existing technology the local adhesion at, for instance grain boundaries, inclusion-matrix interfaces or particle-matrix interfaces in composite materials, can be measured. By locating the site of a microfracture and either deducing the stress change of the source or numerically evaluating the local stress state just prior to fracture, the strength of an interface can be deduced. The coupling of such measurements with systematic variations in thermomechanical treatments offers the possibility of uniquely determining the physical quantities that control adhesion at the microscopic level. The pay off would be materials with scientifically tailored interface properties for optimum performance.

Present fracture measurement methods are unable to truly resolve the dynamics of fracture when cracks are propagating at speeds close to the sound wave velocity. In this velocity regime, acoustic emission, if suitably performed, can observe, without perturbation, dynamic fracture–even if the crack extends no more than one grain diameter. These experiments, while at the fringes of today's expertise, do offer the possibility of providing experimental insights to guide the development of micromechanical modeling of dynamic fracture and even the critical evaluation of modeling predictions.

These opportunities exist for traditional engineering materials. They promise to make valuable contributions to our understanding of material-fracture relations. However, there exists an extra incentive to apply them to the emerging generation of materials-advanced ceramics and composite materials–because these materials rely upon the use of hard, brittle constituents to bestow superior properties. The micromechanisms of failure in these materials promise to be strong sources of acoustic emission and in turn this acoustic emission could prove invaluable in guiding the development of fracture resistance properties.

Just around the corner awaits numerous improvements to the basic measurement methodology of acoustic emission. Developments that would extend the range of sample geometries and material types for which quantitative work can be performed. New analysis methods, based upon frequency domain measurement, are in the offing, and could solve some of the problems that may arise in the application of existing methods to fracture. The incentive for this would be the successful fulfillment of the opportunities discussed above.

References

[1] Wadley, H.N.G., in D.O. Thompson and D.E. Chimenti (eds.), *Review of Progress in QNDE*, Vol. 5, 1986 (to be published by Plenum, New York).

[2] Scruby, C.B., *Defect Characterization and Monitoring by Acoustic Emission*, AERE, Harwell Report Number R11409, Oct. 1984.

[3] Scruby, C.B. and Wadley, H.N.G., *Nucl. Eng., 11*, 1983, pp. 278.

[4] Bill, R.C., Ph.D. Thesis, University of Michigan, 1970.

[5] Fleischman, P., Lakestani, F. and Baboux, J.C., *Mater. Sci. Eng., 29*, 1977, pp. 205.

[6] Kiesewetter, N. and Schiller, P., *Phys. Stat. Sol. (a), 38*, 1976, pp. 569.

[7] Scruby, C.B., Wadley, H.N.G. and Sinclair, J.E., *Phil. Mag. 44:2*, 1981, pp. 249.

[8] Wadley, H.N.G. and Scruby, C.B., *Met. Sci., 12*, 1978, pp. 285.

[9] Rusbridge, K.L., Scruby, C.B. and Wadley, H.N.G., *Mat. Sci. Eng., 59*, 1983, pp. 151.

[10] Scruby, C.B., Wadley, H.N.G. and Rusbridge, K.L., *Mat. Sci. Eng., 59*, 1983, pp. 169.

[11] Scruby, C.B., Jones, C., Titchmarsh, J.M. and Wadley, H.N.G., *Metall. Sci., 15*, 1981, pp. 241.

[12] Simmons, J.A. and Clough, R.B., in J. Hirth and M. Ashby (eds.), *Dislocation Modeling of Physical Systems*, 1981, Pergamon Press, NY.

[13] Wadley, H.N.G., Simmons, J.A. and Turner, C., in D.O. Thompson and D.E. Chimenti (eds.), *Review of Progress in QNDE, 3B*, 1984, Plenum Press, New York.

[14] Breckenridge, F., *J. Acoustic Emission, 1*, 1982, pp. 87.

[15] Simmons, J.A. and Wadley, H.N.G., in D.O. Thompson and D.E. Chimenti (eds.), *Review of Progress in QNDE, 3B*, 1984, Plenum Press, New York.

[16] Simmons, J.A., Turner, C. and Wadley, H.N.G., to be published in *J. Acoust. Soc. Amer.*, 1986.

[17] Hsu, N., private communication.

[18] Hsu, N. and Hardy, S.C., in Y.H. Pao (ed.), *Elastic Waves and Nondestructive Testing of Materials, AMD, 29*, AMSE, 1978, pp. 85.

[19] Wadley, H.N.G., Scruby, C.B. and Shrimpton, G., *Acta Metal., 29*, 1981, pp. 399.

[20] Ohira, Y. and Pao, Y.H., in D.O. Thompson and D.E. Chimenti (eds.), *Review of Progress in QNDE, 5*, 1986, Plenum Press, New York.

[21] Wadley, H.N.G. and Scruby, C.B., *Int. J. Fract.*, 1983, pp. 117.

[22] Achenbach, J.D., Hirashima, K.I. and Ohno, K., *J. Sound and Vibration, 89*, 1983, pp. 523.

APPLICATIONS OF QUANTITATIVE AE METHODS: DYNAMIC FRACTURE, MATERIALS AND TRANSDUCER CHARACTERIZATION

W. Sachse
Department of Theoretical and Applied Mechanics
Cornell University
Ithaca, New York - 14853

Abstract

This paper summarizes the applications of quantitative acoustic emission (AE) measurements for investigation of dynamic fracture processes and for the characterization materials and transducers. The basis of the technique is a measurement system in which two of the three components comprising it (source, structure, sensor) have known temporal and spatial characteristics such that the temporal and spatial features of the third component can be recovered by appropriate signal processing methods. Examples presented in this paper include the characterization of cracks in brittle materials, the evaluation of the frequency-dependent attenuation and dispersion in absorptive composite materials and the determination of the aperture effect of a detecting transducer.

1. Overview

Acoustic emission (AE) signals are the transient elastic waves accompanying the sudden, localized change of stress or strain in a material. The elastic waves propagate through the medium and to the surface of the specimen where they can be detected by one or more sensors. Reliable systems for the detection and processing of AE signals have long been sought to detect and to monitor the formation and growth of cracks in materials. The focus of recent research on quantitative AE techniques has been towards the development of signal processing procedures which can be used to recover from the detected AE signals information about the characteristics of the source of emission, the properties of the medium and the characteristics of the sensor and the signal amplification system. This paper reviews recent developments which have been made by us in each of these areas.

2. Characterization of Crack Sources of AE

A long-sought application of AE technology is to detect, locate, analyze and monitor the nucleation and growth of cracks in materials. In order to critically investigate the feasibility of quantitative AE techniques for doing this, experiments are required in which cracks are reproducibly generated in materials, the emitted signals analyzed and the results of the analysis compared with independently measured crack parameters.

The characteristics of an AE source refer to its location and source type. The latter is specified in terms of its moment tensor components, its spatial distribution, and its temporal characteristics. The procedures by which a source of AE can be located in a structure have been discussed elsewhere [1,2]. In order to recover the temporal characteristics as well as the moment tensor components or the spatial distribution of a source, the effects of the specimen boundaries and the effects of the sensor and auxiliary electronics must be deconvolved from the detected signals. This is discussed in the first part of this paper.

2.1. *Experiments*

A key feature of this work involves the reproducible generation of a crack at a particular source point. This was made possible by using a diamond indenter pressed against a specimen. The choice of indenter shape and size and the indented specimen material determines to a large extent the specifics of the crack formed [3]. The emitted signals are detected with small, broadband piezoelectric transducers of 1.3 mm active area situated at various angles about the crack source at a distance approximately 2h (h - plate thickness) from it. The transducers are connected to preamplifiers giving a signal gain of 80 dB and a detection bandwidth extending from less than 10 kHz to above 2 MHz. Additionally, a capacitive ultrasonic transducer 3.1 mm in diameter was used as a broadband detector of normal displacement signals at the epicentral position. This transducer was connected to a charge amplifier resulting in a detection bandwidth extending from 10 kHz to 10 MHz [4]. The processing of the waveforms detected by a piezoelectric transducer involves an additional step in the deconvolution to remove the characteristics of the transducer from the signals. After amplification, the signals are recorded by an appropriate number of transient waveform digitization recorders operating at a sampling rate of 30 MHz with 10-bit resolution. These were interfaced to a minicomputer-based data acquisition, recording, and processing system. In the measurements to be described, the AE measurement system consisted of nine data recording channels.

With the measurement system described above, Hertzian cone cracks were reproducibly formed by pressing a small spherical indenter onto the surface of a brittle material such as glass. Reproducible AE signals were recorded. As shown in Figure 1, the detected signals verify that the formation of such a crack under these conditions is equivalent to a small vertical force drop occurring at the source point [5]. During unloading from load values above a minimum critical value, so-called lateral vent cracks form, emitting AE signals such as that shown in Figure 2(a) whose features are similar to those of the signal shown in Figure 2(b). The latter is detected during the formation of a penny-shaped crack which can be obtained by pressing a conical indenter onto the surface of a glass plate specimen [6].

The formation of a lateral vent crack obtained by unloading a spherical indenter or a penny-shaped crack generated by loading with a conical indenter is often adequately modelled by a single horizontal dipole acting near to the surface of the specimen. Evaluation of the stress field under the conical indenter, obtained from the Boussinesq solution [7], shows regions of tensile and compressive hoop stresses. The crack generated by a conical indenter appears in the region of tensile hoop stresses whose contours resemble a penny.

2.2. *Theory and Waveform Processing*

The displacement signals u_k detected at a receiver location, $\mathbf{x}$, in a structure from an arbitrary source $f(\mathbf{x}',t)$ located at $\mathbf{x}'$ having source volume V_o can be written as [8]

$$u_k = \int_{V_o} f_j(\mathbf{x}',t) * G_{jk}(\mathbf{x}/\mathbf{x}',t)\,dV \tag{1}$$

where the term G_{jk} represents the dynamic Green's function of the structure. Insight into this formulation can be obtained if one makes a Taylor series expansion of the Green's function of the source point, $\mathbf{x}'$, about any reference point, say the origin, to give [8,9]

$$u_k(\mathbf{x},t) = \int_{V_o} f_j(\mathbf{x}',t) * G_{jk}(\mathbf{x}/0,t)\,dV + \int_{V_o} \mathbf{x}'_i f_j(\mathbf{x}',t) * G_{jk,i}(\mathbf{x}/0,t)\,dV \quad . \tag{2}$$

The first term in this equation, representing the monopole contribution, can be rewritten in the form

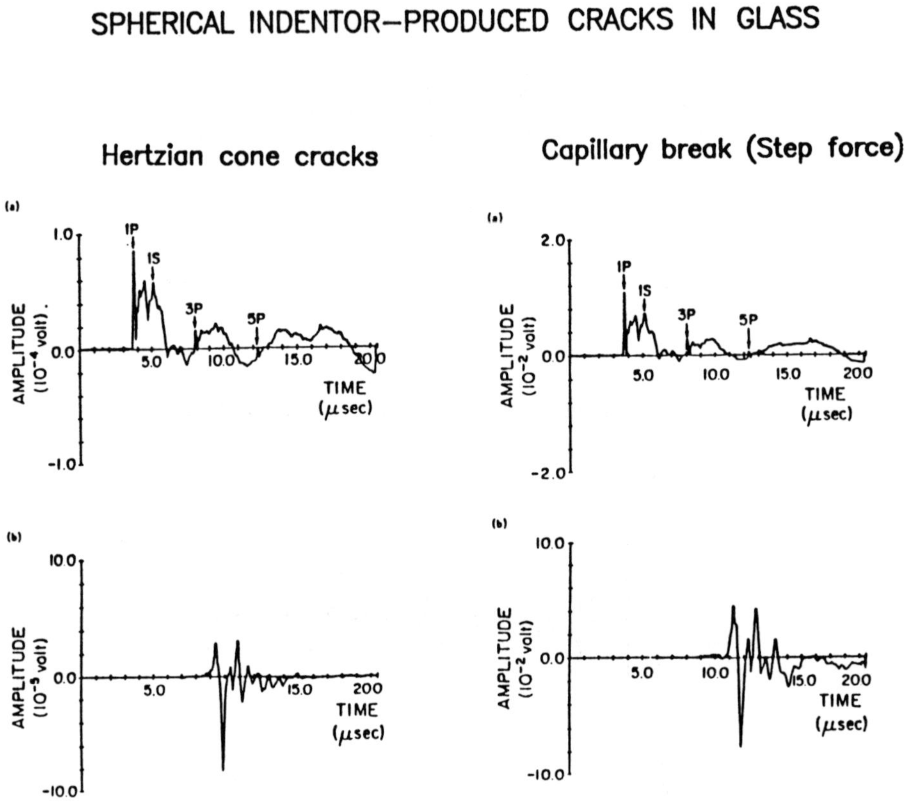

Figure 1: *AE signals in solda-lime glass.*
(a) Spherical indenter-produced, Hertzian cone crack, epicentral signal;
(b) 2-h Off-epicentral signal;
(c) Capillary break, step unloading force signal at epicenter;
(d) 2-h Off epicentral signal.

$$u_k^{(m)} = \{ \int_{V_o} f_j(\mathbf{x}',t)dV \} * G_{jk}(\mathbf{x}/0,t)$$

or more compactly:

$$u_k^{(m)} = S_j(t) * G_{jk}(\mathbf{x}/0,t) \quad (3)$$

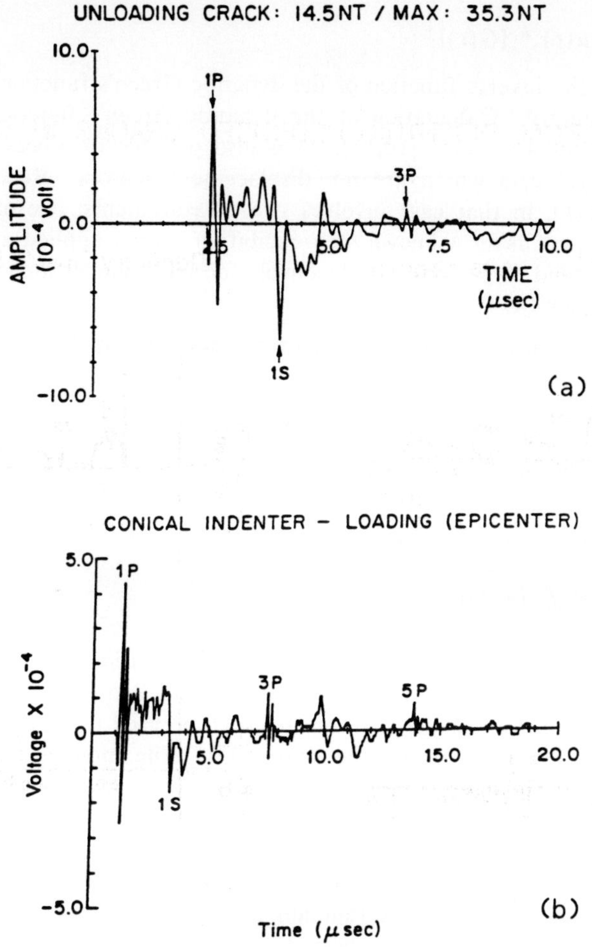

Figure 2: *AE signals at epicenter in soda-lime glass.*
(a) 100μ Spherical indenter, unloading crack signal detected at 14.5 N. during unloading from 35.3 N.
(b) 120° Conical indenter-produced, penny-shaped crack signal.

where $S_j(t) = \int_{V_o} f_j(\mathbf{x}',t)\,dV$.

Based on the results of the experiment shown in Figure 1, the formation of a Hertzian cone crack in a brittle solid is identified as a monopolar source whose time-characteristics can be recovered by inverting Equation (3). If one is dealing only with the time-dependence of the source, then Equation (3) can be inverted to yield for the time function of the source:

$$S(t)=u(t) * [G(t)]^{-1} \tag{4}$$

where $[G(t)]^{-1}$ is the inverse function of the dynamic Green's function appropriate to the testing geometry. Calculation of the dynamic Green's functions has been made by Pao and his associates [10,11]. It is also possible to process the signals detected with transducers which are not displacement sensors. Recovery of the source function, $S(t)$, in that case involves two measurements, one of which is a calibration experiment using a known step excitation source applied at the source position and the second to perform the deconvolution in a manner analogous to that expressed by Equation (4) [12].

The second term in Equation (2), which represents the dipole contribution, can be rewritten as

$$u_k^{(d)}=\{\int_{V_o} \mathbf{x}'_i f_j(\mathbf{x}',t)\,dV\} * G_{jk,i}(\mathbf{x}/0,t)$$

or

$$u_k^{(d)}=M_{ij}(t) * G_{jk,i}(\mathbf{x}/0,t) \tag{5}$$

where $M_{ij}(t)=\int_{V_o} \mathbf{x}'_i f_j(\mathbf{x}',t)\,dV$.

The dynamic Green's function appearing in this equation has been evaluated for various source/receiver geometries by Ceranoglu and Pao [11]. The term, M_{ij}, is called the moment tensor of the source and it can be used to represent any combination of double forces to model the source [1]. The moment tensor can be expressed in terms of the crack's displacement jump and its area in a material [13]. If each of the components of the moment tensor of the source has the same time dependence and the spatial and temporal characteristics of the source are separable, then the moment tensor can be rewritten as a product of the moment tensor magnitudes, M_{ij}, and the source-time function, $S(t)$,

$$M_{ij}=S(t)\cdot m_{ij} \tag{6}$$

In recovering the characteristics of a dipolar source from the measured signals, $u_3(t)$, one uses knowledge of the Green's functions, G, and inverts Equation (6) to recover both $S(t)$ and the components m_{ij}. In the most complicated deformation processes for which the orientation of the crack is arbitrary, all six components of the moment density tensor may need to be evaluated. As of now, this has not yet been done. Here are shown the results obtained for several simpler cases.

For the case described by Equation (6) in which both the time-function and the components of the moment tensor of a source are to be recovered, several processing algorithms have been developed. One of these is based on an algebraic scheme in which the amplitudes of particular ray arrivals in the detected signals are directly related to the amplitudes of the Green's function [14,15]. Once these are determined, the source time-function can be recovered as before. The problems related to this scheme result principally from the amplitude measurement precision requirement made on specific waveform features. An alternative is a double-iterative processing scheme demonstrated by Michaels and Pao [16]. In this, the procedure is initiated with a guess of the moment tensor magnitudes, after which a solution of the source time-function can be obtained by deconvolution. This solution is used, in turn, to evaluate a new set of moment tensor magnitudes with the iteration process continued until the obtained results converge to a final answer. The procedure converges slowly for certain configurations of the source moment tensor. It has been found that in many cases an appropriate initial guess for the double-iterative scheme can be obtained by making use of the results obtained from the algebraic processing scheme.

Evaluation of the moment tensor components of the penny-shaped crack via one of the procedures listed in the previous section often leads to difficulties since the results are invariably initial guess-dependent. The development of a new algorithm by Chang [17] based on a division of the detected signals in the frequency-domain shows great promise to overcome these problems but this algorithm has not yet been tested with actual AE waveforms. In the procedure, the division results in an elimination of the common source-time function, leading to a set of linear algebraic equations which can be solved directly to recover the ratios of the components of the moment tensor, m_{ij}. Once these are found, the source-time function can be recovered by standard time- or frequency-domain deconvolution algorithms.

An alternative approach by which the ratios of the moment tensor components can be recovered involves measurement of the radiated field of the AE signals from the source. The amplitude of the emitted signal measured by a set of transducers at a distance R and at various angles θ from a dipolar source, can be written in terms of the normal displacement amplitude, u_3, which is expected to be of the form

$$u_3(\theta,t)=a_1(t)+a_2(t)\cos^2\theta \tag{7}$$

where θ is the orientation of the normal to the crack in the polar-cylindrical coordinate system. For a single horizontal dipole source, aligned parallel to the crack normal, the term $a_1(t)$ is expected to be near zero. For the three-dipole crack model often adopted (c.f. Scruby [15]), in which the crack is modelled in terms of three orthogonal dipoles of strength, $\lambda+2\mu$, λ and λ, where λ and μ are the Lamè constants, the term $a_1(t)$ is expected to be approximately 0.28. Equation (7) can be rewritten to account for an arbitrary crack orientation relative to the sensor array

$$u_3(\theta,t)=a_1(t)+a_2(t)\cos^2(\theta-\theta_c) \tag{8}$$

where θ_c is the orientation of the normal of the crack relative to the orientation of the sensor array.

2.3. *Results*

By a proper choice of the angular arrangement of sensors around the indenter, eight channels are sufficient to determine the characteristics of the radiated field of the indentation-generated cracks. Each channel of the measurement system was calibrated by breaking a glass capillary 0.05 mm ID and 0.08 mm OD at the location of the crack on the specimen and recording the simultaneous responses. This source is an axisymmetric step excitation normal force [18]. The amplitudes of both the first P- and S-wave arrivals were determined and the corresponding amplitudes in the crack-generated signals were normalized with these calibration amplitudes. The result obtained after one indentation is shown in Figure 3. The solid and dashed lines correspond to the radiation patterns of the P- and S-wave amplitudes, respectively, determined by a non-linear least-square fit of Equation (8) to the measured amplitude data. From the curve fit, the parameters of the radiated field can be extracted. Once these are determined, the measured amplitudes at each transducer location can be reflected across the various symmetry planes to generate image data points as depicted in Figure 4. Inspection of the amplitude ratio, a_1/a_2, shows that many of the tensile cracks produced by the conical indenter are best modelled as resulting principally from a single horizontal dipole. Hence, the three-dipole model of a tensile crack, did not appear to correctly represent the indentation-produced point-cracks in the specimens described here. The orientation of the crack, relative to the array configuration, is also recovered from the fit to Equation (8). The difference between the observed and recovered crack orientation angles in one set of seven indentation experiments was always less than 6° and on average less than 3°.

The radiation field of the signals emitted from lateral vent unloading cracks generated in glass specimens was also measured. As expected, it was found that the formation of multiple cracks is accompanied by a radiation field of AE signals which is of almost uniform amplitude in every direction, differing markedly from the results observed for a single crack.

Using the ratios of the moment tensor components recovered from the radiated field as input data, it is possible to process a displacement waveform at almost any receiver point with the double-iterative processing scheme to reliably recover the individual magnitudes of the moment tensor as well as their time function. By choosing to process an epicentral waveform, one achieves a large signal-to-noise ratio. Figures 5(a) and (b) show one example of a detected epicentral displacement signal and the recovered source-time function for an indentation crack.

An alternate means for generating cracks without the use of indenters and an applied external force involves the use of a rapid, localized cooling of the region near a small scratch on the surface of a brittle solid [19]. The experimental means for doing this is illustrated in Figure 6. If the cooling proceeds symmetrically about the scratch, then the resultant crack will very likely be of pure Mode I type for which moment tensor is simply

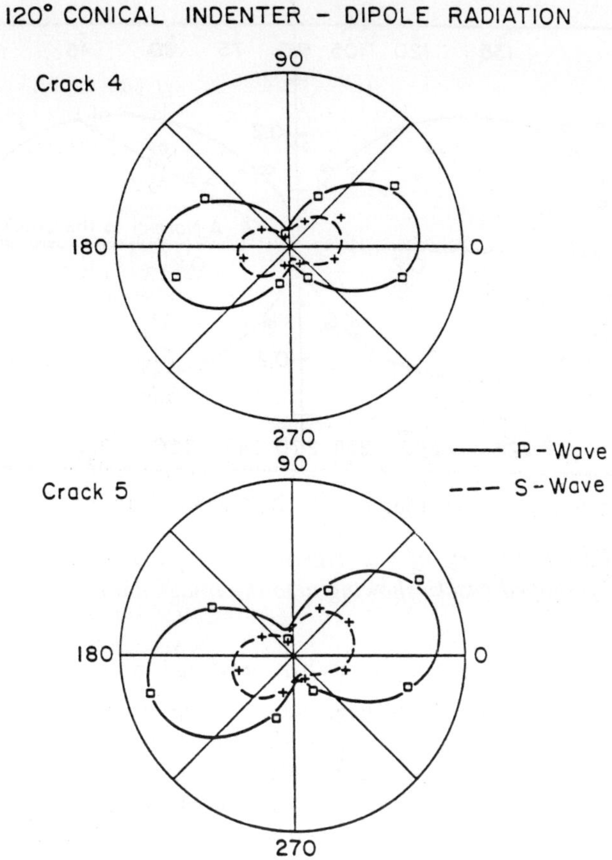

Figure 3: Radiation patterns of the signals accompanying the formation of a penny-shaped crack in glass.

$$\mathbf{M} = \begin{vmatrix} M_{xx} & 0 & 0 \\ 0 & M_{yy} & 0 \\ 0 & 0 & 0 \end{vmatrix} = \begin{vmatrix} D_{11} & 0 & 0 \\ 0 & D_{22} & 0 \\ 0 & 0 & 0 \end{vmatrix} \cdot \hat{S}^d(t) \qquad (9)$$

Here $\hat{S}^d(t)$ represents the normalized source-time functions of the dipolar source. The radiation pattern for such a source is given by

$$u(\theta,t) = d(t)(\cos^2\theta + b(t)) \ . \qquad (10)$$

Radiation patterns having this angular dependence were observed in more than ten

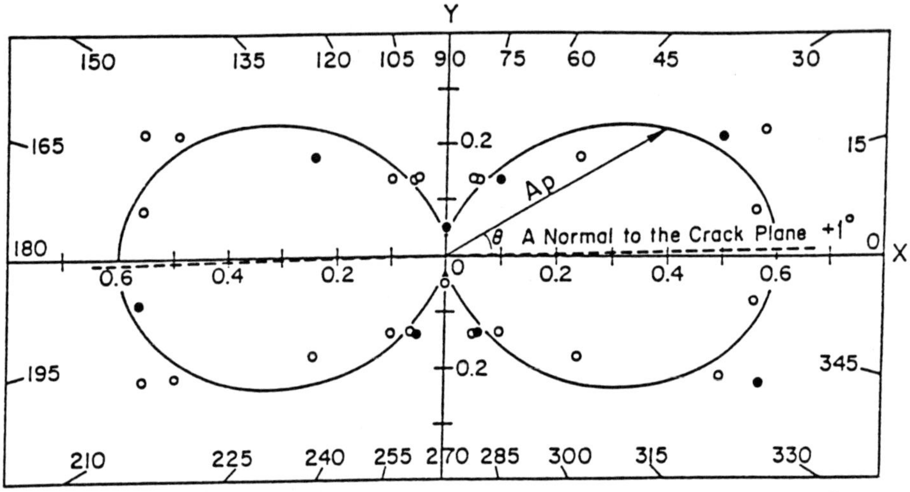

Figure 4: *Radiation pattern of the signals accompanying the formation of a penny-shaped crack, showing also the image data points.*

thermal-crack experiments. A sample result is presented in Figure 7.

As described above, only the relative values of the components comprising the moment tensor can be recovered from measurement of the radiation pattern of the emitted AE signals. Deconvolving the signals at two or more receiver points permits recovery of the source-time function. An alternative procedure to one of the detailed signal deconvolution procedures described in Sect. 2.2, which also appears to be far more robust when processing actual experimental waveforms, has been formulated. In this, the strength of the dipoles used to represent the source of emission can be determined directly from an analysis of the emitted elastic waves [20].

The procedure relies on a comparison between the epicentral displacement signals corresponding to the formation of a crack and a normal force, step unloading, calibration signal. The latter corresponds to a monopolar source, which can be written as

$$V_m(t) = k u_3^m(t) = k F_3 \hat{S}^m(t) * G_{33}(t) \tag{11}$$

where F_3 is the magnitude of the unloading force, $\hat{S}^m$ is the step-like source function of unit magnitude, G_{33} is the appropriate dynamic Green's function and $V_m(t)$ is the voltage of the measured calibration signal. The signal corresponding to the formation of a thermal crack, represented by the dipolar forces, $D_{11} + D_{22}$, is

$$V_d(t) = k u_3^d(t) = k (D_{11} + D_{22}) \hat{S}^d(t) * G_{13,1}(t) \ . \tag{12}$$

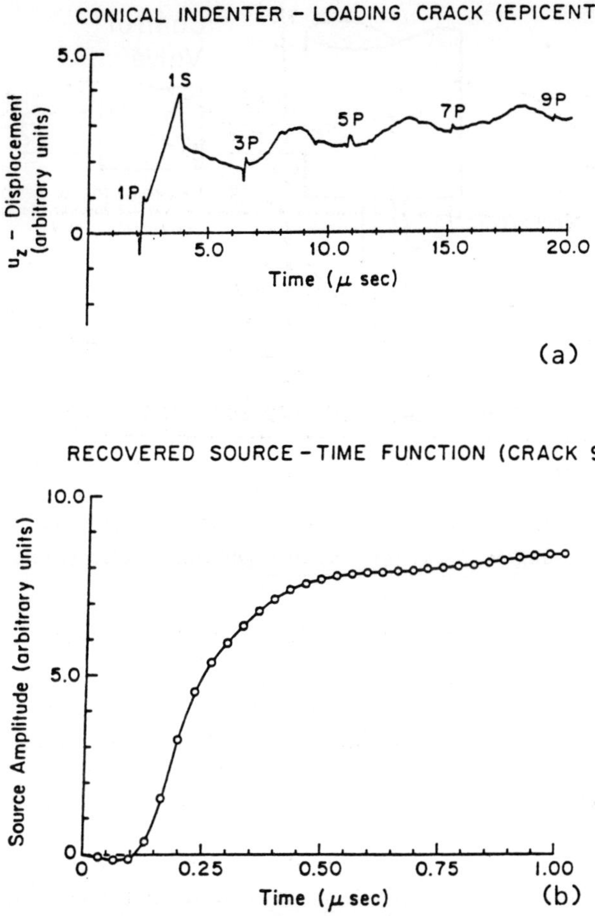

Figure 5: *Penny-shaped crack in soda-lime glass.*
(a) Detected epicentral normal displacement signal;
(b) Recovered source-time function.

As for the calibration source, $\hat{S}^d(t)$ is a step-like source function of unit magnitude and $G_{13,1}(t)$ is the appropriate derivative of the Green's function. Then, for times sufficiently large compared to the risetime of the source, one can form the ratio between these signals to obtain

$$D_{11}+D_{22}=F_3 h \left| \frac{\hat{u}_3^m}{V_m} \right| \cdot \left| \frac{\hat{u}_3^d}{V_d} \right|^{-1} . \qquad (13)$$

where h is the thickness of the plate. This equation is written in terms of $\hat{u}^m$ and $\hat{u}^d$ which represent, respectively, the non-dimensional normalized epicentral displacements corresponding to the step-like source-time functions of unit magnitude

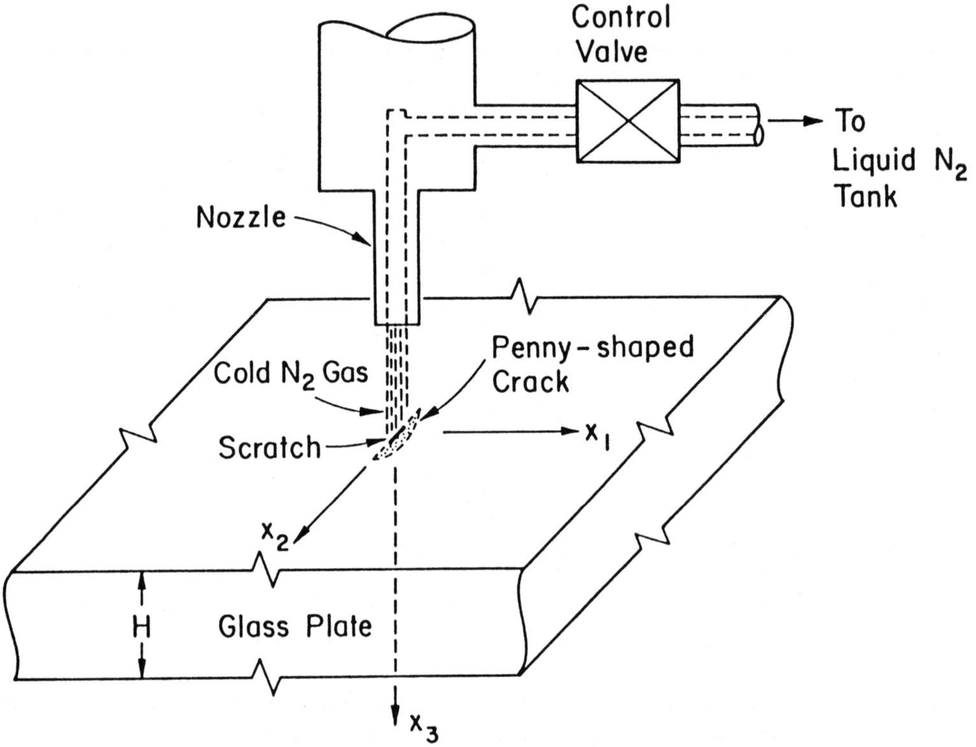

Figure 6: Schematic of the generation of a thermal crack in a glass plate.

of vertical force and dipolar source type.

To determine the values of $D_{11}+D_{22}$, the average value of the detected voltage (or displacement) signals after the arrival of the first P-wave signal is measured. When this result is combined with the moment tensor ratio determined from analysis of the radiation pattern, the explicit values of both D_{11} and D_{22} can be determined. This approach has been demonstrated by analyzing the signals from eleven thermally generated cracks. It was found that for the soda-lime glass specimens, the dipolar source strength normal to the crack plane, D_{11}, ranges from 0.009 N-m corresponding to the formation of cracks ranging from 10^{-5} to 10^{-6} m^2 in extent.

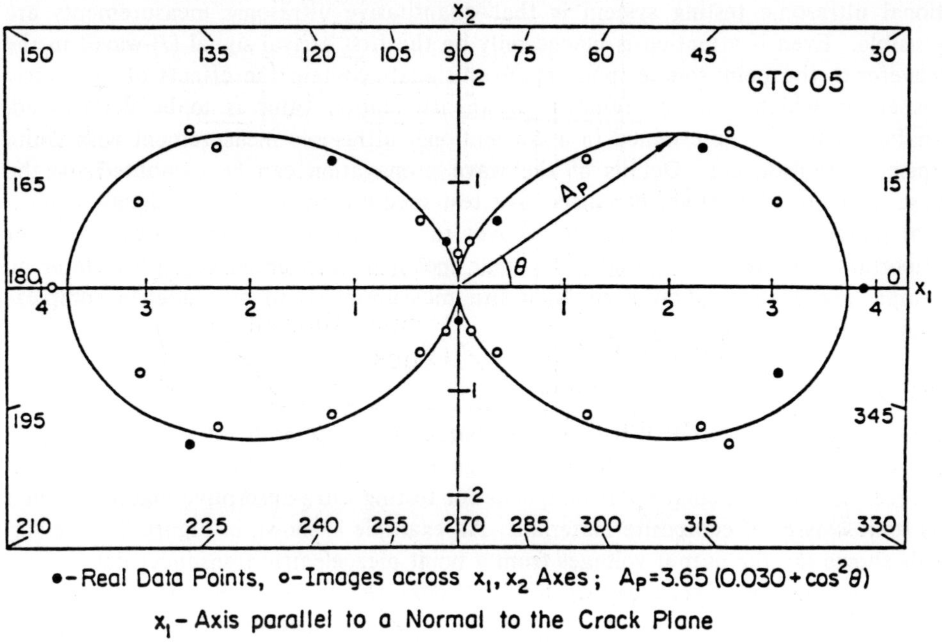

Figure 7: *Radiation pattern of the signals associated with the formation of a thermal crack in glass and the corresponding image data points.*

3. Composite Materials Characterization

In characterizing composite materials, information regarding the spatial variations of composition, configuration (including ply layup and thickness), porosity, fiber/matrix disbonds and cracking and voids or inclusions is sought from a test. Complementing radiographic methods, ultrasonic measurements of absorption and dispersion of a composite material will be required if such materials are to be characterized.

3.1. *Point-Source/Point-Receiver Testing*

The source of AE in a structure can often be properly treated as a point source. While forming the basis of exploration geophysics, the use of such a source for materials inspection and characterization applications is, however, not yet well established. To implement this, a simulated source of emission such as a normal force [18] or a focused laser beam [21] or a spark [22] can be used. The characteristics of the latter two sources may be modelled either by a normal force or a horizontal dipolar source depending on the energy of the excitation, the extent of the interaction region, and the presence of a surface film acting as a constraining layer. By also including a point receiver, a new ultrasonic testing system is formed. The principal advantage of such a point-source, point-receiver system over a conventional ultrasonic testing system is that quantitative ultrasonic measurements are possible. Even if attention is placed only on the first arrival signal (P-wave) in the waveform, the point-source, point-receiver signals contain the effects of geometric dispersion which must be removed if an attenuation value is to be determined, analogous to that determined in a conventional ultrasonic measurement with finite aperture transducers. Details of the wave propagation can be computed for the case of a non-absorptive, non-dispersive test medium by which the signals from a real material can be processed to recover these material characteristics. Another important consideration is that non-planar specimens can be tested with little or no sample preparation. This permits in situ measurements to be made on complex-shaped structural components.

3.2. *Results*

By selecting a point source whose time-characteristics resemble a Heaviside step, a source function is obtained featuring high amplitudes at low frequencies. Such sources can be particularly advantageous for testing ultra-absorptive materials, such as some classes of composite materials. An example is shown in Figure 8. The signals shown are the output voltages from a point piezoelectric transducer detecting a step unloading on a specimen of Kevlar/Epoxy 0.7 inches (1.9 cm) thick with a rubberized coating on one side 5.5 mm thick. In one measurement, the unloading was made on the rubberized coating and in the other, it was on the back side. The oscillations present in the waveforms appear to be the result of reverberations in the rubberized layer. In contrast to these results, the signal-to-noise ratio of conventional ultrasonic measurements made at 1 MHz is so low that it is difficult to detect any signal in either pitch-catch or pulse-echo modes. This observation severely limits the signal processing which can be performed on the signal to recover the properties of the material.

The theoretical normal displacement waveform corresponding to the epicentral testing configuration in a material with no damping was computed with the algorithm developed by Ceranoglu [10,11]. This is shown in Figure 9. Superimposed on this curve is the measured displacement waveform which has been scaled to bring it on scale. In highly absorptive materials it is generally difficult to identify in the displacement signals the arrivals of the individual rays comprising the waveforms. Since the wave arrivals are enhanced in the velocity signals, it is probably advantageous to extract the attenuation from such signals. In an actual test, the load level of the step-excitation must be measured and the corresponding velocity waveform computed for the specific input load-drop. By comparing the amplitudes of particular ray arrivals in the measured and computed waveforms, a determination of the broadband absorption of particular wave modes should be possible, but this has not yet been carried out.

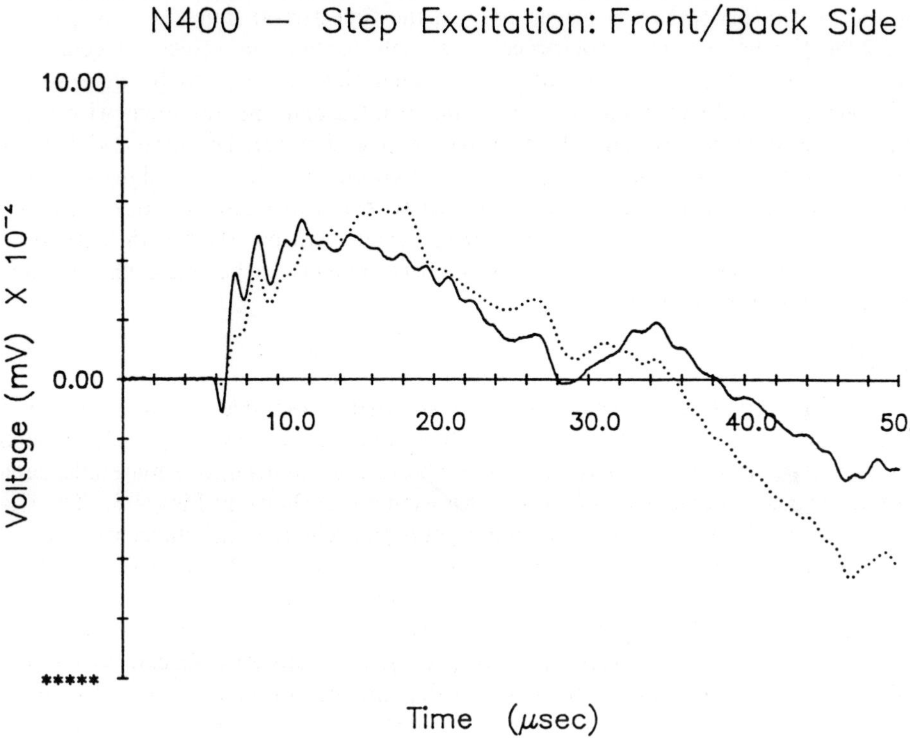

Figure 8: Received voltage signal from a point piezoelectric transducer in a Kevlar/Epoxy laminated specimen. Step excitation source.

The measured waveforms also contain the effects of wave dispersion. We propose that in materials which are not too absorptive, the Fourier phase spectral technique [23], which was developed for recovering the dispersion relation, phase and group velocities, can be modified for use with point sources, provided that the geometric dispersion can be properly accommodated. Weaver [24] has recently shown how to extend Knopoff's [25] epicentral solution for a wave from a point source in a plate to the case in which the wave propagation is through a viscoelastic material. In that case, the wavenumber of the first arrival P-wave is related to the Fourier phase function, ϕ (ω), according to

$$\phi(\omega)=k\,h+(\frac{\pi}{2}+\frac{A}{kh})+\ldots \qquad (14)$$

where

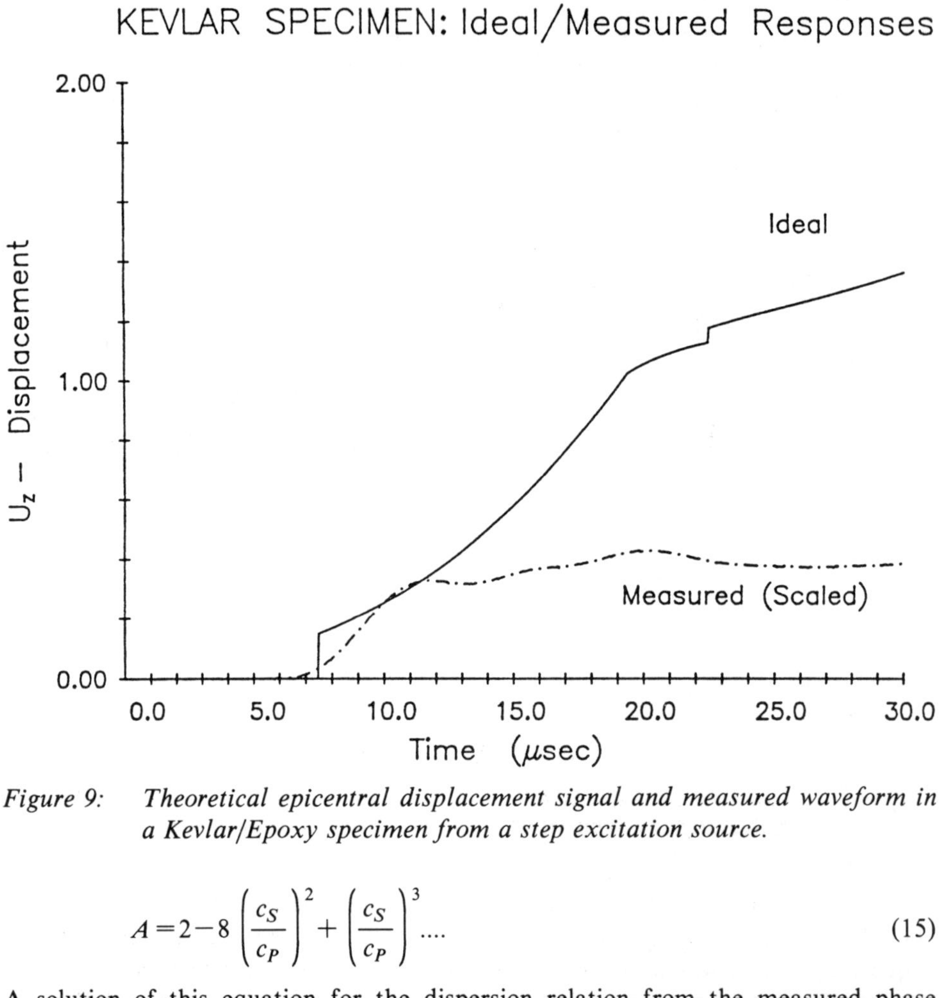

Figure 9: *Theoretical epicentral displacement signal and measured waveform in a Kevlar/Epoxy specimen from a step excitation source.*

$$A = 2 - 8\left(\frac{c_S}{c_P}\right)^2 + \left(\frac{c_S}{c_P}\right)^3 \ldots \tag{15}$$

A solution of this equation for the dispersion relation from the measured phase function can be obtained by an iterative procedure. As for the plane wave situation, the phase velocity, ω/k, and the group velocity, $\partial\omega/\partial k$, are obtained directly once the dispersion relation has been determined. An analogous analysis procedure has also been developed by Weaver for the shear-wave signal in the epicentral waveform [24].

While this algorithm has not yet been implemented for waveforms determined in actual measurement situations, the analysis of a simulated signal has been completed. The test signal corresponds to a displacement signal expected from a low-frequency pulse source in a material whose wavespeed is 0.25 cm/μsec. The synthesized waveform and the windowed portion related to the arrival of the P-wave are shown in Figure 10(a). In Figure 10(b) is the dispersion relation derived from an implementation of Equation (14) while in 10(c) is the derived phase velocity. Although there is an apparent rise of the recovered wave speed at low frequencies, at frequencies above 600 kHz the input wave speed is fully recovered. While this is only a preliminary result, a measurement technique based on phase spectral analysis of point-source ultrasonic signals to recover the frequency characteristics of waves propagating in dispersive materials looks promising.

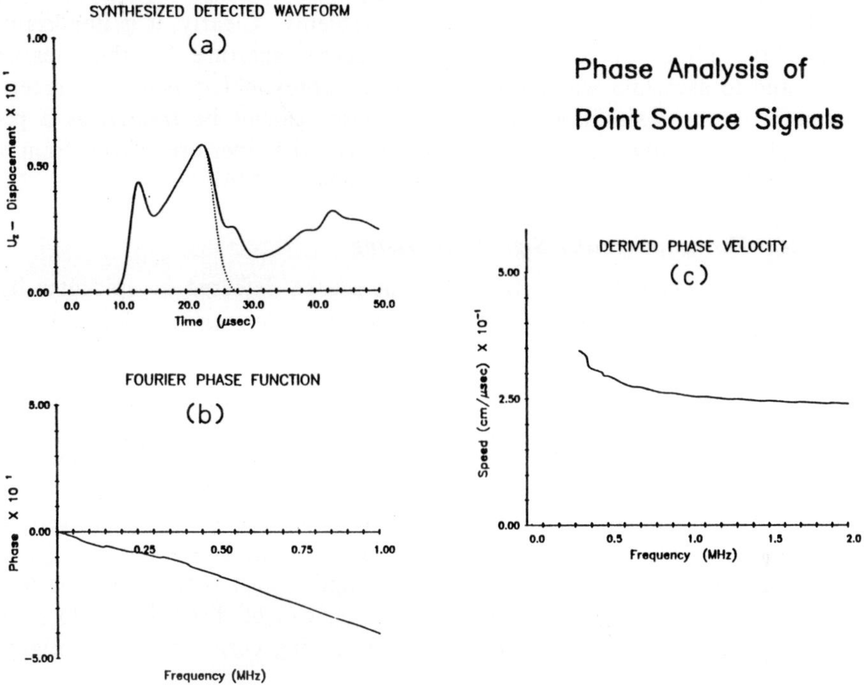

Figure 10: Phase spectral analysis of point-source ultrasonic waveforms.
(a) Synthesized waveform showing windowed portion;
(b) Recovered Fourier phase function;
(c) Derived phase velocity.

4. Sensor Characterization

The smoothing of a received waveform, caused by the finite receiving area of a transducer, is associated with the aperture effect of a transducer. As described in the two previous sections, when quantitative information about a source or a material is to be extracted from the ultrasonic signals, it is advantageous to use a point receiver to detect these signals in a material. However, most of the commercially available ultrasonic transducers have finite receiving areas. For a point receiver, little energy is taken from the measured system, thus minimizing transducer loading effects; but at the same time, the generated output signal is generally also small, hence giving rise to signal-to-noise problems. Clearly, it is important to understand the effect of a finite-size transducer's aperture on the measured waveform and to ascertain when a sensor can be approximated as a point receiver. Furthermore, in those situations in which a sensor cannot be treated as a point receiver, one would like to remove or deconvolve the aperture effect from the detected signals by appropriate signal processing schemes [26].

4.1. *Problem Formulation and Signal Processing*

The vertical displacement signal from a point source detected at epicenter by a finite-aperture transducer follows from Equation (1).

$$u_3(\mathbf{x},t) = \int_{-\infty}^{\infty} s(\tau)d\tau \iint_A \rho_z(\mathbf{x}') G_{33}(\mathbf{x}/\mathbf{x}', t-\tau) \tag{16}$$

Here $\rho_z(\mathbf{x}')$ is the spatial sensitivity function which is a function of position inside the active region of the transducer and $s(t)$ is the source time function. For the case in which the receiver is circularly symmetric with respect to an epicentral point source, it can be represented as a line receiver. The line receiver in general will not possess a uniform sensitivity across its length and it is this variation which is sought. As shown in Figure 11, the line receiver can be discretized into several concentrated point receivers with appropriate weighting constants. The weighting functions of the line receiver after discretization are:

$$W_i = A_i \cdot \bar{\rho}_i \tag{17}$$

where A_i is the area of each discretized ring and $\bar{\rho}_i$ is the average sensitivity over the ring.

After discretizing Equation (16) both in time and space, it can be written in compact matrix form as follows:

$$[G_X(T)]\ [W(X)] * S(T) = U(T) \tag{18}$$

where $S(T)$ is the discretized source-time function for the point source and $W(X)$ is the discretized weighting function. The elements of the Green's functions matrix are a function of both the time and position indices. From this equation one can see that if the Green's functions are the same but shifted only in time, the multiplication between the G-matrix and the spatial weighting function becomes a

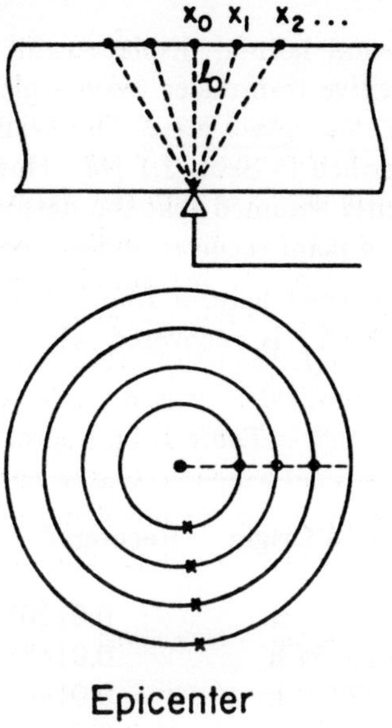

Figure 11: Representation of a transducer into a line receiver and the discretization into concentrated point receivers.

convolution operation. For a linear system, the operation of a shift and a sum corresponds to a convolution. The weighting function, $W(X)$, is then equivalent to a time function $W(T)$. Although the function $W(X)$ is not equi-spaced, $W(T)$ is. Hence, the equivalent spectrum of the transducer aperture is the Fourier transform of $W(T)$. This function is dependent on the geometric arrangement of the source and receiver. The case analyzed to date is for a source directly over the center of a receiving transducer. Although for off-center transducer-source arrangements, $W(T)$ will differ, the approach demonstrated here will still be applicable.

It can be seen from Equation (18) that if the aperture is not too large, only a few receiver points are needed to represent the transducer, hence requiring the computation of only a few Green's functions. Conversely, for a large aperture transducer, many Green's functions must be included in Equation (18). If it is assumed that the Green's functions are position-independent, then only one Green's function, shifted in time, is required. After deconvolution, the recovered time-dependence is the convolution result of $W(T)*S(T)$. For a known aperture, the characteristics of the source, $S(T)$, can be recovered by deconvolving the function $W(T)$. If both $W(T)$ and $S(T)$ are unknown, an iterative deconvolution algorithm must be used to recover them [16,27].

4.2. *Experimental Results*

Figure 12(a) is the experimental normal displacement signal detected by a 0.25 inch (6.4 mm) diameter capacitive transducer when a glass capillary is fractured on the top surface of a 1/4 inch thick glass plate. This transducer is the same as that used for the experiments described in Sect. 2.1 [4]. However, in the earlier deconvolution studies, it was implicitly assumed that the detected signals corresponded to those from a point source and a point receiver testing configuration.

In the measurements described here, a finite aperture is considered and the aperture weighting function $W(X)$ is recovered with the iterative method. The results are listed in Table I.

Table I

Distance to Origin	Recovered $W(X)$
x = 0.0 h	0.01201
x = 0.26154 h	0.01486
x = 0.37292 h	0.01460
x = 0.46043 h	0.01113
x = 0.53590 h	0.00534
x = 0.60386 h	-0.00043

The value of $W(X)$ reduces to almost zero at 0.536 h. That there is a non-zero weighting function for values greater than the transducer radius implies that the transducer has an apparent radius which exceeds the actual size of the transducer. This result can be attributed to the fringing effect associated with the operation of a capacitive transducer. The recovered source-time function is compared in Figure 12(b) with the recovered source-time function obtained if a point-source and point-receiver assumption is made. It is seen that the transducer aperture effect leads to artifacts in the recovered source-time function which are removed if the aperture effect of the transducer is taken into account.

5. Conclusions

The application of quantitative AE measurements for dynamic fracture, materials and transducer characterization studies has been demonstrated. The basis of the technique is a measurement system in which two of the three components comprising it - source, structure, sensor - have known temporal characteristics and spatial characteristics such that features of the third component can be recovered by signal processing methods. Examples presented in this paper have included the characterization of cracks in brittle materials, the evaluation of the frequency-dependent attenuation and wave propagation in ultra-absorptive composite materials and determination of the aperture effect of a detecting transducer.

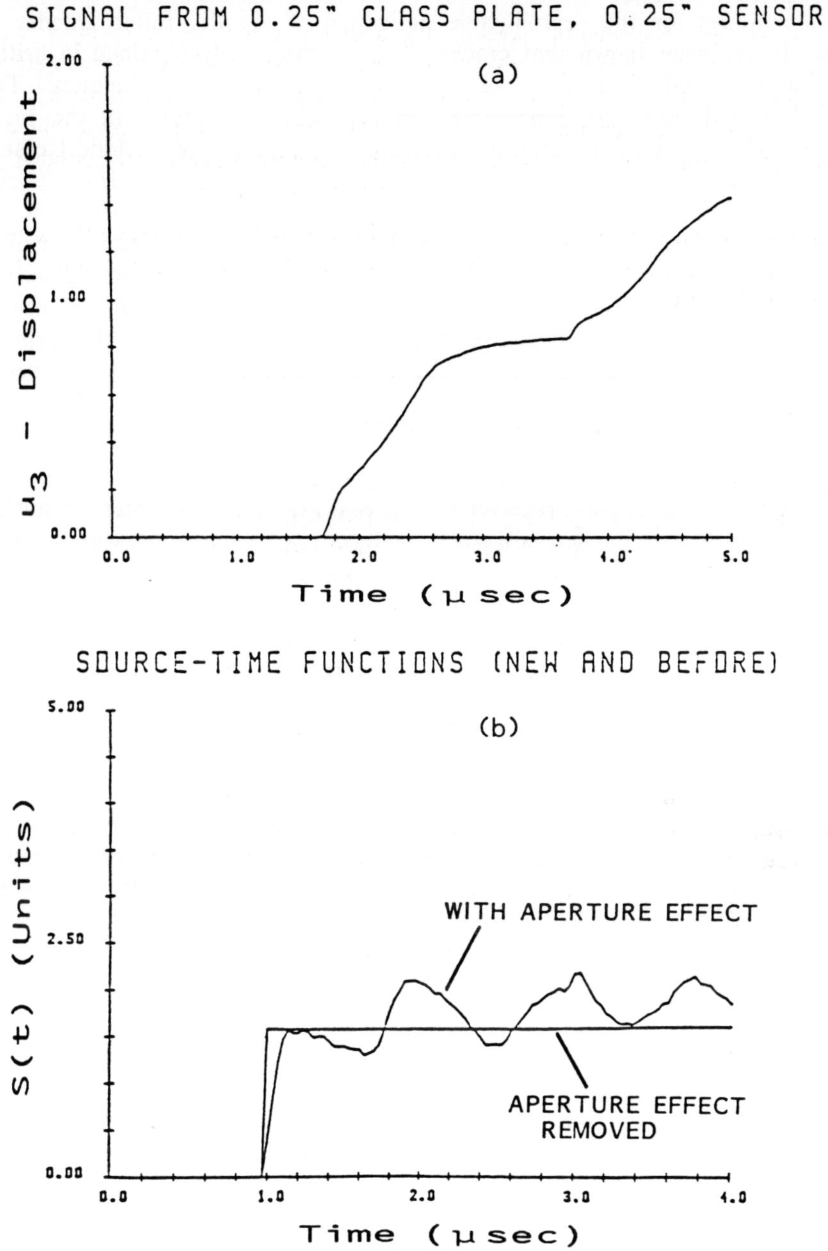

Figure 12: *Finite aperture transducer experimental results.*
(a) Normal displacement waveform at epicenter, step unloading;
(b) Comparison of recovered source time functions of the signal in (a), with and without aperture correction.

Investigation of dynamic fracture processes with quantitative AE methods requires techniques for the reproducible activation of specific failure modes in materials. It has been shown that cracks can be reproducibly obtained in brittle materials such as soda-lime glass via indentation and thermal techniques. The emitted AE signals are very repeatable and they exhibit features in the time-domain and in their radiation patterns which are characteristic of a Mode I crack. It has been shown that the orientation of the crack can be recovered from its AE radiation pattern as can the relative magnitudes of the principal components of the moment density tensor. It was further demonstrated that the dominant contribution to the crack AE source can be modelled by a single force dipole acting normal to the crack plane. An experimental procedure was described which makes use of a step-force calibration signal to recover the actual dipole strength of the forming crack from the emitted signals. It was further demonstrated that the recovered temporal characteristics of the crack source function can provide useful information about the dynamics of the on-going fracture process.

It was also demonstrated with preliminary simulations and actual measurements that a point-source, point-receiver testing configuration with a step excitation source forms the basis of a new ultrasonic materials testing method which shows promise as a means for recovering the frequency-dependent attenuation and dispersion characteristics of ultrasonic waves propagating through an absorptive composite material specimen.

The aperture effect of a receiving transducer was determined by processing the signals detected in a calibration test to recover the transducer's weighting function representing the spatial sensitivity of a transducer. A receiving transducer's aperture must be accounted for if the true time characteristics of the source are to be recovered in the solution to an inverse source problem. If both the weighting function and source time function are unknown in an experiment, an iterative deconvolution algorithm must be used to recover each of them.

Acknowledgments

It is a pleasure to acknowledge the ideas and work of my students and associates, K.Y. Kim, C. Chan, C.P. Chen, S. Howard, W. Pierce and V Smith. Portions of this work were supported by the Naval Research Laboratory, the Office of Naval Research and the National Science Foundation through grants to the Materials Science Center and the College of Engineering at Cornell University. Some of the experiments were facilitated by equipment purchased through DoD Equipment Grant N00014-83-G-0200 from the Office of Naval Research.

References

[1] Pao, Y.H., "Theory of Acoustic Emission", in *Elastic Waves and Non-destructive Testing of Materials*, Y.H. Pao, Ed., AMD-Vol. 29, Am. Soc. Mech. Engrs., New York (1978), pp. 107-128.

[2] Sachse, W. and Pao, Y.H., "Locating and Characterizing Sources of Acoustic Emission", *Quant. NDE in the Nuclear Industry*, R.B. Clough, Ed., Am. Soc. Metals, Metals Park, Ohio (1982), pp. 326-331.

[3] Lawn, B.R. and Wilshaw, T.R., "Review Indentation Fracture: Principles and Applications", *J. Mater. Sci.*, 10, 1049-1081 (1975).

[4] Kim, K.Y. and Sachse, W., "Self-Aligning Capacitive Transducer for the Detection of Broadband Ultrasonic Displacement Signals", *Rev. Sci. Instrum.*, 57, 264-267 (1986).

[5] Kim, K.Y. and Sachse, W., "Characteristics of Acoustic Emission Signals of Hertzian and Unloading Cracks in Glass", *J. Appl. Phys.*, 55 (8), 2847-2856 (1984).

[6] Kim, K.Y. and Sachse, W., "Characteristics of Acoustic Emission Signals from Indentation Cracks in Glass", in *Progress in Acoustic Emission, II*, M. Onoe, K. Yamaguchi and H. Takahashi, Eds., Jap. Soc. Non-destr. Inspect., Tokyo (1984), pp. 163-172.

[7] Sneddon, I.N., "Boussinesq's Problem for a Rigid Cone", *Proc. Cambridge Phil. Soc.*, 44, 492-507 (1948).

[8] Backus, G., "Interpreting the Seismic Glut Moments of Total Degree Two or Less", *Geophys. J.R. astr. Soc.*, 51, 1-25 (1977).

[9] Stump, B.W. and Johnson, L.R., "The Determination of Source Properties by the Linear Inversion of Seismograms", *Bull. Seism. Soc. Am.*, 67, 1489-1502 (1977).

[10] Pao, Y.H., Gajewski, R.R. and Ceranoglu, A.N., "Acoustic Emission and Transient Waves in an Elastic Plate", *J. Acoust. Soc. Am.*, 65 (1), 96-105 (1979).

[11] Ceranoglu, A.N. and Pao, Y.H., "Propagation of Elastic Pulses and Acoustic Emission in a Plate: Part I. Theory; Part II. Epicentral Response; Part III. General Responses", *ASME J. Appl. Mech.*, 48, 125-147 (1981).

[12] Michaels, J.E., Michaels, T.E. and Sachse, W., "Applications of Deconvolution to Acoustic Emission Signal Analysis", *Materials Evaluation*, 39 (11), 1032-1036 (1981).

[13] Aki, K. and Richards, P., *Quantitative Seismology*, Vol. 1, W.H. Freeman and Co., San Francisco (1980), Chapter 3.

[14] Chen, C.P., "Acoustic Emission Source Characterization of Fatigue Cracks in 7075-T6 Aluminum Alloy Thin Plate", Ph.D. Dissertation, Cornell University, Ithaca, NY (1985).

[15] Scruby, C.B., "Quantitative Acoustic Emission Techniques", *Research Techniques in Nondestructive Testing*, Vol. 8, R.S. Sharpe, Ed., Academic Press (1985), 141-210.

[16] Michaels, J.E. and Pao, Y.H., "The Inverse Source Problem for an Oblique Force on an Elastic Plate", *J. Acoust. Soc. Am.*, 77 (6), Jun. 1985, pp. 2005-2011.

[17] Chang, C. and Sachse, W., "Frequency Division Method for AE Source Characterization", *J. Acoust. Soc. Am., 75* (5), May 1986, pp. 1307-1316.

[18] Hsu, N.N., Simmons, J.A. and Hardy, S.C., "An Approach to Acoustic Emission Signal Analysis - Theory and Experiment", *Materials Evaluation*, 35, No. 10, pp. 100-106, 1977.

[19] Savage, J.C. and Mansinha, L., "Radiation from a Tensile Fracture", *J. of Geophys. Res.*, 68, 6345-6358 (1963).

[20] Kim, K.Y. and Sachse, W., "Characteristics of an Acoustic Emission Source from a Thermal Crack in Glass", *Intl. Journal of Fracture.* (In Press).

[21] Dewhurst, R.J., Hutchins, D.A., Palmer, S.B. and Scruby, C.B., "Quantitative Measurements of Laser-Generated Acoustic Waveforms", *J. Appl. Phys.*, 4064-4071 (1982).

[22] Sachse, W., "Characteristics of Spark-generated Ultrasonic Signals", MSC Report 5149, Materials Science Center, Cornell University, Ithaca, NY (November, 1983).

[23] Pao, Y.H. and Sachse, W., "On the Determination of Phase and Group Velocities of Dispersive Waves in Solids", *J. Appl. Phys.*, 48, 4320-4327 (1978).

[24] Weaver, R.L., "Frequency Dependence of Generalized Ray Arrivals at Epicenter in a Viscoelastic Plate". (In preparation).

[25] Knopoff, L., "Surface Motions of a Thick Plate", *J. Appl. Phys.*, 24, 661-670 (1958).

[26] Chang, C. and Sachse, W., "Separation of Spatial and Temporal Effects in an Ultrasonic Transducer", in *Review of Progress in Quantitative Non-destructive Evaluation*, Vol. 5A D.O. Thompson and D.E. Chimenti, Eds., Plenum Press, New York (1986), pp. 139-143.

[27] Chang, C. and Sachse, W., "Analysis of Elastic Waves Signals from an Extended Source in a Plate", *J. Acoust. Soc. Am.*, 77 (4), 1335-1341 (1985).

DEFECT CHARACTERIZATION BY SCATTERING METHODS

FLAW CHARACTERIZATION BY ULTRASONIC SCATTERING METHODS

J.D. Achenbach
The Technological Institute
Northwestern University
Evanston, Illinois

Abstract

Some aspects of flaw characterization by ultrasonic scattering methods are briefly reviewed. Recent results on the effects of crack face interactions and anisotropy are indicated. A frequency-domain inverse method which yields the crack-opening volume is discussed. The last section comments on the computation of relevant stress intensity factors directly from processed ultrasonic measurements.

1. Introduction

Analytical modeling of ultrasonic wave scattering provides valuable quantitative information for methods to detect and characterize cracks. Even though both the geometrical configuration and the process of ultrasonic wave propagation must be simplified to accommodate an analytical approach, the characteristic features of the scattering phenomenon can be maintained. Results obtained by the modeling approach aid in the design of efficient testing configurations, as well as in the interpretation of experimental results.

In the frequency domain, the exact formulation of elastodynamic scattering problems results in the statement of a set of singular integral equations. For scattering by cracks, the unknown quantity in the integral equations is the crack-opening displacement. Generally, the integral equations must be solved numerically. Two methods can be used to obtain the set of singular integral equations: Fourier transform techniques and Green's function methods. For complicated configurations only the Green's function method provides an effective approach.

Section 2 of this paper presents a brief summary of relevant elastodynamic theory. A well-known representation integral expresses the scattered field in terms of the crack-opening displacement. An expression for the electrical signal strength in terms of the crack opening displacement and a traction term integrated over the surface area of the crack is also included. Recent results for the direct scattering problem are briefly reviewed in Section 3.

Most of the available theoretical work has concentrated on the perfect mathematical crack. Deviations from this idealized case, such as effects of crack face interactions have been discussed in Section 4. Section 5 presents some recent results for reflection by a flaw plane of periodically spaced microcracks. Another potentially important effect, that of material anisotropy, has been discussed in Section 7.

Whereas a substantial body of literature has become available for the direct problem of scattering by cracks, relatively little has been published on the inverse problem. A method based on inverse time-domain ray tracing, which has yielded some interesting results, is discussed in Section 8. The rudiments of a novel inverse method to characterize cracks from data in the frequency domain have been discussed in Section 9.

An ultimate goal of quantitative non-destructive evaluation is to obtain information on the residual strength of materials and components. Information on the location, size, shape and orientation of cracks provides input for the calculation of critical strength parameters such as the stress-intensity factor. Section 10 briefly comments on the computation of relevant stress intensity factors directly from processed ultrasonic measurements.

Recent review papers on ultrasonic QNDE which include substantial discussions of scattering of ultrasonic waves by cracks are those of Fu [1] and Thompson [2]. Interesting practical applications have been discussed by Coffey and Chapman [3].

2. Theory

Two fundamental results of steady-state time-harmonic elastodynamic theory are needed to develop the equations which govern the scattering of time-harmonic waves by inhomogeneities in elastic solids. They are the basic singular solution and the reciprocal identity. The basic singular solution, usually referred to as the Green's vector, is (see e.g. [4])

$$\mu u_{i;k}^{G}(\mathbf{x}-\mathbf{X})=\frac{1}{k_T^2}\frac{\partial^2}{\partial x_i \partial x_k}[-G(k_L R)+G(k_T R)]+G(k_T R)\delta_{ik}\,, \qquad (2.1)$$

where L and T refer to longitudinal and transverse waves, respectively, and

$$G(k_\alpha R)=\frac{\exp(ik_\alpha R)}{4\pi R}\ , \quad \alpha=L,T \tag{2.2}$$

$$k_L=\frac{\omega}{c_L}\ , \quad c_L^2=(\lambda+2\mu)/\rho\ , \tag{2.3a,b}$$

$$k_T=\frac{\omega}{c_T}\ , \quad c_T^2=\mu/\rho\ , \tag{2.4a,b}$$

and

$$R=|\mathbf{x}-\mathbf{X}|\ . \tag{2.5}$$

In the usual manner the term $exp(-i\omega t)$ has been omitted. Equation (2.1) represents the displacement component at position $\mathbf{x}$ and in direction x_i, due to a unit time-harmonic force in direction x_k, applied at position $\mathbf{X}$.

The reciprocal identity relates two time-harmonic elastodynamic states of the same frequency, which are labeled by the superscripts A and B, in the following manner, see [4],

$$\int_V (f_i^B u_i^A - f_i^A u_i^B)dV=\int_S (T_i^A u_i^B - T_i^B u_i^A)dA \tag{2.6}$$

Here V and S represent the volume and the bounding surface(s) of the body, and f_i and T_i are the components of body forces and surface tractions.

Now let us consider scattering of an incident wave by a crack. The components of the incident field are $u_i^{in}(\mathbf{x})$. The scattered field is defined as the difference between the total field and the incident field:

$$u_i(\mathbf{x})=u_i^{in}(\mathbf{x})+u_i^{sc}(\mathbf{x})\ . \tag{2.7}$$

The components of the corresponding stress tensors are related by

$$\tau_{ij}(\mathbf{x})=\tau_{ij}^{in}(\mathbf{x})+\tau_{ij}^{sc}(\mathbf{x}) \tag{2.8}$$

Usually the crack is assumed to be free of surface tractions; the conditions on the crack faces A^+ and A^- then are

$$\text{on } A^+ \text{ and } A^-:\ \tau_{ij}^{sc}n_j=-\tau_{ij}^{in}n_j \tag{2.9}$$

where n_j are the components of the normals to the crack faces. An application of the reciprocal identity (2.6), where solution A is taken as the basic singular solution and solution B is the scattered field, yields the following well-known representation for the scattered field

$$u_k^{sc}(\mathbf{x})=-\int_{A^+}\tau_{ij;k}^{G}(\mathbf{x}-\mathbf{X})[u_i^{sc}(\mathbf{X})]n_j^+\,dA(\mathbf{X}) \tag{2.10}$$

Here $\mathbf{n}^+$ is directed from A^- to A^+ and $\tau_{ij;k}^G(\mathbf{x}-\mathbf{X})$ is the stress corresponding to the basic singular solution given by Equation (2.1), while $[\mathbf{u}^{sc}(\mathbf{X})]$ defines the crack-opening displacement

$$[\mathbf{u}^{sc}(\mathbf{X})] = \mathbf{u}^{sc}(\mathbf{X}+\varepsilon\mathbf{n}) - u^{sc}(\mathbf{X}-\varepsilon\mathbf{n}) \tag{2.11}$$

Equation (2.10) is an exact representation of the scattered field.

For quantitative non-destructive evaluation, an alternative reciprocal relation has been derived by Auld [5]. Consider a two transducer system, where transducer I, with power P, produces the incident field, and transducer II is the receiver. The ratio of received electrical signal strength over incident signal strength is denoted by Γ. Auld's formula gives the change of Γ due to scattering by an imperfection. For a crack we have

$$\delta\Gamma = [(E_{II})_{crack} - (E_{II})_{no\ crack}]/E_I = \frac{-i\omega}{4P}\int_{A^+}\tau_{ij}^{II}[u_i^I]n_j\,dA \tag{2.12}$$

In Equation (2.12), E_I and E_{II} are the strengths of the electrical signals in transducers I and II, respectively; τ_{ij}^{II} is the stress induced by transducer II in the absence of the crack, and $[u_i^I]$ is the crack-opening displacement generated by the exciting transducer I.

Equation (2.12) has considerable appeal since it directly relates the scattered field to an electrical signal. On the other hand, Equation (2.10) may be used equally well, particularly if motion is measured by other methods, for example, laser interferometry. It should, however, be noted that in both cases the unknown quantity is the crack-opening displacement. Hence, in the direct problem the COD must first be computed, while in the inverse problem the COD can conceivably be the result of the inversion procedure. Considerable numerical effort has been directed towards the calculation of COD's for the direct problem. The inverse problem in the frequency domain is still in an embryonal stage.

3. The Direct Problem

The stresses corresponding to Equation (2.10) can be obtained by substituting $u_k^{sc}(\mathbf{x})$ into Hooke's law. Then by letting $\mathbf{x}$ approach A^+, the stresses should satisfy (2.9). The resulting equations are integral equations for $[u_i^{sc}(\mathbf{x})]$. These equations require careful handling, because the integrands contain singularities. The general system of equations has been presented by Budiansky and Rice [6].

For simple geometries, such as penny-shaped cracks and two-dimensional slits, it is convenient to use Fourier transform techniques to derive a governing system of dual integral equations. These kinds of equations can be reduced to a single singular integral equation, which can then be solved numerically [4].

In recent years numerous results have become available for fields generated by scattering of ultrasonic waves by cracks. Solutions for two-dimensional configurations in unbounded bodies have been discussed by Achenbach, et al. [4], who also listed earlier references. The three-dimensional case of scattering of a plane wave by a penny-shaped crack has been investigated by Krenk and Schmidt [7], and Martin [8]. The more difficult configurations of a surface-breaking and a subsurface crack have been considered by Mendelsohn et al. [9], and Achenbach and

Brind [10], for the in-plane problem. Three-dimensional scattering by a surface-breaking crack has been analyzed by Angel and Achenbach [11]. For the surface breaking crack comparisons between the analytical results of [9] and [11] and experimental data have been presented by Yew et al. [12], Dong and Adler [13] and Vu and Kinra [14]. Experimental results for reflection of a surface wave by a sub-surface crack oriented normal to the free surface were obtained by Khuri-Yakub et al. [5], who observed very satisfactory agreement with the theoretical results of reference [10]. There are of course also numerical studies, principally by the T-matrix method. Typical of these is the work of Visscher [16].

In recent work [17], analytical modeling has been improved by including a number of aspects typical of practical configurations. Thus, it was taken into account that the transmitting and receiving transducers are often coupled to the specimen by placement in a water bath. The transmission of the signal from the water to the solid requires careful investigation. It was also taken into account that only part of the crack may be illuminated by the incident signal. The relation between the scattered field and the electrical signal recorded by the receiving transducer was made part of the overall approach by directly relating this signal to the crack-opening displacement which is generated by the incident wave. The crack-opening-displacement was computed by the use of the frequency-domain methods developed earlier. The process of modeling from the initial electrical pulse to the measured electrical pulse included conversion to the time-domain by the use of the fast Fourier transform method.

The complexity of the exact method of solution has stimulated the development of approximate approaches. In two well-known approximate theories, which are valid at low and high frequencies, respectively, a form of $[u^{sc}(\mathbf{X})]$ is postulated and $\mathbf{u}^{sc}(\mathbf{x})$ is subsequently computed by (2.10). At low frequencies the static crack-opening-displacement can be substituted in the integral, to give the so-called quasi-static scattering theory. At very high frequencies the geometrical elastodynamic field on the insonified crack face can be used as the crack-opening displacement. Subsequent evaluation of (2.10) produces the "physical elastodynamics" or Kirchhoff approximation to the scattered field.

At high frequencies a totally different approximation is presented by ray theory. In ray theory it is assumed that disturbances propagate along straight or curved rays, and that the interaction of rays with inhomogeneities follows simple geometrical rules which can be established on the basis of solutions to canonical problems. If the rules are known, then rays can be traced and (in principle) the signals that propagate along all rays passing through a point of observation can be superimposed to yield the complete field. The geometrical aspects of ray theory have intuitive appeal, and they are relatively simple. From the mathematical point of view, ray theory gives an expansion which has asymptotic validity with respect to "high frequency" or "small time after arrival of a disturbance".

The scattering of a bundle of rays by a crack-like flaw follows relatively simple rules. At sufficiently high frequencies, diffraction at certain points on the crack edge, which have been called the "flash points", produces the dominant part of the scattered field. The flash points emit bundles of diffracted waves which propagate towards a point of observation. The basic theory has been presented in reference [4].

4. Crack Face Interactions

The results of the preceding sections are for a perfect mathematical crack, that is, an infinitesimally thin crack, with smooth faces. The effects of crack face interaction are ignored in considerations of the perfect mathematical crack. However, the failure processes that result in a crack generally produce rough crack faces. Once a crack has been opened and the crack faces have undergone the slightest relative sliding displacement, it must be expected that the crack will never completely close again due to incompatibility of the rough crack faces. Under subsequent loading conditions the faces of the crack generally may not be free of surface tractions, as is assumed for a perfect mathematical crack, nor will there be perfect contact between the crack faces. Unless the crack faces are completely separated due to a state of prestress, a complicated interaction between the crack faces is to be expected, which will be different for opening and closing of the crack on the one hand, and relative sliding of the crack faces on the other. For a slightly closed crack, the interaction of the crack faces will be a non-linear process which will depend strongly on the magnitudes of the tractions transmitted across the contacting crack faces.

The effects of interaction between contacting crack faces are of particular interest in studies of crack detection and crack characterization by the use of the specular reflection and scattering of ultrasonic waves. A perfect mathematical crack acts as a perfect screen for reflection and scattering. A crack with interacting crack faces may be a poor reflector, and thus difficult to detect and characterize. In recent work [18] it has been assumed that for the purpose of computing the fields of stress and deformation elsewhere in the body, the interaction between the upper and lower faces of the flaw plane can be described by appropriate relations between the tractions and displacement across a perfectly flat surface. This surface may be considered as the median plane of the actual flaw surface. Let us consider a two-dimensional configuration with a flaw plane in the plane $y=0$, as shown in Figure 1. In the analytical model we consider averaged tractions and averaged displacement discontinuities per unit area, with respect to coordinates in the flaw plane. The averaged tractions are continuous, which implies that at $y=0$:

$$\sigma_y^+ = \sigma_y^- = \sigma_y^* \ , \ \sigma_{yx}^+ = \sigma_{yx}^- = \sigma_{yz}^* \ , \tag{4.1a,b}$$

where the + and - signs refer to the upper and lower sides of the flaw plane. It should be noted that (4.1a,b) include the conditions for a perfect mathematical crack, which are that the stresses vanish. In the present model for a flaw plane with rough faces, (4.1a,b) will be supplemented by relations between the stresses σ_y^*,

σ^*_{yx}, and the displacement discontinuities $[v]$ and $[u]$, respectively.

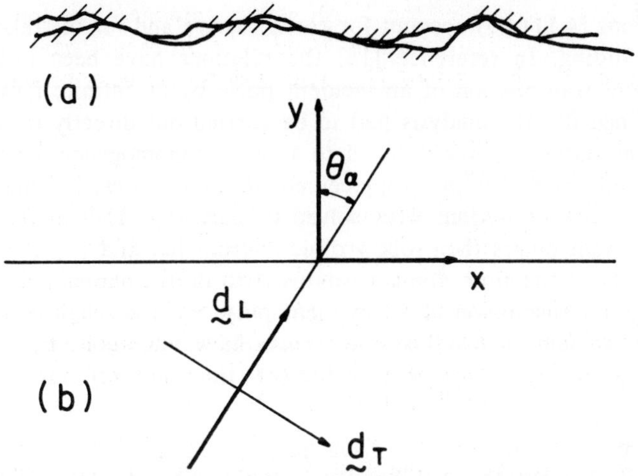

Figure 1: Schematic depiction of a flaw plane

We will first consider the opening mode of the flaw plane. It is reasonable to assume that in the unloaded state, i.e., when $\sigma^*_y=0$, the faces will be slightly separated: $[v]=\Delta>0$. In the closing mode we have $\sigma^*_y<0$, and the required stress will increase rapidly as $[v]\to 0$. In fact we assume that an infinite interface stress σ^*_y is required to close the flaw plane completely $[v]=0$, i.e., to completely flatten out the roughness of the faces. The opening displacement cannot be negative since that would imply overlap of the faces. To separate the faces, $[v]>\Delta$, a slight resistance has to be overcome. The behavior described here can be represented by the relation

$$\sigma^*_y = T\frac{[v]-\Delta}{[v]} \tag{4.2}$$

Here T is the maximum tensile force (very small) that can be transmitted across the faces. Equation (4.2) represents a nonlinear spring. It takes into account different behavior for opening and closing.

We have assumed that separation of the flaw-plane faces is independent of sliding of the faces. The opposite can, however, not be assumed. The resistance to sliding depends very much on the magnitude of σ^*_y. When $\sigma^*_y>0$ there will be very little resistance. A convenient relation between σ^*_{yx} and $[u]$ is that we require $|\sigma^*_{yx}|\leq S$, and

$$[\dot{u}]=\dot{\sigma}^*_{yx}/C \text{ for } |\sigma^*_{yx}|<S\ ,\ C=\text{ constant}\ ,\ S=S_0\exp(-\alpha\sigma^*_y). \tag{4.3}$$

Note that the critical value S depends on σ^*_y. The relation that we have assumed satisfies the condition that S is small for $\sigma^*_y>0$, while S increases rapidly as σ^*_y becomes negative, i.e., as the faces are pressed together. The key point to observe is the possibility of displacement hysteresis. The parameters T, Δ, C, α, and S_0 have to be determined from the crack geometry and from experimental data.

Equations (4.1)-(4.3) account for crack closure and for the related resistance to crack-face sliding. In reference [18] the relations have been used to investigate reflection and transmission of an incident pulse by an infinite flaw plane. Because of the nonlinearity, the analysis had to be carried out directly in the time domain. The problem statement was reduced to a set of inhomogeneous nonlinear ordinary differential equations for the displacement discontinuities, $[u]$ and $[v]$, across the flaw plane. These equations were solved numerically. The reflected and transmitted displacement pulses then follow directly from $[u]$ and $[v]$. Next the Kirchhoff approximation in the time domain was used to derive expressions for the specular reflection and transmission of an incident pulse by a crack with interacting crack faces. Both incident longitudinal and transverse waves were considered. The loss of specular reflection as compared to a perfect (traction-free) crack was exhibited by specific example.

In Figures 2a and 2b we have plotted the far-field amplitudes for wave incidence on a crack. The discontinuity in U_L is related to the rapid closure of the crack. The dashed curves in Figure 2 are the same amplitudes for the stress free crack. Also plotted in Figures 2c and 2d are the magnitudes of the Fourier transforms of U_L and U_T.

5. Flaw Plane of Periodically Spaced Microcracks

In reference [19], the interaction of elastic waves with a planar array of periodically spaced cracks of equal length was investigated. Normal incidence of both longitudinal and transverse waves was considered. The geometry is shown in Figure 3. As some distance from the plane of the cracks, the displacement fields appear as the superposition of a finite number of reflected and transmitted plane-wave propagating modes. The reflection and transmission coefficients corresponding to the propagating mode of order zero were computed for various values of the frequency and the ratio of the crack length to the crack spacing. The results are comparable to those obtained for scattering of elastic waves by surfaces with periodic roughness. The results of reference [19] can also be brought into perspective with investigations for the reflection of elastic waves by a flaw plane, when the flaw plane is viewed as a region where separation zones and contact zones can both exist, and friction may be significant over the contact zones. As discussed in Section 4, Achenbach and Norris [18] have proposed a set of non-linear boundary conditions to account for the separation and friction effects. Other boundary conditions, of a linear type, are discussed by Thompson and Fiedler [20]. In reference [19] it is assumed that the separation zones are not affected by the action of the incident wave. The displacements as well as the stresses are continuous over the plane regions in between the cracks. Hence, the configuration models an interface with predetermined separation zones and very rough contact zones.

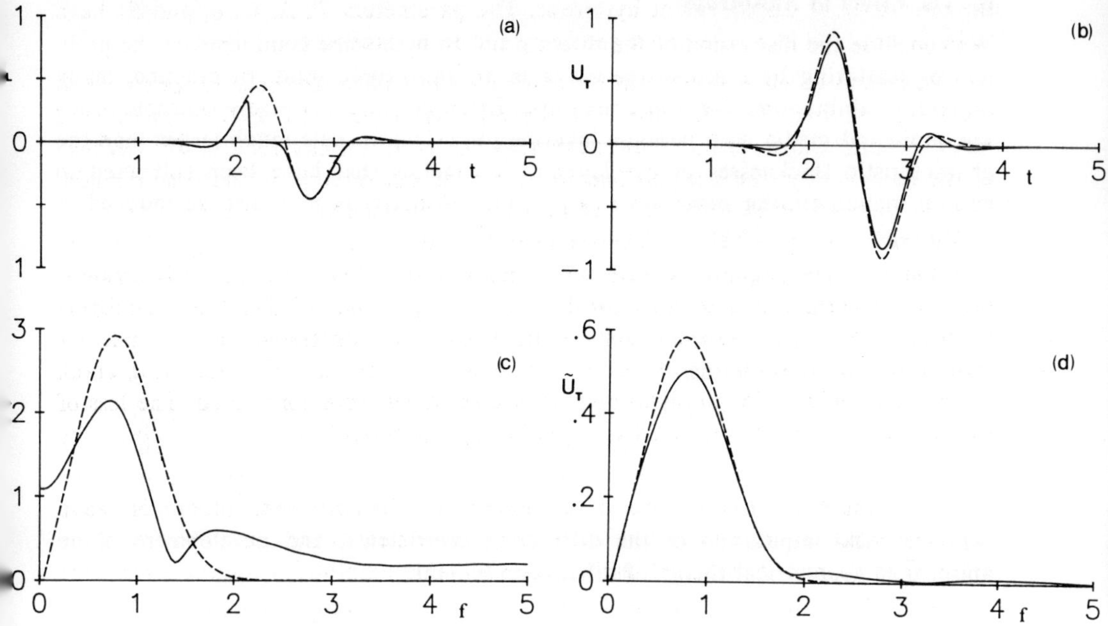

Figure 2: Far-field amplitudes for normally incident L- and T-waves, (a) and (b), respectively; (c) and (d): spectra of (a) and (b).

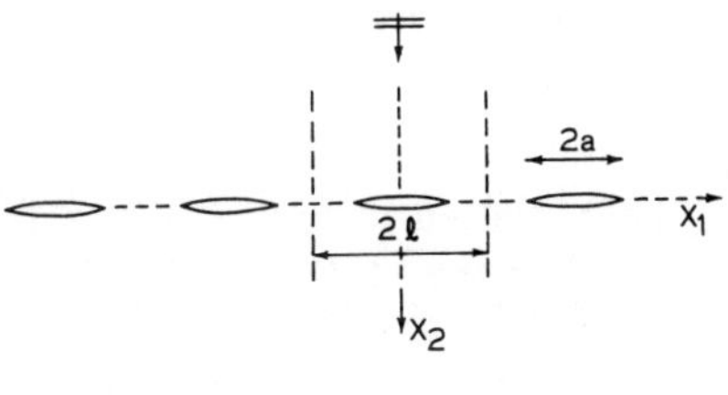

Figure 3: Array of cracks

Curves for the reflection coefficients versus $2\ell/\lambda_T = \omega\ell/\pi c_T$ are shown in reference [19]. In reference [19] the exact theory has also been compared with two approximate theories. The first theory is a static limit to the dynamic theory. The second one is inspired by Thompson and Fiedler's paper [20], where the array of cracks is replaced by an equivalent spring-layer. Both approximations agree with the exact theory at low frequencies.

6. The Effect of Anisotropy

We conclude the discussion of the direct problem with some comments on the problem of scattering by a stress-free crack in an anisotropic solid. In practice, many materials exhibit some form of anisotropy. Examples of transversely isotropic solids are laminated media with periodic layering (on a length scale much larger than the characteristic thicknesses of the layers), and metals that have been subjected to certain manufacturing processes (e.g., rolling). Anisotropy may also be induced in an isotropic solid by large prestressing or straining.

The number of known solutions for scattering problems in anisotropic solids is much smaller than for isotropic elasticity. The present discussion, which is based on reference [21], is concerned only with the direct scattering from cracks in transversely isotropic solids.

We first consider a semi-infinite crack located in a plane of symmetry of a transversely isotropic material. The incident wave motion is confined to the plane which is normal to the crack edge. Therefore, the scattering geometry is two dimensional. This problem can be viewed as a canonical problem whose solution is necessary for the computation of the diffraction coefficients, and development of an anisotropic geometrical theory of diffraction (GTD).

Now, let a plane quasi-longitudinal wave of unit amplitude be incident at angle θ with the plane of the crack, on a crack of finite length D which is located in a plane of symmetry. The diffracted quasi-longitudinal field in the back scattered direction can then be computed by using the diffraction coefficient for each tip of the crack individually. Superposition of the two crack-tip diffractions gives a high-frequency approximation to the back scattered or pulse-echo signal. It is assumed that D is of the same order as the wavelength.

Taking the origin at the center of the crack, the far field amplitude $A \equiv |(r/D)^{1/2}\mathbf{u}(r,\theta)|$ has been plotted in Figure 4 as a function of the dimensionless frequency $D\omega/\sqrt{b}$. We note that $\omega/\sqrt{b}$ is the longitudinal wave number in the direction normal to the slit ($\theta=\pi/2$). The isotropic result ($a/d=4$) is also plotted in Figure 4 for comparison. Because b is the same for the two examples, the effect of anisotropy disappears at normal incidence. The parameters a, b and d are defined in reference [21]. Figure 4 shows that the anisotropy is important, especially at angles of incidence near grazing.

7. Inverse Problem – Ray Methods

In the direct problem the incident wave and the geometrical configuration are known. For a given point of observation the positions of the flash points on the crack edge can then be determined by geometrical considerations, and the scattered field can subsequently be determined by direct ray tracing. If the geometrical configuration is unknown, but information is available on the diffracted field, an inverse ray tracing procedure can be attempted to determine the flash points on the crack edge from which diffracted signals have emanated.

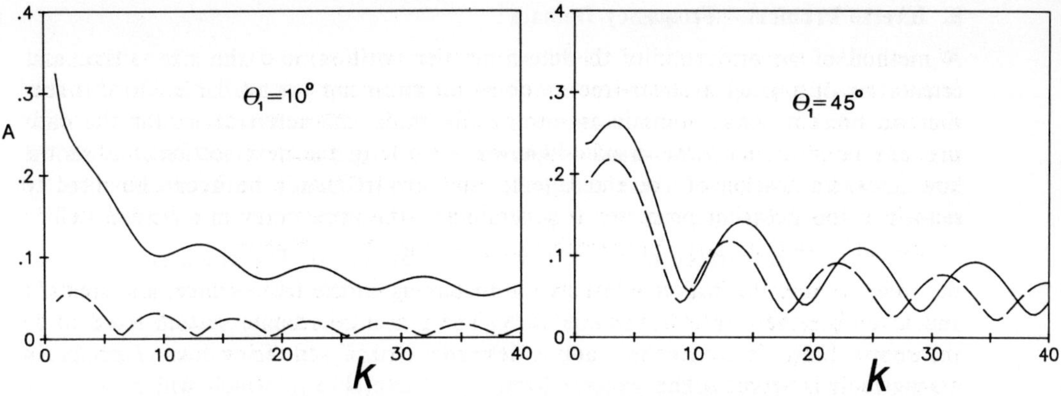

Figure 4: Backscattered amplitude from a finite crack. Dashed line is the isotropic result; $k=D\omega/\sqrt{b}$.

In recent papers [22-24] two analytical methods have been developed to map the edge of a crack by the use of data for diffraction of elastic waves by the crack-edge. These methods are based on elastodynamic ray theory and the geometrical theory of diffraction, and they require as input data the arrival times of diffracted ultrasonic signals. The first method maps flash points on the crack edge by a process of triangulation with the source and receiver as given vertices of the triangle. By the use of arrival times at neighboring positions of the source and/or the receiver the directions of signal propagation, which determine the triangle, can be computed. This inverse mapping is global in the sense that no *apriori* knowledge of the location of the crack edge is necessary. The second method is a local edge mapping which determines planes relative to a known point close to the crack edge. Each plane contains a flash point. The envelope of the planes maps an approximation to the crack edge. The material containing the crack was taken as a homogeneous, isotropic, and linearly elastic solid. More recently, extensions to include anisotropy of the material have been given in reference [23]. Mathematical details and a fairly detailed error analysis can be found in reference [22]. References [22-24] also include applications of the methods to synthetic data. It is of particular interest that the local mapping technique allows for an iteration procedure whereby the result of a computation suggests an improved choice of the base point which in the subsequent iteration yields a better approximation to the crack edge. A comparison with experimental data has been given in reference [24].

8. Inverse Problem – Frequency Domain

A method of general validity to determine the position, and the size, shape and orientation of a crack from frequency-domain data is not yet available. In this section we present some preliminary results for crack characterization, for the case that the orientation of the crack is known *apriori*. In the next section it is shown how an extrapolation of these results to the zero-frequency limit can be directly related to the stress-intensity factor generated by the same crack in a tension field.

The flat free surface of a body is defined by $x_2=0$. The crack is parallel to the free surface. We will consider four transducers on the free surface: a source (I) and three receivers (II, III and IV). Relative to a coordinate system fixed to I, the approximate center of the crack is defined by $(0, \bar{x}_2, 0)$. Transducer I produces a longitudinal wave of the general form $u_2=U\exp(ik_Lx_2)$, which will generate a crack opening displacement $[u_2^I]$ only. Using Auld's formula, Equation (2.12), we obtain for the pulse-echo configuration

$$\delta\Gamma_1=C_IV^{(o)}\,, \tag{9.1}$$

where

$$C_I=\frac{\omega^2}{4P}\frac{U}{c_L}(\lambda+2\mu)e^{ik_L\bar{x}_2}\,, \tag{9.2}$$

and $V^{(o)}$ is the crack opening volume:

$$V^{(o)}=\int_A[u_2^I(\xi_1,\xi_2)]d\xi_1d\xi_2 \tag{9.3}$$

In (9.3) the origin of the local coordinates ξ_1 and ξ_2, which are in the plane of the crack, is taken at the geometrical center of the crack.

Next, we consider a pitch-catch configuration with transducer I as the source, and transducer II as the receiver. Suppose relative to transducer II, the center of the crack is defined by $(\bar{x}_1,\bar{x}_2,\bar{x}_3)$, where $\bar{x}_2$ is the same unknown as above. The stress τ_{22}^{II}, which is the only stress component that will enter Equation (2.12), is then of the form

$$\tau_{22}^{II}=(\lambda+2\mu d_2p_2)ik_LUe^{ik_L\bar{\mathbf{x}}\cdot\mathbf{p}} \tag{9.4}$$

where

$$p_i=d_i=\bar{x}_i/r \tag{9.5a,b}$$

$$r=(\bar{x}_i\bar{x}_i)^{1/2} \tag{9.6}$$

Equation (9.4) may be rewritten as

$$\tau_{22}^{II}=(\lambda+2\mu d_2p_2)ik_LUe^{ik_Lr}e^{ik_L(\xi_1p_1+\xi_2p_2)}$$

This expression can subsequently be substituted in Equation (2.12), to yield $\delta\Gamma_{II}$ in terms of integrals of the form

$$I=\int_A e^{ik_L(\xi_1 p_1+\xi_2 p_2)}[u_2^I(\xi_1,\xi_2)]d\xi_1 d\xi_2 \tag{9.8}$$

A useable inverse method can be obtained if it can be assumed that

$$k_L\xi_1 p_1 \ll 1 \quad , \quad k_L\xi_2 p_2 \ll 1 \tag{9.9}$$

In that case the exponentials may be expanded, and if two terms are retained only, we find

$$\delta\Gamma_{II}=C_{II}V^{(o)}+ik_L C_{II}(p_1 V^{(11)}+p_2 V^{(12)}) \tag{9.10}$$

where $V^{(o)}$ is defined by (9.3), and

$$C_{II}=\frac{\omega^2}{4P}\frac{U}{c_L}(\lambda+2\mu d_{2}p_2)e^{ik_L r} \tag{9.11}$$

$$V^{(11)}=\int_A \xi_1[u_2^I(\xi_1,\xi_2)]d\xi_1 d\xi_2 \tag{9.12}$$

$$V^{(12)}=\int_A \xi_2[u_2^I(\xi_1,\xi_2)]d\xi_1 d\xi_2 \tag{9.13}$$

Equation (9.1) corresponds to two equations (real and imaginary parts) for the three unknowns: $\bar{x}_2$ and the real and imaginary parts of $V^{(o)}$. Similarly Equation (9.10) corresponds to two equations for seven unknowns: $\bar{x}_2$ and the real and imaginary parts of $V^{(o)}$, $V^{(11)}$ and $V^{(12)}$. Together (9.1) and (9.10) yield four equations for seven unknowns. It is, however, apparent that another 2 receivers on the free surface will produce another set of four equations of the form (9.10) without the introduction of additional unknowns. Hence transducers *I*, *II*, *III* and *IV* will produce eight equations for seven unknowns. A general method to solve such an overdetermined set of equations can be reduced to a nonlinear least-squares problem. Details for a similar calculation have been discussed by Achenbach, Sotiropoulos and Zhu [25].

9. The Stress-Intensity Factor

It is well known that for a crack in a tension field of magnitude σ_o, the crack opening volume is related to the stress-intensity factor by the relation

$$\int_A [u_2(\xi_1,\xi_2)]dA=\frac{2(1-\nu^2)}{3E}\sigma_o\int_S \rho_c\left(\frac{K_I}{\sigma_o}\right)^2 ds \tag{10.1}$$

where S is the edge of the crack, see e.g., reference [26]. As shown by Budiansky and Rice [27], Equation (10.1) provides the basis for a direct calculation of K_I from the zero-frequency limit of the ultrasonic measurement of the crack opening volume $V^{(o)}$. From reciprocity we have

$$\int_A [u_2(\xi_1,\xi_2)]d\xi_1 d\xi_2=\frac{\sigma_o}{U(\lambda+2\mu)}\lim_{\omega\to 0}\frac{V^{(o)}}{ik_L} \tag{10.2}$$

Substitution of (10.2) into (10.1) relates the zero-frequency limit of an ultrasonic measurement directly to the stress-intensity factor for the same crack. An approximate solution of Equation (10.1) can be found in reference [27].

Acknowledgment

The work presented here was carried out in the course of research sponsored by the Office of Naval Research (Contract N00014-85-K-0401).

References

[1] Fu, L.S., "Mechanics aspects of NDE by sound and ultrasound", *Applied Mechanics Reviews, 35*, 1982, 1047-1057 pp.

[2] Thompson, R.B., "Quantitative ultrasound nondestructive evaluation methods", *J. Appl. Mech.*, (50th Anniversary Issue), *50*, 1983, 1191-1201 pp.

[3] Coffey, J.M. and Chapman, R.K., "Application of elastic scattering theory for smooth flat cracks to the quantitative prediction of ultrasonic defect detection and sizing", *Nucl. Energy, 22*, 1983, 319-333 pp.

[4] Achenbach, J.D., Gautesen. A.K , McMaken, H., *Ray Methods for Waves in Elastic Solids – With Applications to Scattering by Cracks*, Pitman, Boston, 1982.

[5] Auld, B.A., "General electromechanical reciprocity relations applied to the calculation of elastic wave scattering coefficients", *WAVE MOTION, 1*, 1979.

[6] Budiansky, B. and Rice, J.R., "An integral equation for dynamic elastic response of an isolated 3-D crack", *WAVE MOTION, 1*, 1979, 187-192 pp.

[7] Krenk, S. and Schmidt, H., "Elastic wave scattering by a circular crack", *Phil. Trans. R. Soc. Lond. A308*, 1982, 167-198 pp.

[8] Martin, P.A., "Diffraction of elastic waves by a penny-shaped crack", *Proc. R. Soc. Lond. A378*, 1982, 262-285 pp.

[9] Mendelsohn, D.A., Achenbach, J.D. and Keer, L.M., "Scattering of elastic waves by a surface-breaking crack", *WAVE MOTION, 2*, 1980, 277-292 pp.

[10] Achenbach, J.D. and Brind, R.J., "Scattering of surface waves by a subsurface crack", *J. of Sound and Vibration, 75*, 1981, 43-56 pp.

[11] Angel, Y.C. and Achenbach, J.D., "Reflection and transmission of obliquely incident Rayleigh waves by a surface-breaking crack", *J. Acoust. Soc. Am., 75*, 1984, 313-319 pp.

[12] Yew, C.H., Chen, K.G. and Wang, D.L., "An experimental study of interaction between surface waves and a surface-breaking crack", *J. Acoust. Soc. Am., 75*, 1984, p. 189.

[13] Dong, R. and Adler, L., "Measurements of reflection and transmission coefficients of Rayleigh waves from cracks", *J. Acoust. Soc. Am., 76*, 1984, 1761-1763 pp.

[14] Vu, B.Q. and Kinra, V.K., "Diffraction of Rayleigh waves in a half-space. I. Normal edge crack", *J. Acoust. Soc. Am., 77*, 1985, 1425-1430 pp.

[15] Khuri-Yakub, B.T., Shui, Y., Kino, G., Marshall, D.B. and Evans, A.G., "Measurement of surface machining damage in ceramics", in *Review of Progress in Quantitative Nondestructive Evaluation, Vol. 3*, D.O. Thompson and D.E. Chimenti (Eds.), Plenum Press, New York, 1984, 229-238 pp.

[16] Visscher, W.M., "Elastic wave scattering by a surface-breaking or sub-surface planar crack. II. Three dimensional geometry", *J. Appl. Phys., 57*, 1985, 1538-1550 pp.

[17] Achenbach, J.D., Lin, W. and Keer, L.M., "Mathematical modeling of ultrasonic wave scattering by sub-surface cracks", *Ultrasonics, 24*, 1986, 207-215 pp.

[18] Achenbach, J.D. and Norris, A.N., "Loss of specular reflection due to nonlinear crack-face interaction", *J. of Nondestructive Evaluation, 3*, 1982, 229-239 pp.

[19] Angel, Y.C. and Achenbach, J.D., "Reflection and transmission of elastic waves by a periodic array of cracks", *J. Appl. Mech., 52*, 1985, 33-41 pp.

[20] Thompson, R.B. and Fiedler, C.J., "The effects of crack closure on ultrasonic scattering measurements", in *Review of Progress in Quantitative NDE, Vol. 3*, Plenum, New York, 1984, 207-216 pp.

[21] Norris, A.N. and Achenbach, J.D., "Elastic wave diffraction by a semi-infinite crack in a transversely isotropic material", *Q.J. Mech. and Appl. Math., 37*, 1984, 565-580 pp.

[22] Norris, A.N. and Achenbach, J.D., "Mapping of a crack edge by ultrasonic methods", *J. Acoust. Soc. Am., 72*, 1982, 264-272 pp.

[23] Norris, A.N., "Inverse ray tracing in elastic solids with unkown anisotropy", *J. Acoust. Soc. Am., 73*, 1983, p. 421.

[24] Achenbach, J.D., Norris, A.N., Ahlberg, L. and Tittmann, B.R., "Crack mapping by ray methods", *Review of Progress in Quantitative Nondestructive Evaluation, Vol. 2A*, D.O. Thompson and D.E. Chimenti, (Eds.), Plenum Press, New York, 1982, 1097-1116 pp.

[25] Achenbach, J.D., Sotiropoulos, D.A. and Zhu, H., "Characterization of cracks from ultrasonic scattering data", submitted for publication.

[26] Budiansky, B. and O'Connell, R.J., "Elastic moduli of a cracked solid", *Int. J. Solids and Structures 12*, 1976, 81-97 pp.

[27] Budiansky, B. and Rice, J.R., "On the estimation of a crack fracture parameter by long wavelength scattering", *J. Appl. Mech., 45*, 1978, 453-454 pp.

EXPERIMENTAL RESULTS OF ULTRASONIC SCATTERING FROM DISCONTINUITIES

Laszlo Adler
Department of Welding Engineering
The Ohio State University
Columbus, Ohio

Abstract

Recent developments of materials characterization by ultrasonic waves are summarized by presenting a systematic approach to discontinuity analysis.

1. Introduction

Figure 1 presents a systematic approach to ultrasonic evaluation of material structures which may be a weld or a joint, for example:

- Locate all flaws
- Characterize each flaw (determine size, shape, orientation and composition)
- Characterize the material (determine the elastic properties, grain size, surface roughness, etc.)
- Evaluate the seriousness of the flaws' presence (using fracture mechanic techniques).

In order to determine the presence (or absence) of weld discontinuities, an ultrasonic image is produced. Defects will be delineated as areas of increased ultrasonic-echo return. For flaws, large compared with the beam dimensions, the ultrasonic image shows the extent of the defect. Flaws, on the order of and smaller than the ultrasonic wavelength, scatter the incident sound beam. Some of this energy returns to the transducer and appears on the image as a weakly scattering

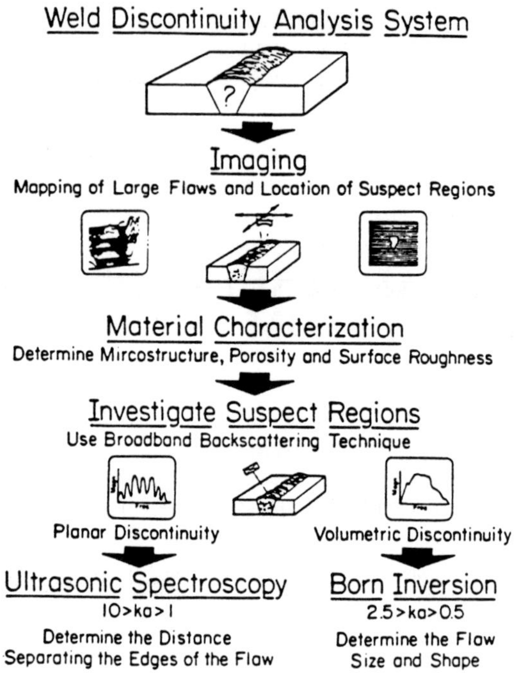

Figure 1: Systematic procedure for ultrasonic nondestructive evaluation

region in the material. Regions such as these are flagged for later investigation.

Next, the characteristics of the weld metal are measured. The grain structure and roughness of the ultrasonic beam entrance surface will have an effect on all subsequent measurements, and so must be determined. In addition, the mechanical properties of the material surrounding any flaws will profoundly affect the strength of the weld. Also, regions of the weld containing dense porosity or clouds of inclusions will lower weld strength. Determinations of the concentration and size distribution of voids or inclusions should permit calculations of estimated weld mechanical properties.

One now returns to the suspect regions in the welded structure, namely areas of increased echo return in the ultrasonic image. A broadband ultrasonic wave is directed toward the suspect region and the backscattered signal is then processed to obtain a magnitude spectrum (amplitude versus frequency). If the spectrum shows deep and periodic modulation, the suspect region contains a planar discontinuity (e.g. crack). If, however, the spectrum is relatively smooth, the flaw is volumetric (e.g. pore). Characterization of the defect requires further signal processing. Discrimination of planar from volumetric discontinuities permits the appropriate processing algorithm to be selected.

Ultrasonic spectroscopy (using cepstral analysis) is utilized for determining the size of crack-like flaws. Upon processing, the distance (along the ultrasonic beam path) is calculated which separates the near and far edges of the flaw. Interrogation of the defect from a number of angles may be used to obtain the flaw's size and shape.

Volumetric defects are characterized by "Born inversion" processing. This algorithm returns both a line-of-sight estimate of flaw radius and the cross-sectional area of the flaw (along the ultrasonic beam path). "Observation" of the defect at a number of "look angles" discloses its shape and dimensions.

Once all flaws are characterized, data tabulating:

- Coordinates of all flaws
- Shape and size of all defects
- Material elastic properties (Young's and shear moduli, Poisson's ratio)
- RMS surface roughness
- Concentration and size distribution for areas containing clouds, of porosity, or inclusions

can be output for evaluation using fracture mechanics techniques.

2. Description of System Elements

2.1. *Imaging*

The purpose of ultrasonic imaging is to locate regions in the weld that contain flaws. A focused ultrasonic beam is directed toward the weld. Changes in material elastic properties or density will cause reflection and scattering of the incident beam. A portion of this echo energy is intercepted by the receiving transducer. The receiver output is recorded with the spatial coordinates of the scattering region (Figure 2). If the entire volume of the weld is scanned, one builds up a 3-dimensional map of echo amplitude versus position.

2.2. *Display Modes*

Display of echo information is normally presented one plane at a time. The image is termed a *C-scan* (echo amplitude at constant depth) if the image plane is parallel to the sample surface (x-y plane). It is a *B-scan* (display brightness indicates echo amplitude) if any other plane (e.g. y-z, y-z or oblique) is presented. It is also possible to display the echo strength as a brightness-modulated three-dimensional isometric presentation (Figure 3). Echo amplitude is shown as display height as a function of two orthogonal spatial coordinates. Because of the ease in implementing this isometric display format, it was chosen for all our B- and C-scans.

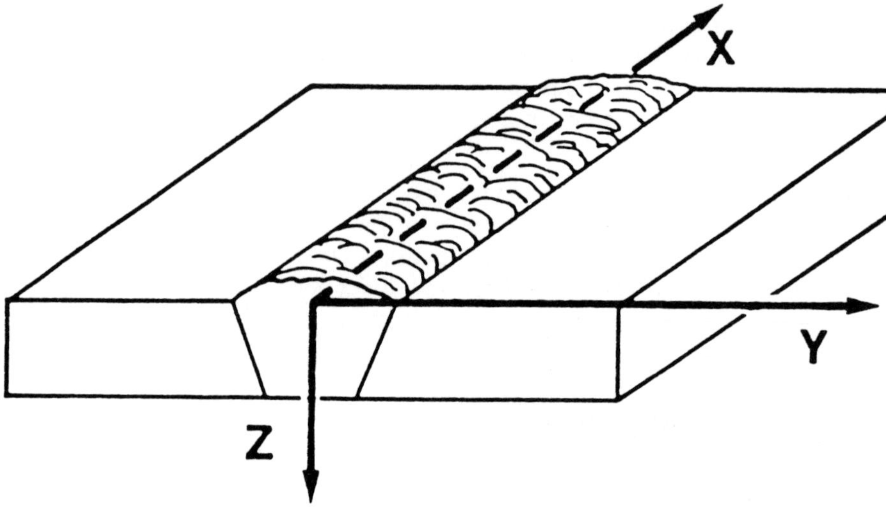

Figure 2: Assignment of coordinates to a weld

2.3. *Image Optimization*

The scattering amplitude measured for a particular element in the weld is dependent on

- the ultrasonic wave mode (longitudinal, shear, surface wave)
- the frequency of the incident wave
- the incident angle, and
- the spatial distribution of the incident beam.

Each of these factors may be adjusted such that flaw detectibility is optimized. Analytical, as well as, experimental work is needed in this area.

2.4. *Image Processing*

It is possible to accentuate the presence of flaws in an ultrasonic image by postprocessing. A method, gradient processing, for enhancing edges in an image is introduced here. Let the echo amplitude as a function of spatial position be represented by $f(x,z)_1$ (B-scan display mode). Then the gradient is defined as the vector

$$\mathbf{G}[f(x,z)] = \begin{bmatrix} \frac{\delta f}{\delta x} \\ \frac{\delta f}{\delta z} \end{bmatrix} . \tag{1}$$

The gradient vector points in the direction of the maximum rate of change of $f(x,z)$ and its magnitude gives the maximum rate of change of $f(x,z)$ per unit distance in the direction of **G**. For manipulation of the digital image, the magnitude of

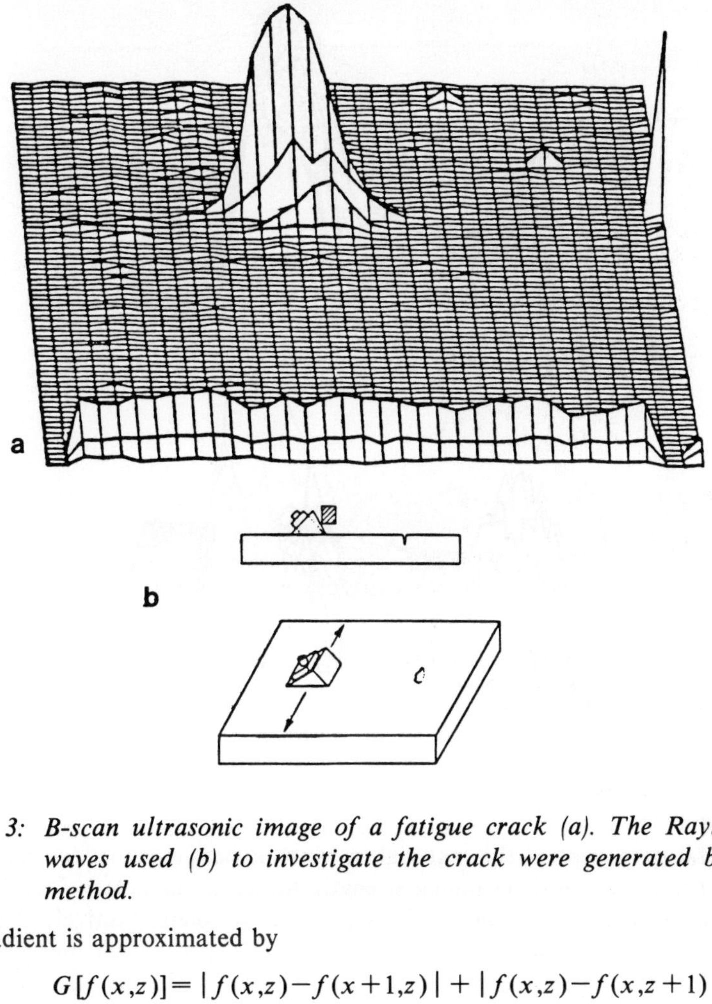

Figure 3: B-scan ultrasonic image of a fatigue crack (a). The Rayleigh surface waves used (b) to investigate the crack were generated by the wedge method.

the gradient is approximated by

$$G[f(x,z)] = |f(x,z) - f(x+1,z)| + |f(x,z) - f(x,z+1)| \; . \tag{2}$$

The gradient of the ultrasonic image is displayed if the gradient is above a threshold value. If the gradient is below threshold (presumably the case for pixels representing unflawed regions in the weld), the image pixel is set to zero. Flaws in the processed image stand out more clearly (Figure 4b) than in the original ultrasonic image (Figure 4a). Research is needed for adaptive threshold setting, other methods of image processing and automated flaw recognition schemes.

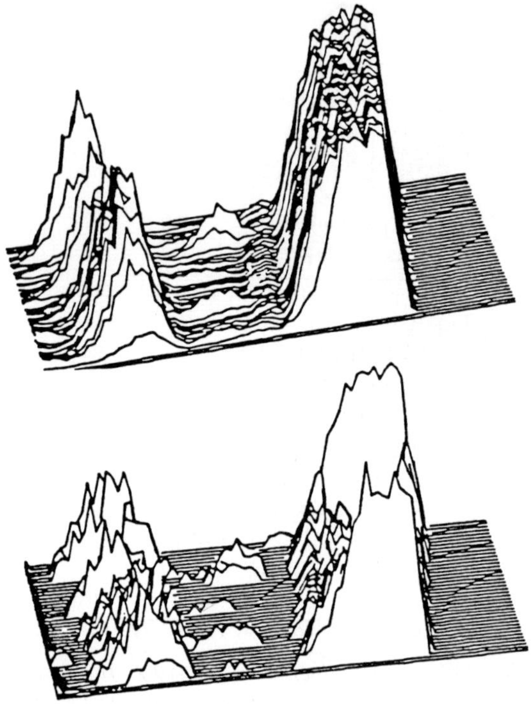

a b

Figure 4: Ultrasonic B-scan images of a specimen containing multiple defects. Image before (a) and after (b) gradient processing.

3. Materials Characterization

The mechanical properties of the material in which a flaw is embedded may be as important as defect size in determining strength. Material characteristics which are of importance are listed in Table 1. It is possible to nondestructively determine many of these properties from ultrasonic velocity and attenuation measurements. Both analytical and experimental studies have been made to determine how frequency-dependent velocity and attenuation measurements may be used to infer the concentration and size of pores occurring in dense clouds.

Inhomogeneities (such as pores and inclusions weaken the structural components in which they occur. It is important to nondestructively determine the size distribution and concentration of pores or inclusions. A multiple scattering theory was developed for treating wave propagation through inhomogeneous material. Matrix and second phase elastic properties inclusion concentration and size distribution are used as input. Dispersion and frequency-dependent attenuation are calculated using theory.

Table 1: Material Properties

Tensile Modulus
Shear Modulus
Tensile Strength
Shear Strength
Bond Strength
Hardness
Surface Finish
Impact Strength
Fracture Toughness
Anisotropy
Microstructure
Grain Size
Porosity, Void Concentration
Phase Composition
Hardening Depth
Residual Stress
Heat Treatment Profile
Fatigue Damage

3.1. *Ultrasonic Wave Propagation in Cast Iron-Graphite Composite*

The same multiple scattering treatment applied to porous media may also be used to analyze the problem of an ultrasonic wave traversing a material containing compact flake and nodular graphite, and also specimens of gray iron are studied. Because the graphite in nodular iron appears mostly as spheres, it was thought the multiple scattering theory could be used with only minor changes to include the elastic properties of the graphite nodules.

The properties of the cast iron matrix were estimated from ultrasonic velocity measurements on a sample of 1045 steel. This type of steel was chosen because it has approximately the same relative percentages of ferrite and pearlite. Table 2 lists pertinent material properties for the matrix (iron) and inclusions (graphite nodules). Elastic moduli were calculated from the velocities.

Table 2: Properties of the Constituents in Nodular Cast Iron

			Velocities (km/sec)
Material	Longitudinal	Transverse	Density (g/cm^3)
iron (1045 steel)	5.8	3.1	7.2
graphite	3.2	1.8	2.2

Values for the matrix material, iron, and the inclusions, graphite spheres were input to the multiple scattering theory to determine the phase velocity and attenuation as a function of frequency. Narrowband and broadband ultrasonic experiments were performed on carefully prepared specimens of cast iron. Figure 5 plots theoretically-determined attenuation and experimental measurements. The shape of the curves are identical; however, there is an offset. We ascribe this to incorrect assumptions of the wave speed in the graphite and to the fact we assumed a uniform size of graphite nodules, when there is actually a distribution of sizes.

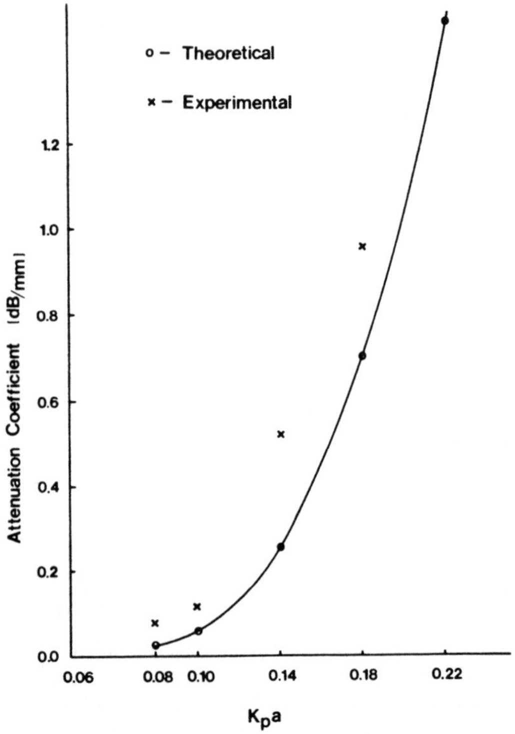

Figure 5: Attenuation of ultrasonic waves in nodular cast iron as a function of frequency (o - multiple scattering theory and × - experiment).

3.2. *Surface Roughness*

The condition of the surface through which an ultrasonic beam enters the material can have a profound effect on the frequency content of the pulse, the angle of the refracted energy and on the spatial distribution of the beam in the solid. Both imaging and defect characterization techniques will be affected by the surface finish on the weld under evaluation. Much of the unreliability of contact ultrasonic testing arises from multipath effects and trapped air due to roughness on the sample surface.

We have begun studies aimed at nondestructively inferring parameters describing surface roughness (rms roughness and correlation length for randomly-rough surfaces and periodicity and peak to valley height for periodic surfaces). Surface roughness may be measured ultrasonically from the angular or frequency dependence of wave scattering.

Ultrasonic measurement of rms roughness (h) may be carried out in the following manner. An ultrasonic transducer, operated in the pulse-echo mode, is used to record the intensity $<I>$ of backscattered ultrasonic compressional waves. The experiment is repeated for a smooth-surfaced sample of the same material (giving I_o).

If low frequencies are used, the backscattered intensity is approximated as

$$<I> \simeq I_o e^{-k^2 h^2} . \tag{3}$$

where k is the wave number of the ultrasound in the liquid bath. The roughness, h, may be estimated from

$$h \simeq \left(\frac{\Delta dB}{-17.37\, k^2} \right)^{1/2} \tag{4}$$

where ΔdB is the relative amplitude of the waves backscattered from the rough surface compared to that of the smooth surface. The advantages of the ultrasonic measurement of surface roughness are

- it is nondestructive,
- determinations may be made rapidly, and
- whereas mechanical profilometers measure roughness only along the line of stylus traverse, the ultrasonic method averages over the entire insonified area (be it small or large).

4. Defect Characterization

Ultrasonic images produced in the first stage of our evaluation disclosed the extent of large flaws and identified weakly scattering regions as areas potentially containing smaller defects. For a fracture mechanics evaluation, the size and shape of these small discontinuities is required (in addition to their location, as determined from the ultrasonic image). Flaw characteristics may be determined from ultrasonic scattering data; however, the defect must first be classified as volumetric (3-dimensional, e.g. pore or inclusion) or planar (2-dimensional, e.g. crack-like).

The classification of defect type is carried out by identifying discriminatory features in the backscattering frequency spectrum. The procedure below is followed.

1) A broadband ultrasonic pulse is directed toward the suspect region in the weld.
2) The backscattered signal from the region is sampled, digitized and stored in computer memory.
3) A system normalization signal is acquired by sending the incident pulse through an unflawed region in the material toward a perfect reflector (polished solid-air surface).
4) Perturbing effects of the data acquisition system and material intervening between the ultrasonic transducer and the defect are removed by deconvolving the defect signal by the signal from the reference reflector.
5) The magnitude spectrum is displayed.
   - The defect is classified as planar if the spectrum has deep, periodic modulation (explanation given later).
   - The defect is classified as a volumetric flaw if the spectrum is relatively smooth.
6) If there is some ambiguity in classifying the defect type (5), then the suspect region is interrogated from a number of angles.
   - A planar discontinuity will return large backscattering signals for orientations in which the incident ultrasonic beam is directed normal to the plane of the flaw.
   - Amplitude of signals returned from a volumetric discontinuity depend less strongly on angle.
7) If the discontinuity is planar, an inversion scheme termed ultrasonic spectroscopy is followed.
8) If a volumetric flaw is to be characterized, the Born inversion algorithm is utilized.

The Born inversion procedure is summarized here.

4.1. *Born Inversion*

As an ultrasonic wave strikes a volumetric flaw (Figure 6), the cross-sectional area encountered increases as the wave propagates. When the wave is just incident on the defect (time, t_1), the area encountered is minimal. As the wave moves onward (time t_2, then t_3,...) the area encountered increases (A_2,A_3,...). The cross-sectional of the flaw reaches a maximum when the wavefront reaches the center of the flaw. The area encountered decreases thereafter. The Born inversion algorithm returns the area function A_i versus t_i. From this, a line-of-sight estimate of flaw radius may be calculated.

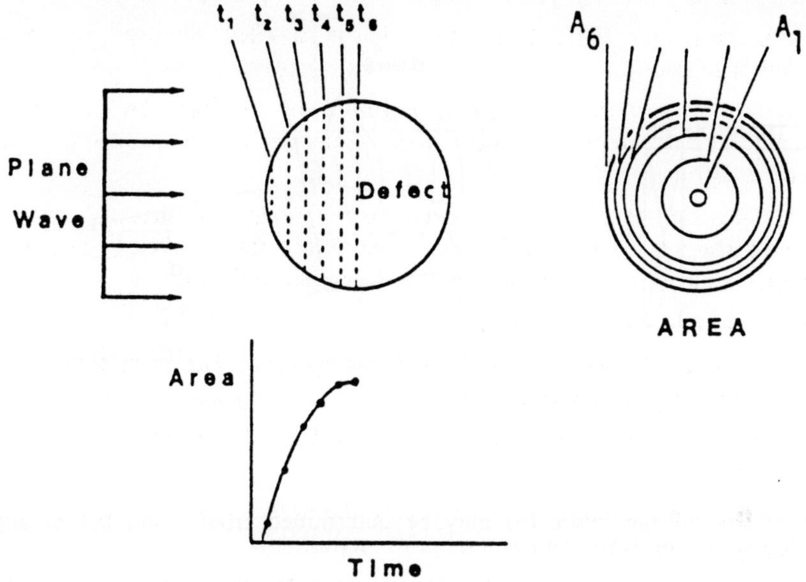

Figure 6: Plane wave encountering a scatterer. (a) position of wavefront at times t_i. (b) area of the scatterer encountered at times t_i and (c) area function.

4.2. *Ultrasonic Spectroscopy*

Ultrasonic spectroscopy applied to defect characterization was pioneered by Adler, et.al. [1,2,3]. The fundamental ideas underlying the technique are summarized here (details may be found in the book by Fitting and Adler, [4]).

An ultrasonic wave directed toward a planar discontinuity will be in part specularly reflected from it, but also scattering will occur from the near and far edges (flashpoints [5]) of the flaw (Figure 7). If the echo is transformed to the frequency domain (via the Fast Fourier Transform, FFT), the time spacing of the signals may be determined.

Consider a single signal $y(t)$ which has a spectrum $Y(2\pi f)$. The magnitude spectrum of two such signals, separated in time by $2t_o$, has been shown [6] to be $|2\cos 2\pi f t_o|\,|Y(2\pi f)|$. That is the spectrum is modulated, and the period is determined by the time separation (Figure 8). The spacing (Δf) of the frequency minima may be used to determine the time separation (Δt) of the ultrasonic echoes:

$$\Delta f = 1/2t_o = 1/\Delta t \ . \tag{5}$$

If the wave speed is known, then the separation of the near and far flaw edges may be calculated and the flaw dimensions determined from

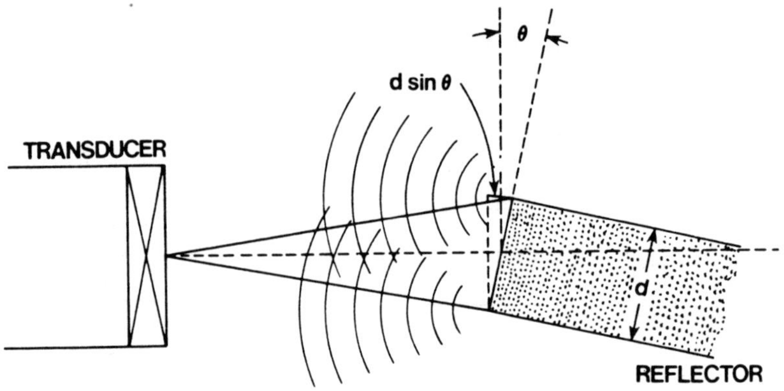

Figure 7: Ultrasonic waves scattered from the edges of a planar reflector.

$$d = \frac{c}{2\Delta f \sin\theta} \,. \tag{6}$$

The orientation of the crack (θ) may be determined from a number of angular measurements.

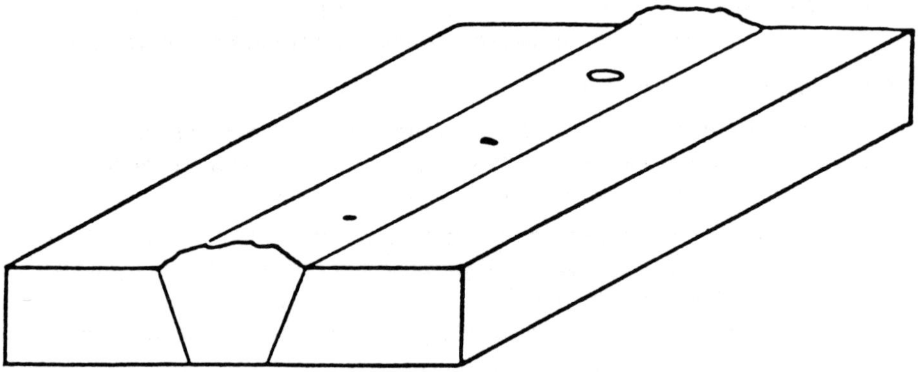

Figure 8: Weld sample #2 containing planar discontinuities 1/16″, 1/8″ and 1/4″ in diameter.

5. Examples of Ultrasonic Nondestructive Weld Evaluation

As a test of our systematic method for evaluating welded structures (Figure 1 and previous descriptions), we began with two welds containing well-characterized defects.

Weld sample #2 was carefully prepared to contain only planar discontinuities. Three artificial flaws, nominally 1/16″, 1/8″ and 1/4″ in diameter were incorporated into the weld along its center line (Figure 8). The plane of the defects was parallel to the sample surface. Both the top and bottom surfaces of the specimen were ground flat.

5.1. *Imaging*

A 5 MHz focused transducer (approximately 6″ focal distance, 1″ diameter) was scanned in a water bath above the weld. The water bath length was adjusted to focus the ultrasonic beam approximately at the depth the discontinuities were thought to occur. Spacing between transducer positions (along the weld) was 5 mm.

- The ultrasonic waveform returning from the weld is sampled, digitized and displayed (Figure 8).
- The signal is full-wave rectified (Figure 8).
- An estimate of the envelope of the waveform is determined (Figure 8).
- The number of positions along the weld where the transducer will be located is input (Figure 9).
- A file name identifying the image data to be stored is given (Figure 9) and
- The depth (time) range to be used for image display is defined (using the graphic terminal's cursor) (indicated by the vertical lines in Figure 10).

5.2. *Ultrasonic Spectroscopy*

Broadband ultrasonic pulses from the 15 MHz unfocused transducer were coupled to the weld through a 6″ water path. Backscattering signals were acquired for each flaw at 0-, 10- and 20-degrees refracted angles in the solid. The spectra computed from these signals was plotted versus frequency. The absence of deep modulation at 0-degrees indicates the plane of each flaw is approximately parallel to the surface of the weld. The average spacing of minima in the spectra was determined (Table 3) and equation (6) used to calculate the flaw diameter.

Table 3: Average spacing of minima in the backscattring spectra flaws in weld #2

ID	10-degree Measurement	20-degree Measurement
A	9 MHz	4 MHz
B	4.7 MHz	2.3 MHz
C	2.55 MHz	1.13 MHz

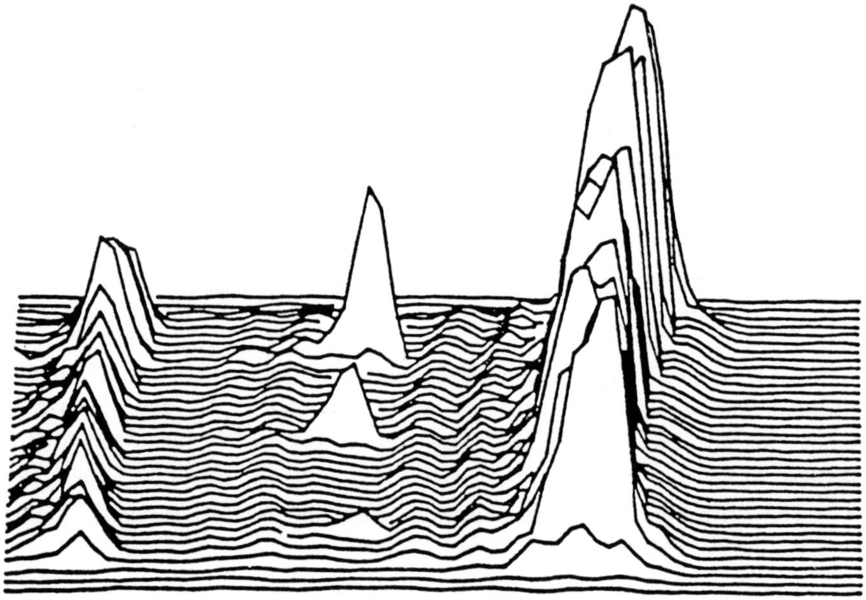

Figure 9: *B-scan ultrasonic image of a weld containing three planar discontinuities (1/16″, 1/8″ and 1/4″ in diameter).*

5.3. *Nondestructive Evaluation Summary for Weld #2*

Longitudinal Wave Speed:	5939 m/sec
Shear Wave Speed:	3627 m/sec
Young's Modulus:	2.566
Shear Modulus:	8.219×10^{10}
Poisson's Ratio:	0.283

Suspect Regions:	ID	X	Y	Z
	A	2.0 cm	0	7.5
	B	7.5 cm	0	7.5
	C	12.5 cm	0	7.5

Flaw Type:	ID	Type
	A	planar
	B	planar
	C	planar

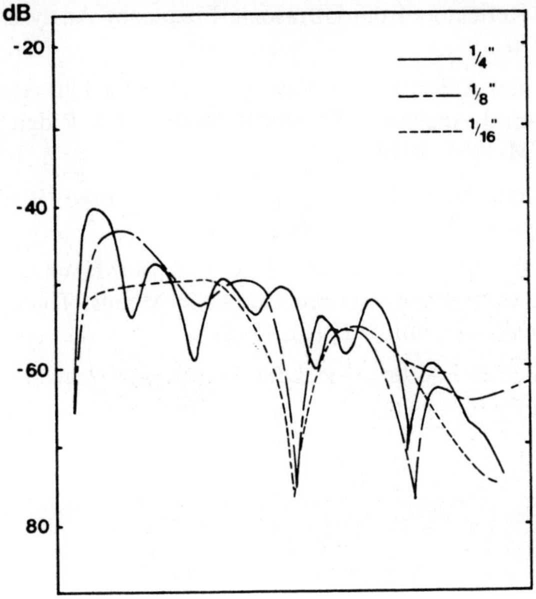

Figure 10: Backscattering spectra from suspect regions in weld #2.

Flaw Orientation:	ID	Angle with Respect to Surface
	A	0 °
	B	0 °
	C	0 °

Flaw Dimension: (assume circularly symmetric)	ID	10 ° Measurement	20 ° Measurement
	A	1.8 mm	2.06 mm
	B	3.5 mm	3.6 mm
	C	6.36 mm	7.3 mm

Compare these ultrasonically determined results with the actual position and dimensions below:

ID	X	Y	Z	Diameter (mm)
A	1.8 cm	0	7.5 cm	1.6 mm
B	7 cm	0	8 cm	3.2 mm
C	12.05 cm	0	8 cm	6.35 mm

References

[1] Whaley, H.L. and Adler, L., "Flaw Characterization by Ultrasonic Frequency Analysis", *Mat. Eval.*, 29 (8), 1971.

[2] Whaley, H.L. and Adler, L., "Model for the Determination of the Size and Orientation of Reflectors from Ultrasonic Frequency Analysis", *J. Acoust. Soc. Am.*, 48, 1, 1970.

[3] Whaley, H.L. and Adler, L., "A New Technique for Ultrasonic Flaw Determination by Spectral Analysis," Technical Memo, Oak Ridge National Laboratory, ORNL-TM-3056, 1970.

[4] Fitting, D.W. and Adler, L., *Ultrasonic Spectral Analysis for Nondestructive Evaluation*, Plenum Press, New York, 1981.

[5] Achenbach, J.D., Adler, L., Lewis, D.K. and McMaken, H., "Diffraction of Ultrasonic Waves by Penny-Shaped Cracks in Metals: Theory and Experimentation", *J. Acoust. Soc. Am.*, 66, (6), 1979.

[6] Simpson, W.A., "A Fourier Model for Ultrasonic Frequency Analysis", *Mat. Eval.*, 34, (12), 1976.

FRACTURE MECHANICS AND QNDE

APPLICATION OF QNDE TO RETIREMENT-FOR-CAUSE ANALYSIS

Jann N. Yang
School of Engineering and Applied Science
The George Washington University
Washington, D.C.

Abstract

The retirement-for-cause (RFC) is a procedure under which structural components are inspected periodically. The component is retired from service only when it has incurred quantifiable damage detected by the NDE system. The application of NDE results to RFC analysis is described in detail, in particular the importance of dealing with the NDE uncertainty. The RFC analysis methodology is summarized, and the sensitivity of structural reliability and life-cycle-cost with respect to the capability of NDE systems is examined. Numerical examples are given to demonstrate the importance of RFC life management.

1. Introduction

Most rotor disks of gas turbine engines are limited by the low-cycle fatigue life (LCF) that is defined as the number of cycles necessary to initiate a crack approximately 0.03 in. (0.76mm) long. The statistical distribution of LCF life, which is obtained for a given set of loading conditions (stress/strain, time, temperature) involves tremendous dispersion. The ± 2 standard deviation bounds may span two orders of magnitude in LCF life. Traditionally, the disk design life is determined from the distribution of LCF life at an occurrence rate of 1 in 1000 disks. It is at this design life that all LCF-limited disks are retired from service. This procedure has been very successful in preventing the catastrophic failure in service. However, in retiring 1000 disks because one may fail, the remaining 999 unfailed disks are not utilized. Recently, the concept of retirement-for-cause (RFC) has been proposed (Refs. [1-12]) for screening one bad part and certifying the remaining 99.9% for additional safe utilization of the life capacities of engine disks.

Under retirement-for-cause life management, each rotor disk is inspected periodically. When a crack is detected during inspection, the disk is retired; otherwise, it is returned to service until the next inspection maintenance. This procedure could be repeated until the disk has incurred detectable damage (crack), at which time it is retired for that reason (cause). Retirement-for-cause is, then, a methodology under which an engine disk would be retired from service when it has incurred quantifiable damage, rather than because an analytically determined design life had been exceeded. As such, the NDE plays an essential role in the RFC life management.

One important task in the application of RFC procedures is the determination of the inspection interval. When the inspection interval τ is too long, the structural reliability will be too low. On the other hand, if the inspection interval is too short, the cost of inspection or replacement will be too high and the down time may be unacceptable. Hence, an optimal inspection interval should be determined such that the life-cycle-cost of engine components is minimum and a high level of the component reliability is maintained. To accomplish such an objective, RFC analysis methodologies have been attempted recently (Refs. [4-9]), which are capable of estimating the probability of component failure, the percentage of replacement during each inspection maintenance, the life-cycle-cost, etc.

The NDE system plays an important role not only in the RFC life management, but also in many areas of practical importance, including the design of critical components, the product quality assurance in manufacturing, the monitor of expected structural damage/deterioration, the avoidance of sudden failures from unknown flaws in service, the life extension of aging components, etc., [13-14]. Frequently, new aircraft systems may not be deployed before the design life of aging aircraft. Under this circumstance, the life extension of aging aircraft becomes a necessity, and the NDE system plays a key role.

It is the purpose of this paper to (i) describe the applications of NDE results to the RFC analysis of structural components under scheduled inspection maintenance, (ii) summarize the RFC methodology to demonstrate the effect of the NDE capability, and (iii) examine the sensitivity of structural reliability and life-cycle-cost savings with respect to the capability of NDE systems. Numerical examples are given to demonstrate the importance of NDE systems in the RFC life management.

2. Application of NDE Results to RFC Analysis

Current NDE techniques are not capable of repeatedly producing correct indications when applied to flaws of the same length. The chance of detecting a given crack depends on many factors, such as the location, orientation, and shape of the flaw, materials, inspectors, inspection environments, etc. As a result, the probability of detection (POD) for all cracks of a given length has been used in the literature to define the capability of a particular NDE system in a given environment. Methods for establishing a POD curve using experimental test results and POD curves for some NDE systems are available, for instance, in Refs. [15-17] and references therein.

In some applications, such as the damage tolerance specifications, the NDE system is required to be capable of detecting a specified crack length, a_{NDE}, with a 90% detection probability and a 95% confidence level for a particular structural detail. a_{NDE} is referred to as the inspection limit. Such a deterministic approach, however, does not provide statistical information regarding the structural reliability, the expected (average) repair or replacement rate, etc.

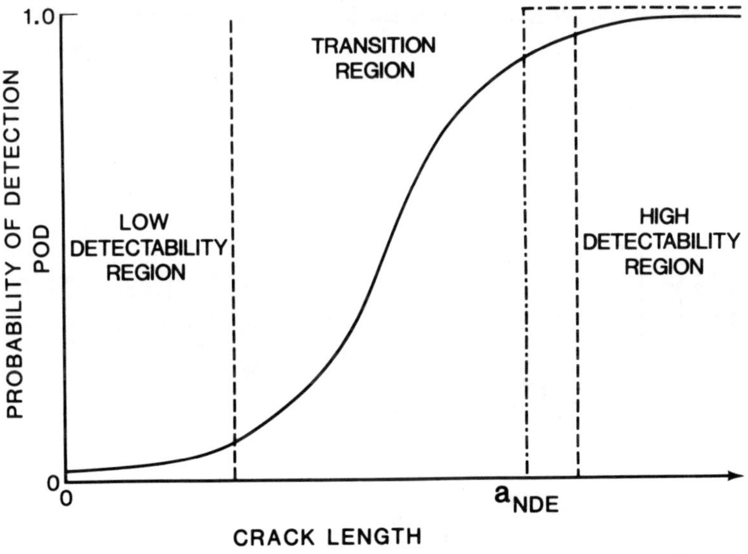

Figure 1: Typical experimental result for POD versus crack length

A typical POD curve usually consists of three regions: (i) low detectability region, (ii) transition region, and (iii) high detectability region as shown in Figure 1. It follows from Figure 1 that the NDE consists of two types of wrong indications: (i) failure to give a positive indication in the presence of a crack length greater than a_{NDE}, referred to as Type I error, and (ii) given a positive indication when the crack length is smaller than a_{NDE}, referred to as Type II error. The Type I error allows the components containing a crack length longer than a_{NDE} to return to service, thus greatly increasing the potential safety hazard. For safety critical components, Type I error is of primary concern. Type II error rejects good components and, hence, increases the replacement rate and the life-cycle-cost. For the retirement-for-cause life management in which the life-cycle-cost is an important criterion, both Type I and Type II errors are of practical concern. As such, the POD curve associated with 50% rather than 95% confidence level should be used. It is obvious then that the ideal inspection capability of an NDE system is a unit step function as shown in Figure 1 by the dashed-dot curve in which both Type I and Type II errors are zero. Unfortunately, such an ideal NDE system may be far from reality. Methods for reducing either Type I error or Type II error or both using multiple inspections procedures have been proposed and described in details in Refs. [18-19].

In applications the POD curve represents the go/no-go situation. For example, the level of NDE signal (or voltage) is preselected, and the component is rejected (no-go) if the magnitude of the inspection signal (or voltage) exceeds the preselected level; otherwise the component is accepted (go) and returned to service. In other words, there are only two outcomes after inspection, i.e., reject (no-go) or accept (go). Consequently, when a component is either rejected or accepted, there is no further information regarding the possible flaw size in the inspected components. This approach may not be the optimal procedure in extracting the information for each component being inspected.

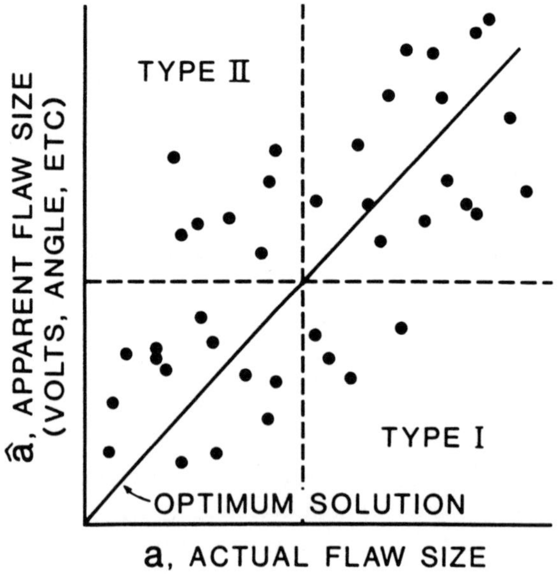

Figure 2: Apparent flaws size "â" versus actual flaw size "a"

The second approach in presenting the NDE capability is the plot of the apparent flaw size, "$\hat{a}$", versus the actual flaw size "a" as shown in Figure 2. The apparent flaw size "$\hat{a}$" is estimated from the inspection outcomes of the specimen, whereas the actual flaw size is determined precisely by breaking that specimen. Figure 2 contains more information than Figure 1. In fact the POD curve can be derived from the "$\hat{a}$" versus "a" data by drawing a horizontal line in Figure 2 through a given level of "$\hat{a}$".

The advantage of using the "a" versus "$\hat{a}$" information for the NDE system is as follows. On the basis of the inspection result, "$\hat{a}$", for an individual component, the next inspection interval for that component may be adjusted. Further, "$\hat{a}$" provides information to statistically up-date the crack size during the next inspection, leading to a better determination of the crack size in that component. Such advantage can be capitalized for structures involving small population size, such as bridges and dams. For a fleet of large number of gas turbine engine disks or aircraft components, however, it becomes impractical to schedule different inspection interval for different individual component. Moreover, the tracking system for all disks using Baysian statistical up-date procedures may become too cumbersome. While the information for "$\hat{a}$" versus "a" is a better representation for the capability of a NDE system, it is rarely used in the analysis. Likewise, unlike the POD curve, such information is quite limited in the literature.

To establish the POD curve from experimental test results, it is convenient to assume a functional form. Furthermore, the functional form for the POD curve is convenient for the RFC analysis. Two possible functional forms have been used (Refs. [17-18],[7-8]), i.e.,

$$POD(a)=\frac{\exp(\alpha^*+\beta^* lna)}{1+\exp(\alpha^*+\beta^* lna)} \quad ; \quad 0<a<\infty \tag{1}$$

in which $POD(a)$ is the probability of detecting a crack size "a", and α^* and β^* are constants defining the NDE capability. Equation (1) is referred to as the log odd function.

Another *POD* function, referred to as the Weibull function can also be used;

$$\begin{aligned} POD(a)&=0 \quad ; \quad a<\varepsilon \\ &=1-\exp\left\{-\left(\frac{a-\varepsilon}{\beta}\right)^{\alpha}\right\} \quad ; \quad a\geq\varepsilon \end{aligned} \tag{2}$$

in which ε is the crack size below which the crack cannot be detected by the NDE system. In Equation (2), α and β are constants, representing the bandwidth and the central location of the POD curve, respectively.

For an ideal NDE system, the POD curve is a unit step function at a_{NDE}, i.e., $POD(a)=0$; for $a<a_{NDE}$; $POD(a)=1$ for $a\geq a_{NDE}$. Such an ideal *POD* curve can be obtained from the Weibull function by setting $\varepsilon=a_{NDE}$, $\beta\rightarrow 0$ and $\alpha\rightarrow\infty$.

Hence, the Weibull function given by Equation (2) includes the ideal *POD* curve as a special case.

The uncertainty of crack detection is reflected by the bandwidth of the *POD* curve; the wider the bandwidth, the more uncertainty and variability. When a *POD* curve is wide-banded, the Type II error (rejecting good components) may be serious and the replacement rate may be too high. To reduce the replacement rate, an engine disk may be inspected by two NDE systems using the intersection rule described as follows [18,19,8]. When a component is inspected and accepted by the first NDE system, it is returned to service; otherwise, it is further inspected by the second NDE system. Some actions may be taken for those components rejected by the first NDE system before being inspected by the second NDE system, for instance, cleaning, polishing, or replicating of the rejected critical locations. The component is replaced only if it is rejected by both NDE systems; otherwise it is returned to service. The second NDE system may be identical to the first one (i.e., the first NDE system is used twice) or it may be a redundant system to be used only for those components rejected by the first NDE system. However, the detection capability of the second NDE system should be much better than that of the first one in order to return to service a high percentage of good components rejected by the first system. The surface preparation of the critical location, such as cleaning, polishing, etc., will increase the NDE capability significantly. The resulting probability of crack detection using two NDE systems can be easily derived (Refs. [18],[7-8]).

When two NDE systems are used for inspection, the cost of inspection by the second NDE system may be quite expensive. This is because special preparation should be made for the rejected disks, such as cleaning, polishing, etc. As a result, in addition to the average percentage of replacement, the average percentage of components inspected by the second NDE system (or rejected by the first NDE system), referred to as the average percentage of *K*-account, should be estimated in the RFC analysis.

3. RFC Analysis Methodology

The RFC analysis essentially deals with the reliability and life-cycle-cost optimization of fatigue-critical components under scheduled inspection maintenance. Important elements in the RFC analysis, such as the initial fatigue quality, crack growth damage accumulation, service load spectra, NDE, etc., involve considerable statistical uncertainty and variability. These statistical uncertainties should be accounted for using the probabilistic approach.

3.1. *Initial Fatigue Quality (IFQ)*

The initial fatigue quality (IFQ) defines the initially manufactured state of a component or details, such as bolt holes, rim holes, cooling air holes, etc., prior to service. It is well known that defects exist not only at the surface of critical locations but also within the material volume, including voids, inclusions, flaws, etc. Other defects may be introduced during material processing (e.g., forging), manufacturing and assembling processes (e.g., hole drilling), etc. In general the initial fatigue quality (IFQ) can be represented by the statistical distributions of either the material defect sizes prior to service, or the time to crack initiation, or both.

Let a_0 be a reference crack size beyond which the linear elastic fracture mechanics is applicable. Then, the initial flaws size, $a(0)$, can be divided into two populations; one with a flaw size larger than a_0, and the other with a flaw size smaller than a_0. For the first population, the statistical distribution has been shown to fit the lognormal distribution well,

$$f_{a(0)}(x) = \frac{\log e}{\sqrt{2\pi x}\,\bar{\sigma}_1} \exp\left\{ -\frac{1}{2} \left[\frac{\log x - \mu_1}{\bar{\sigma}_1} \right]^2 \right\} \quad ; \quad x \geq a_0 \tag{3}$$

in which $f_{a(0)}(x)$ is the probability density function of $a(0)$, and $\bar{\mu}_1$ and $\bar{\sigma}_1$ are, respectively, the mean value and standard deviation of $\log a(0)$, which should be determined from test results. The crack size in this population starts to propagate immediately after the component is introduced into service.

The crack size in the second population, i.e., $a(0) \leq a_0$, involves an incubation period, referred to as the time to crack initiation (TTCI), that is defined as the time (or cycles) required to initiate a crack size a_0. TTCI is a statistical variable involving a large variability. The probability density function of TTCI, denoted by $f_T(t)$, can be represented by the lognormal distribution,

$$f_T(t) = \frac{\log e}{\sqrt{2\pi t}\,\bar{\sigma}_T} \exp\left\{ -\frac{1}{2} \left[\frac{\log t - \bar{\mu}_T}{\bar{\sigma}_T} \right]^2 \right\} \quad ; \quad t \geq 0 \tag{4}$$

in which $\bar{\mu}_T$ and $\bar{\sigma}_T$ are the mean value and standard deviation, respectively, of log TTCI.

For some materials, such as aluminum, titanium, waspaloy, astroloy, etc., the initial fatigue quality can be represented by TTCI alone when the reference crack size a_0 is large. For some powder metallurgy superalloys, such as IN100, however, the referenced crack size can be defined to be rather small such that some initial defects may be larger than a_0. Under this circumstance, the initial fatigue quality should be defined by both the distribution of $a(0)$ given by Equation (3) for $a(0) \geq a_0$ and that of the time to crack initiation given by Equation (4) for $a(0) \leq a_0$ [8].

3.2. *Statistical Crack Propagation*

When the crack size is greater than the reference crack size a_0 the linear elastic fracture mechanics can be used to predict the crack growth damage accumulation in service. Under spectrum loading, the crack growth rate equation can be approximated as follows [8].

$$\frac{da(t)}{dt}=QS^V a^b(t) \tag{5}$$

in which $da(t)/dt$ = crack growth rate, Q, V and b are constants, and S is the "normalized" maximum stress in the design spectrum, in the sense that $S=1$ represents the nominal design spectrum.

Under service environments, the crack growth rates of structural details exhibit considerable statistical variability. Recently, efforts have been made to randomize the deterministic crack growth rate equation given by Equation (5) in order to account for the variability. The lognormal crack growth rate model seems to be very reasonable (Refs. [21-24],[7],[8])

$$\frac{da(t)}{dt}=ZQa^b(t) \tag{6}$$

in which $Z=H_1H_2H_3H_4S^V$ is a random variable accounting for different contributions to the scatter of the crack growth rate. H_1,H_2,H_3,H_4 and S are random variables; H_1 = materials variability, H_2 = variability of crack geometry, such as aspect ratio, stress intensity factor, etc., H_3 = variability of crack modeling, e.g., two-dimensional crack model, H_4 = variability of crack growth damage due to one equivalent cycle, and S = variability of stress resulting from variabilities of the service loads, temperature profile, etc.

Assuming all the random variables H_1,H_2,H_3,H_4 and S to be lognormal with a median value equal to unity (i.e., taking values around unity), the random variable Z is lognormal with a unit median value. Thus, the deterministic equation, Equation (5), represents the median crack growth behavior, whereas the statistical variability of the crack growth rate is taken care of by the lognormal random variable Z, Equation (6), with the probability density function $f_Z(z)$ given by

$$f_Z(z)=\frac{\log e}{\sqrt{2\pi}z\,\overline{\sigma}_z}\exp\left\{-\frac{1}{2}\left[\frac{\log z}{\overline{\sigma}_z}\right]^2\right\} \quad ; \quad z\geq 0 \tag{7}$$

where $\overline{\sigma}_z$ is the standard deviation of $\log Z$, which should be estimated from the statistical variability of random variables H_1,H_2,H_3,H_4 and S [7-8].

3.3 *Reliability and Percentage of Replacement for Structural Components under Scheduled Inspection Maintenance*

Given the distribution of the initial fatigue quality and the POD curve for each NDE system, the average percentages of replacement and K-account during production inspections can be computed [8]. The statistical distribution of the crack size can be derived as a function of served life using the distributions of the initial fatigue quality and the statistical crack growth model. Meanwhile the statistical distribution of the crack size in service is subjected to modification during each inspection and replacement maintenance. The following quantities relevant to the RFC analysis can be computed (see Refs. [7-8]); (i) the average percentage of replacement during the jth inspection maintenance, $R(j)$, for $j=1,2,\ldots$, (ii) the average percentage of K-account during the jth inspection maintenance, $K(j)$ for $j=1,2,\ldots$, and (iii) the probability of structural failure in the jth service interval, $p(j)$, for $j=1,2,\ldots$. The quantities described above can be obtained using either the analytical/numerical solution [7,8] or the method of Monte Carlo simulation [9].

3.4 *Cost/Risk Analyses*

One objective of the RFC analysis is to determine the optimal inspection interval τ such that either the life-cycle-cost (LCC) is minimum or the return of investment (ROI) is maximum, and at the same time a high level of structural reliability in service is maintained. The LCC and ROI models may be rather complicated. However, for the preliminary retirement-for-cause demonstration, a cost/risk analysis for each component is considered herein for simplicity.

Let DL = the conventional design life determined from the distribution of LCF life at an occurrence rate of 1 in 1,000 disks; T_D = the extended engine design life under RFC life management; $m_D = (T_D/DL)-1$ = the total number of replacements for a component in $(0,T_D)$, based on the conventional design; τ = the service inspection interval; and $n_D=T_D/\tau$ = the total number of inspection intervals in $(0,T_D)$.

Without adjusting the interest rate for simplicity, the expected (or average) RFC life-cycle-cost saving for one component in the engine design life, T_D, can be expressed as

$$EC=m_D C_r+C_f p(DL)-\left[\sum_{j=1}^{n_D} C_f p(j)+C_1 n_1+\sum_{j=1}^{n_1} C_2 K(j)+\sum_{j=1}^{n_1} C_r R(j)\right] \tag{8}$$

where $n_1=n_D-1$, C_r = the cost of one component (e.g., disk); C_f = the cost of consequence of failure (it may be the cost of one engine or one airplane); $p(DL)$ = the probability of failure of a component in $(0,DL)$ based on the conventional design; $p(j)$ = the probability (or average percentage) of failure of a component in the jth service interval, which is a function of the inspection interval τ; C_1 = the cost of inspecting one component using the first NDE system, including the cost of maintaining the system; and C_2 = the cost of inspecting one component using the second NDE system, including the cost of maintaining the system.

On the right-hand side of Equation (8), the first two terms denote the expected cost for a component in $(0,T_D)$ using the conventional design procedure. All bracketed terms indicate the expected cost for a component under the RFC life management, including the expected (average) costs of failure, inspections, and replacement. The optimal inspection interval τ is determined by maximizing Equation (8) subjected to the constraint that the probability of failure of a component in the design service life T_D should be smaller than an acceptable value P_a^*, i.e., $p(T_D) \leq P_a^*$.

4. Demonstrative Example

The third-stage turbine disk of the TF-33 gas turbine engine is considered for demonstrative purposes. The material of the engine disk is Incoloy 901 and 10 bolt holes are the critical locations where cracks may develop [1]. Currently, the design life for such a disk is 2,500 cycles, i.e., the disk is discarded after DL = 2,500 cycles of service life. Under RFC life management, the design life is extended to T_D = 12,500 cycles.

Available test results indicate that the initial fatigue quality can be represented by the distribution of time to crack initiation (TTCI) alone with the reference crack size a_0 = 0.03 in (0.76 mm). The median life and the coefficient of variation of TTCI are, respectively, $\overline{T}$ = 4,000 cycles and V_T = 75%, from which the log normal parameters in equation (4) are computed as $\overline{\mu}_T = log\overline{T}$ = 3.6 and σ_T = 0.29. From the spin-pit test results under a nominal loading spectrum ($S=1$), the average propagation life from a_0 = 0.76 mm to the critical crack size a_c = 0.335 in. (8.51 mm) is approximately 11,500 cycles [1]. The resulting crack growth data ($S=1.0$) are best fitted by Equation (5) with the results; $Q=4.22\times10^{-4}$ and $b=1.2914$. The coefficient of variation of the crack growth rate is V_Z = 30%, and hence, the lognormal parameter for Z is σ_z = 0.1276, Equation (7). The parameter V appearing in Equation (5) is approximately 1.0469 [7].

To investigate the effect of NDE capability on the RFC analysis results, four POD curves shown in Figure 3 are considered. These four POD curves are expressed by the log odd function, Equation (1), with the following parameter values: α^* = 55.28 and β^* = 16.4 for POD No. 1; α^* = 66.6 and β^* = 23.4 for POD No. 2; α^* = 28.94 and β^* = 11.73 for POD No. 3; α^* = 13.44 and β^* = 3.95 for POD No. 4.

Without inspection maintenance, the cumulative probability of failure is computed as a function of service life. The result is shown in Figure 4 by a solid curve designated as curve 0. It is observed that the cumulative probability of failure increases as the service life increases.

Suppose the POD curve for the first NDE system is given by Curve 1 of Figure 3. Since such a POD curve is narrow-banded, the second NDE system is not necessary. Under the scheduled inspection maintenance, the cumulative probabilities of failure are presented in Figure 4 by solid curves designated by 1-4. The

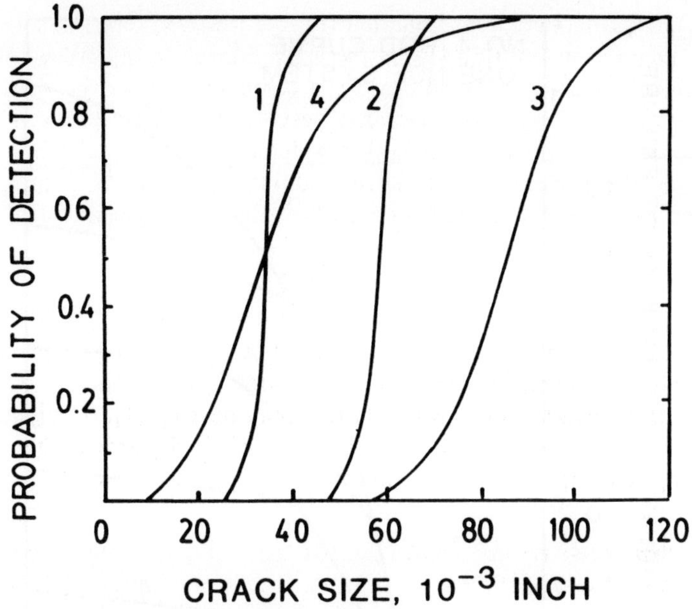

Figure 3: Various POD curves for NDE system

number shown on each curve (or curve number) denotes the number of inspection maintenances performed in the service life of 12,500 cycles. For example, curve 3 in Figure 4 represents the cumulative probability of failure in which three service inspection maintenances are performed in 12,500 cycles, i.e., the inspection interval is 3,125 cycles. Figure 4 clearly demonstrates that the probability of failure in service is reduced significantly by the application of scheduled inspection maintenance. Furthermore, the proability of failure decreases as the number of inspection maintenances in 12,500 cycles increases.

During each inspection and replacement maintenance, a certain percentage of disks will be replaced, and it can be computed statistically. The total average percentage of disk replacement within the service life of 12,500 cycles is shown in Table 1 for different inspection intervals. It is observed from Table 1 that the total average percentage of replacement increases as the number of inspection maintenances increases. It is mentioned that the total percentage of replacement is 400% based on the conventional design.

Let the POD curve of the first NDE system be the one shown by curve 2 in Figure 3. Although such an NDE system is not as capable as the previous one, the POD curve is narrow-banded, and hence the second NDE system is not employed. The cumulative probabilities of failure are displayed by solid curves in Figure 5 for different inspection intervals. The total average percentage of replacements in the service life of 12,500 cycles is presented in Table 1. It can be observed from Figures 4 and 5 that, with the same inspection interval τ, the probability of failure using POD curve 2 is much higher than that using POD curve 1. However, the

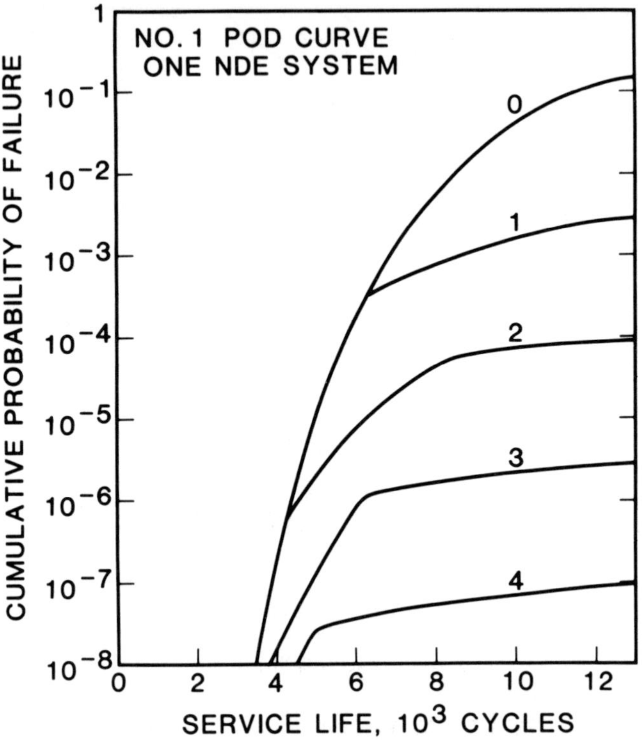

Figure 4: Cumulative proability of failure as function of service life using POD curve 1

Table 1: Total average percentage of replacement and K-account in 12,500 cycles*

No. of Inspections	Inspection Interval, Cycles	POD Curve					
		1	2	3	4	4	4
		One NDE System				Two NDE Systems	
1	6250	65.4	21.2	5.9	62.2	19.8	62.2*
2	4166	95.2	50.9	25.6	100.7	49.5	116.1*
3	3125	108.9	63.3	38.6	127.3	62.0	157.1*
4	2500	120.6	69.8	46.4	149.9	68.7	191.7*
5	2083	129.1	73.7	51.4	170.3	72.7	224.3*
6	1785	135.8	76.4	54.9	189.4	75.5	255.9*
7	1562	141.2	78.6	57.6	207.9	77.6	285.4*
8	1388	147.5	80.2	59.6	226.0	79.3	318.2*

* Percentage of K-Account

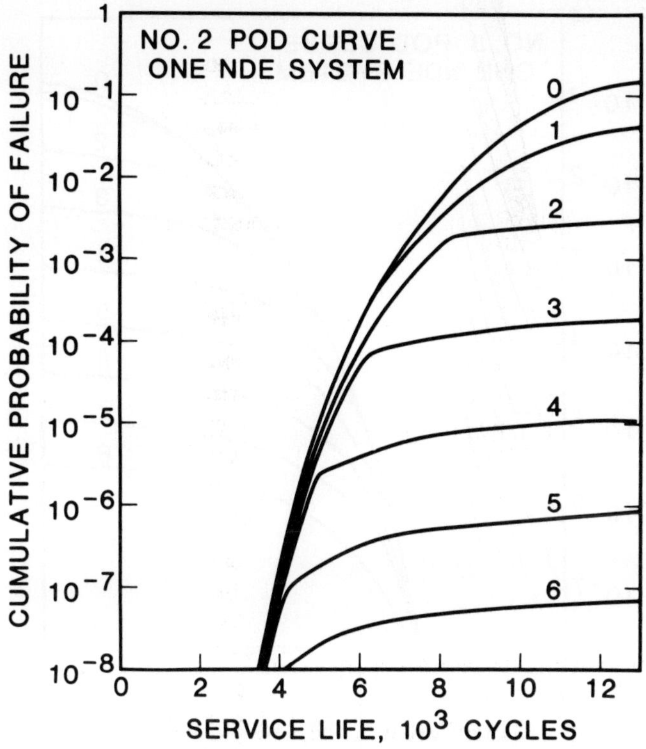

Figure 5: Cumulative probability of failure as function of service life using POD curve 2

total percentage of replacements is smaller with the application of POD curve 2. In other words, a higher reliability is achieved through a higher percentage of replacement.

Consider POD curve 3 as shown in Figure 3 for the first NDE system. By use of one NDE system, the cumulative probabilities of failure are shown in Figure 6. Although the capability of such an NDE system is not outstanding, it is still possible to substantially reduce the failure probability using a shorter inspection interval τ. The total average percentage of replacement is shown in Table 1 for different inspection intervals. As expected, the total average percentage of replacement is much smaller than those associated with POD curves 1 and 2.

Finally, POD curve 4 shown in Figure 3 is considered for the first NDE system. With a single NDE system, the cumulative probabilities of failure are displayed in Figure 7. The total average percentage of replacement in 12,500 cycles is shown in Table 1. As observed from Figure 3, POD curve 4 is wide-banded, involving considerable statistical uncertainty in crack detection. Hence, the second NDE system is employed in which the POD curve is considered to be a unit step function at the crack size $a = \varepsilon = 0.06$ in., which corresponds to a 90% detection

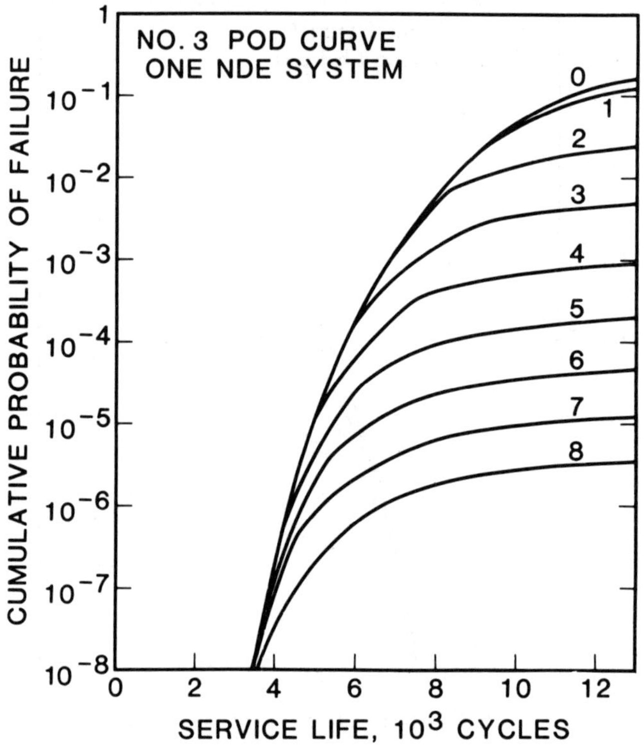

Figure 6: Cumulative probability of failure as function of service life using POD curve 3

probability for the first NDE system. The cumulative probabilities of failure have been computed but not displayed because of space limitations. It is mentioned that the failure probability using two NDE systems is slightly higher than that shown in Figure 7, in which only one NDE system is employed. The average percentage of replacements as well as the average percentage of K-account are given in Table 1. A comparison between the total average percentages of replacement shown in Table 1 indicates that the second NDE system is capable of significantly reducing the total percentage of replacement.

For narrow-banded POD curves, it is observed from Table 1 and Figures 4-6 that a higher improvement of structural reliability is usually accompanied by a higher replacement rate. Furthermore, the efficiency of the NDE system is determined not only by the central location of the POD curve but also by the bandwidth. When the bandwidth of the POD curve is large, a second NDE system is needed to reduce the replacement rate. However, the application of the second NDE system should be balanced by the cost of K-account.

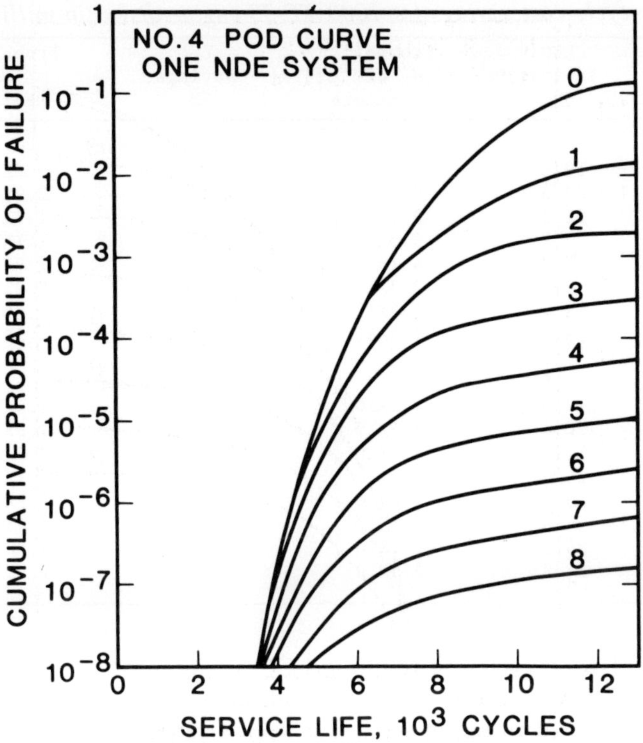

Figure 7: Cumulative proability of failure as function of service life using POD curve 4

Based on the conventional design, there will be four replacements in T_D = 12,500 cycles, i.e., $m_D = (T_D/DL) - 1 = 4$. The following cost values are assumed: the cost of one disk is C_r = 7500 dollars; the cost of inspecting one disk is C_1 = 200 dollars; and the cost of failure of one disk, including the possible failure of the entire aircraft, is C_f = 2,000,000 dollars. The probability of failure $p(j)$ in each service interval and the average percentage of replacement, $R(j)$, during each inspection maintenance have been computed [7-8]. By use of only one NDE system the average percentage of K-account is zero.

With the application of the POD curve 1 shown in Figure 3, the expected life-cycle-cost savings for a fleet of 1000 disks under various inspection intervals have been computed using Equation (8). The results are presented in Table 2 along with the cumulative probability of failure, $p(T_D)$, in the engine design life of 12,500 cycles. The number of disks in a fleet of 1000 expected to fail in service is equal to the cumulative probability of failure $p(T_D)$ shown in Table 2 multiplied by 10^3.

It is observed from Table 2 that a significant life-cycle-cost (LCC) saving can be achieved by the application of RFC procedures. The optimal number of inspection maintenances in 12,500 cycles appears to be 2.0, i.e., the optimal inspection

Table 2: Life-cycle-cost savings for 1000 TF-33 engine disks (in million dollars)

No. of Inspections	LCC Savings, \$M	Probability of Failure	LCC Savings, \$M	Probability of Failure
	POD curve 1		POD curve 2	
1	19.1	3.0×10^{-3}	-49.0	3.9×10^{-2}
2	22.3	9.4×10^{-5}	20.0	2.9×10^{-3}
3	21.2	2.8×10^{-6}	24.3	1.7×10^{-4}
4	20.2	9.5×10^{-8}	23.9	1.1×10^{-5}
5	19.3	4.1×10^{-9}	23.5	7.9×10^{-7}
6	19.0	2.3×10^{-10}	23.1	6.8×10^{-8}
7	18.4	1.7×10^{-11}	22.7	6.7×10^{-9}
8	18.0	1.5×10^{-12}	22.4	7.5×10^{-10}
	POD curve 3		POD curve 4	
1	-217.0	1.2×10^{-1}	- 1.3	1.4×10^{-2}
2	-20.4	2.4×10^{-2}	18.4	1.8×10^{-3}
3	17.4	4.6×10^{-3}	19.3	2.8×10^{-4}
4	24.0	8.8×10^{-4}	17.9	5.1×10^{-5}
5	24.8	1.8×10^{-4}	16.2	1.2×10^{-5}
6	24.6	4.3×10^{-5}	14.6	2.6×10^{-6}
7	24.3	1.1×10^{-5}	13.0	6.7×10^{-7}
8	23.9	3.3×10^{-6}	11.5	1.8×10^{-7}

interval is $\tau = 4167$ cycles. The resulting LCC saving is 22.3 million dollars and the probability of disk failure is 9.4×10^{-5}, indicating 0.094 disks out of a fleet of 1000 disks are expected to fail in service. If such a reliability level for the engine disk is not acceptable, one can choose three inspection maintenances that result in a LCC saving of 21.1 million dollars. The corresponding probability of disk failure is 2.8×10^{-6} (or 0.0028 disk out of a fleet of 1000 will fail in service).

Consider the POD curves 2, 3, and 4 shown in Figure 3. It is mentioned that POD curves (Nos. 1, 2, 3, and 4) are not in increasing capability. The results of the life-cycle-cost saving are presented in Table 2, along with the probability of failure $p(T_D)$. It is observed from Table 2 that there is always an optimal number of inspection maintenance (or optimal inspection interval) at which the LCC saving is maximum. A trend observed is that the optimal number of inspection maintenances increases as the capability of the NDE system is reduced. For instance, POD curve 3 is poorer than POD curve 2, and POD curve 2 is poorer than POD curve 1. The optimal number of inspection maintenances using POD curve 3 is 5, whereas that using POD curve 2 is 3. Another observation is that the LCC saving seems to be insensitive to the number of inspection maintenances in the design life. This is beneficial for decision-making, because it allows for more flexibility in choosing the inspection interval τ without severe penalty. It is further observed from Table 2 that the life-cycle-cost savings are comparable among all of the NDE systems considered. These observations hold because the capability of all the NDE systems considered above are quite good as will be discussed later. The trends observed above may not be true if the input cost data are changed.

5. Conclusions and Discussions

It is demonstrated that the scheduled inspection maintenance can reduce significantly the probability of failure of fatigue-critical components in service. The improvement of structural reliability is sensitive to the capability of the NDE system employed. When appropriate NDE systems are used, the application of retirement-for-cause life management is shown to result in a substantial life-cycle-cost (LCC) saving over the conventional design system. This is accomplished through a significant reduction of part replacement. Furthermore, in the RFC analysis, the uncertainty of the NDE system, as reflected by the bandwidth of the POD curve, plays an important role in determining the necessity of multiple inspection procedures. In fact, one important direction of improving the NDE system for applications to RFC life management is the reduction of the POD bandwidth. The crack detection uncertainty comes from many sources among which the human factor and inspection facilities seem to be quite important. Consequently, intensive efforts have been made to establish computer-controlled fully automatic NDE systems in order to reduce the effect of human factor to a minimum in the entire inspection operations [10].

From the system standpoint, one question of practical importance is whether the current state of NDE technology or capability can satisfy the specified goal of the particular RFC or inspection maintenance requirements. Such a question can be answered once the RFC analysis methodology is established. This is because the RFC methodology takes into account the uncertainty and capability of NDE systems, and a sensitivity study can be conducted to determine the direction in which the improvement of the NDE system is needed. The question will be difficult to resolve without carrying out the RFC analysis.

The following observations from our experience with the development of RFC methodology are interesting [7-8]. Let a_{90} be the crack size corresponding to a 90% detection probability with 50% confidence level. When the propagation life, denoted by N_p, from the reference crack size a_0 to the critical crack a_c is long, the NDE capability is sufficient, if a_{90} is of the same order of a_0, as long as the bandwidth of the POD curve is not large. However, if N_p is short, a_{90} should be smaller than a_0; otherwise little benefit may be derived from RFC procedures. For instance, the four POD curves shown in Figure 3 result in no RFC benefit if they are used for an advanced engine disk discussed in Ref. [8]. Under this circumstance, the propagation life N_p should be increased by reducing the reference crack size a_0 along with the improvement of NDE capability [8].

References

[1] Hill, R.J., Reimann, W.H. and Ogg, J.S., "A Retirement-for-Cause-Study of an Engine Turbine Disk", AFWAL-TR-81-2094, November, 1981.

[2] Annis, C.G., Jr. *et al.*, "Gas Turbine Engine Disk Retirement-for-Cause; An Application of Fracture Mechanics and NDE", *Journal of Engineering for Power*, ASME, Vol. 103, January, 1981, pp. 198-200.

[3] Harris, J.A., *et al.*, "Engine Component Retirement-for-Cause", Contract F33615-80-C-5160, Government Product Div., Pratt & Whitney Aircraft Group, West Palm Beach, Florida.

[4] Besuner, P.M., *et al.*, "Cost/Risk Analysis for Disk Retirement: Vol. 1", FAA-82-3-8, Failure Analysis Associates, Palo Alto, CA, 1983.

[5] Besuner, P.M., *et al.*, "Analysis Methods for F100 Engine Component Retirement for Cause", FAA-81-12-5, Failure Analysis Associats, Palo Alto, CA, 1982.

[6] Reimann, W.H., "Retirement-for-Cause of Engine Components", in *USAF Engine Structural Durability and Damage Tolerance*, AIAA Professional Study Series, Seattle, WA, June-July, 1983.

[7] Yang, J.N. and Chen, S., "Fatigue Reliability of Gas Turbine Engine Components Under Scheduled Inspection Maintenance", paper AIAA 84-0850, May, 1984; *Journal of Aircraft*, AIAA, Vol. 22, May, 1985, pp. 415-522.

[8] Yang, J.N. and Chen, S., "An Exploratory Study of Retirement-for-Cause for Gas Turbine Engine Components", AIAA Paper 84-1220, June 1984; *Journal of Propulsion and Power*, AIAA, Vol. 2, No. 1, January-February, 1986, pp. 38-49.

[9] Watkins, T., Jr., and Annis, C.G., Jr., "Engine Component Retirement-for-Cause: Probabilistic Life Analysis Technique", Air Force Wright Aeronautical Laboratories, AFWAL-TR-85-4075, June, 1985.

[10] "Manufacturing Technology for Nondestructive Evaluation (NDE) System to Implement Retirement-for-Cause (RFC) Procedures for Gas Turbine Engine Components", Contract No. F22615-81-C-5002, System Research Laboratories Inc., NDE System Division, Dayton, Ohio.

[11] Ko, R.T., "RFC Phase I eddy Current Reliability Test Results", paper to be presented at the Symposium on Review of Progress in Quantitative NDE, to be held August, 1986.

[12] Bernes, A.P., "Analysis of the RFC/NDE System Performance Evaluation Experiments", paper to be presented at the Symposium on Review of Progress in Quantitative NDE, to be held August, 1986.

[13] Lincoln, J.W., "Risk Assessment of an Aging Military Aircraft", *Journal of Aircraft*, Vol. 22, No. 8, August, 1985, pp. 687-691.

[14] Forney, D.M., "Develpment of Improved NDE to Meet Structural Integrity Requirement," in *Proc. 1985 aircraft Structural Integrity Program (ASIP) Conf.*, Dayton, Ohio, November, 1985, pp. 375-421.

[15] Packman, P.F., *et al.*, "Reliability of Flaw Detection by Nondestructive Inspection, *ASM Metal Handbook*, Vol. 11, 8th Edition, Metals Park, Ohio, 1976, pp. 214-224.

[16] Yee, B.G.W., *et al.*, "Assessment of NDE Reliability Data", NASA CR-134991, NASA Lewis Research Center, Cleveland, Ohio, 1976.

[17] Berens, A.P. and Hovey, P.W., "Evaluation of NDE Reliability Characterization," AFWAL-TR-81-4160, Air Force Wright Aeronautical Laboratories, WPAFB, November, 1981.

[18] Yang, J.N. and Donath, R.C., "Improving NDE Capability Through Multiple Inspections with Application to Gas Turbine Engine", Air Force Wright Aeronautical Laboratories Technical Report, AFWAL-TR-82-4111, WPAFB, August, 1982.

[19] Yang, J.N. and Donath, R.C., "Improving NDE Capability Through Multiple Inspections with Applications to Gas Turbine Engine", in *Review of Progress in Quantitative Nondestructive Evaluation", Vol. 2A, Plenum Press, New York, 1983, pp. 69-78.*

[20] Manning, S.D. and Yang, J.N., "USAF Durability Design Handbook: Guidelines for the Analysis and Design of Durable Aircraft Structures", AFWAL-TR-83-3027, Air Force Wright Aeronautical Laboratories, Wright-Patterson Air Force Base, OH., January, 1984.

[21] Yang, J.N., Salivar, G.C. and Annis C.G., "Statistical Modeling of Fatigue Crack Growth in a Nickel-Based Super-Alloy", *Journal of Engineering Fracture Mechanics*, Vol. 18, No. 2, June, 1983, pp. 257-270.

[22] Yang, J.N. and Donath, R.C. "Statistical Fatigue Crack Propagation in Fastener Holes Under Spectrum Loading", *Journal of Aircraft*, AIAA, Vol. 20, December, 1983, pp. 1028-1032.

[23] Yang, J.N., *et al.*, "Stochastic Crack Propagation in Fastener Holes", *Journal of Aircraft*, AIAA, Vol. 22, September, 1985, pp. 810-817.

[24] Yang, J.N., Hsi, W.H., Manning, S.D. and Rudd, R.L., "Stochastic Crack Growth Models for Applications to Aircraft Structures", in *Probabilistic Fracture Mechanics and Reliability*, Chapter IV, ASTM-STP, January, 1986.

ELASTIC WAVE INTERACTIONS WITH PARTIALLY CONTACTING INTERFACES: STATUS OF THEORY AND APPLICATION TO FATIGUE CRACK CLOSURE CHARACTERIZATION

R.B. Thompson, O. Buck and D.K. Rehbein
Ames Laboratory
Iowa State University
Ames, Iowa

Abstract

The present state of theoretical understanding of elastic wave interactions with partially contacting interfaces is reviewed. Included are discussions of exact solutions for planar interfaces consisting of periodic arrays of strip cracks; numerical solutions to selected two-dimensional problems involving bounded, partially contacting cracks; and approximate analytical results intended for the reduction of data obtained in particular experimental configurations. Uses of the latter in addressing the inverse problem of deducing characteristics of fatigue crack contact distributions from ultrasonic scattering measurements is then addressed, followed by a discussion of the use of this information in the prediction of the effects of closure on fatigue crack growth.

1. Introduction

The scattering of elastic waves from cracks is important in nondestructive evaluation for two reasons. First, the presence of the scattered signals provides a means for detecting cracks. Second, various features of the signals, including their variations with time and transducer position, provide information useful in determining the sizes of the cracks.

Traditionally, the crack is viewed as a pair of stress free surfaces, separated by an infinitesimal distance and bounded by an edge or tip. This leads to a precise definition of the theoretical scattering problem [1] and to a number of procedures for interpreting the experimental data [2]. A central feature of both is the idea that the scattering from a crack is dominated by phenomena occurring near the tip and associated with the dynamic stress singularity that is created in that region when the crack is illuminated by an elastic wave.

Cracks in real materials, particularly fatigue cracks, do not fully correspond to this model. In particular, it is widely believed that many contacting asperities may exist in the region near the tip [3]. Associated with these contacts are decreases in the crack reflectivity, increases in its transmissivity, and changes in the nature of the signals diffracted from the region near the crack tip [4]. These conditions may have a deleterious effect on the ability to detect and size the crack with ultrasonic nondestructive evaluation techniques.

It is also well known that the contacts influence the growth of the cracks under cyclic loading, and the term "crack closure" has been used to describe the retardation of crack growth associated with their existence [3,5]. A number of theories have recently emerged which describe these effects. It is therefore of interest to study the scattering of elastic waves from such cracks and to seek inverse solutions for obtaining the closure parameters from experimental data.

This paper reviews recent developments. Emphasis is placed on the areas in which mechanics plays a major role. Included are sections on experimental approaches to study the scattering from closed cracks, the present status of theory, and recent results in inverting experimental information to obtain crack parameters.

2. Ultrasonic Studies of Crack Closure

It is commonly accepted in the nondestructive testing literature that large compressive stresses near crack tips will cause a decrease in the reflections of signals from these tips and an increase in the signals transmitted past the crack [6,7]. One such result is illustrated in Figure 1 [8]. Under sufficient loads, cracks have been reported to become invisible. The use of ultrasonic scattering to study the closure phenomena, itself, has been less widespread. Buck et al. used Rayleigh transmission measurements, as illustrated in Figure 2a, in one of the early experimental demonstrations of the existence of closure [9]. By deducing an apparent crack depth from the transmission coefficient and plotting this versus applied load, they were able to deduce a closure stress intensity, defined as that stress intensity which would fully open the crack to its tip. For a number of material systems, e.g., 7075 and 2024 aluminum, subject to a wide range of humidities, they found that the crack growth rate could be predicted from a single, modified Paris Law

$$\frac{da}{dN} = A(K_{\max} - K_{closure})^m \qquad (1)$$

where $K_{\max}$ is the maximum stress intensity factor and $K_{closure}$ is the closure stress intensity factor defined above. The quantity in parentheses is known as the effective stress intensity factor, K_{eff}.

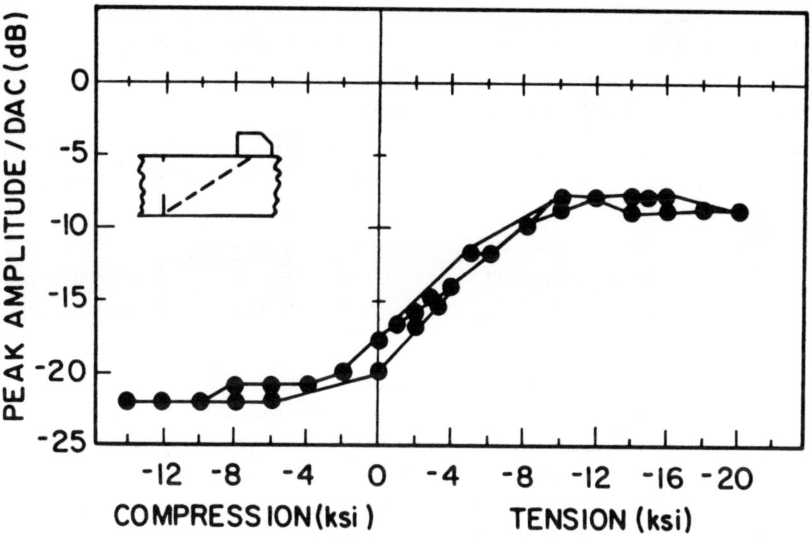

Figure 1: Influence of stress on the detectability fo fatigue cracks (after Ref. [8])

More recent experiments can be divided into three regimes of wavelength λ, crack depth a, and contact separation S. Golan has studied the diffraction of signals from large fatigue cracks grown in compact tension specimen, as shown in Figure 2b [10]. He interprets his data in terms of arrivals from discrete contacts and thus is in the regime $a > S > \lambda$. By applying torque balance arguments to the load dependence of the diffracted signals, he estimates the closure force on each contact region.

Resch et al. have studied the reflection of Rayleigh waves from small cracks in cantilever beam fatigue specimens, as illustrated in Figure 2c. In this regime of $\lambda > a > S$, they have studied a variety of closure phenomena [11] similar to those described above.

The present authors have been concerned with the third possible regime, in which $a > \lambda > S$ [12]. As illustrated in Figure 2d, measurements of transmitted and diffracted signals have been made in the compact tension specimen geometry. Using the scattering theory described in Section 3, preliminary efforts have been

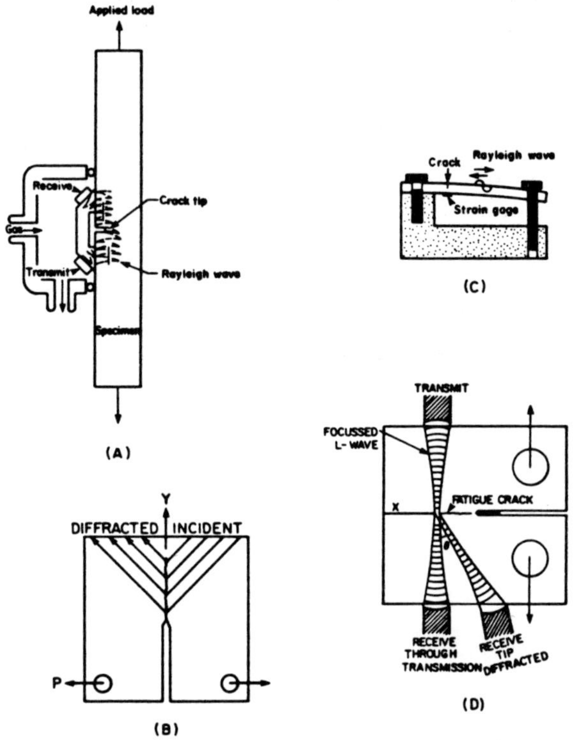

Figure 2: Geometries for ultrasonic studies of fatigue crack closure. (A) Rayleigh wave transmission experiments in part through crack specimen (after Ref. [9]). (B) Shear wave diffraction experiments in compact tension specimen (after Ref. [10]), $a>S>\lambda$. (C) Rayleigh wave reflection experiments in cantilever fatigue specimen (after Ref. [11]), $\lambda>a>S$. (D) Longitudinal wave transmission and diffraction experiments in compact tension specimen (after Ref. [12]), $a>\lambda>S$.

made to deduce the crack parameters, as described in Section 4.

3. Models for Elastic Wave Scattering from Contacting Crack Surfaces

The theory for the elastic wave interaction with partially contacting surfaces, and its application to fatigue crack problems, is partially developed. Included are exact solutions for two-dimensional periodic problems, numerical results for some other two-dimensional cases, and approximate analytical formulae intended for the interpretation of experimental data in the long wavelength regime.

One of the first theoretical examinations of the partial contact problem was performed by Kendall and Tabor [13], who were interested in characterizing the degree of contact of surfaces as a part of tribological studies. They developed formulae for the long wavelength reflection and transmission of longitudinal plane waves normally incident on an interface whose sides were joined by isolated, circular contacts. Their results were extended to the interaction of ultrasonic waves with fatigue crack surfaces joined by asperities by Haines [14]. Thompson et al. [15-17] further generalized the results to a number of different interface topographies through the use of effective boundary conditions. They postulated that the long wavelength interaction of an elastic wave, propagating with the phase variation $\exp[j(\omega t - kz)]$, with the interface could be described through the use of a set of effective boundary conditions

$$\sigma_{3i}^{+} = \sigma_{3i}^{-} \tag{2a}$$

$$\kappa_{ij}(u_j^{+} - u_j^{-}) = \sigma_{3i}^{+} \, . \tag{2b}$$

$\sigma_{3i}^{\pm}$ is the stress on either side of the interface, $u_j^{\pm}$ is the displacement on either side of the interface, and κ_{ij} may be thought of as a generalized distributed spring joining the two sides of the interface. As discussed by Baik and Thompson [17], known static elasticity solutions for a variety of interface geometries can be used to evaluate κ_{ij}. For normal incidence, use of Equation (2) and the one-dimensional elastic wave equation leads to the plane wave reflection R and transmission T coefficients

$$R = j\alpha/(1 + j\alpha) \tag{3a}$$

$$T = 1/(1 + j\alpha) \tag{3b}$$

where

$$\alpha = \pi f \rho v / \kappa \, , \tag{3c}$$

f is the frequency, ρ and v are the material density and wave speed, and κ is the appropriate interface stiffness or spring constant.

The accuracy of this approximation, and its limits, have been explored by Angel and Achenbach [18,19], who developed exact solutions for the transmission and reflection of ultrasonic waves at a planar interface consisting of a periodic array of strip contacts and cracks. Their solutions show that the spring model provides accurate predictions at long wavelengths, as expected. In addition, the exact solution predicts a number of interesting resonance phenomena when the wavelength becomes comparable to the structure of the interface. These are not included in the spring model.

The above solutions all involve plane wave interactions with unbounded interfaces. Numerical solutions have been employed by Visscher (method of optimal truncation) and Punjani and Bond (finite difference method) to treat a bounded, two-dimensional crack containing a discrete set of contacts. Visscher considered the interaction of a straight-crested Rayleigh wave with a surface breaking crack [20,21]. The method involves an expansion of the crack opening displacement in terms of a set of localized functions on the crack surface. This specification of the COD then defines the scattered fields. By sequentially omitting every other localized function, starting from the tip, the effects of an increasing number of equally spaced contacts can be simulated. Figure 3a presents the modulus of the Rayleigh-Rayleigh backscattering amplitude versus k_s and the number of contacts in a crack represented by 20 localized functions. Here k_s is the shear wave number and the crack depth is taken to be unity. Figure 3b presents the corresponding result for the scattering cross-section of the SV wave scattering downward, into the material.

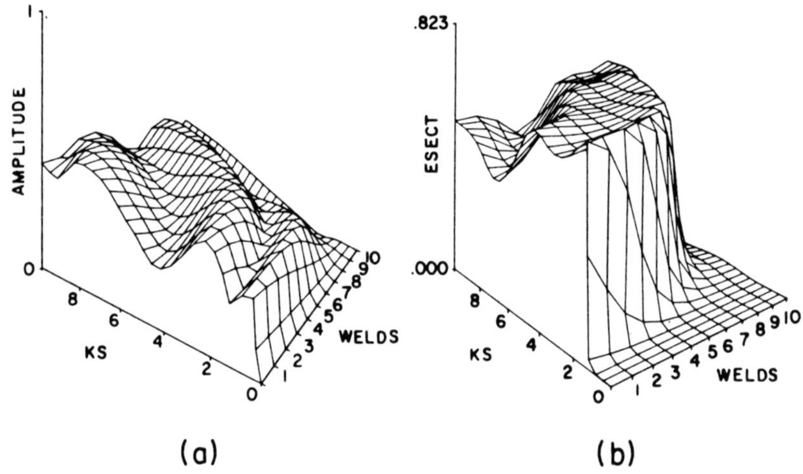

Figure 3: Rayleigh wave interaction with a surface breaking crack as a function of number of equally spaced contacts, starting from tip (after Refs. [20,21]). (a) Backscattred Rayleigh wave. (b) Downscattered SV wave.

Punjani and Bond consider the interaction of an incident longitudinal plane wave with the tip of a two-dimensional crack in an otherwise unbounded medium [22]. Figure 4 presents a finite difference computation of the scattered displacement field, as it would exist at one instant of time, after a crack has been illuminated at 45° by a SV wave. The left hand side shows the response of an ideal crack, while the right hand side shows that of a crack with partially closed regions distributed near its tip. The increase in the transmitted signal in the shadow region, and the greater magnitude and structure of the diffracted waves, are evident.

Thompson et al. have developed approximate models for the interpretation of measurements of the spatial and frequency dependence of signals in the geometry shown in Figure 2d [16,23,24]. To do so, they have further generalized the previously described spring model to approximately include the effects of a spatially varying partial contact, combined with the effects of the transducer beam patterns. The basis for the generalization is the reciprocity theory of Auld [25],

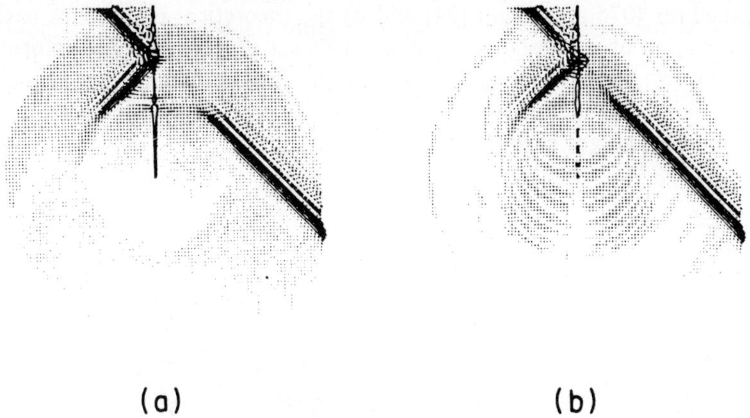

Figure 4: 45° SV wave interaction with a semi-infinite crack (after Ref. [22]). (a) ideal crack; (b) partially contacting crack.

$$\delta\Gamma = \frac{j\omega}{4P}\int_A (u_i^R \tau_{ij}^T - u_i^T \tau_{ij}^R) n_j \, dA \tag{4}$$

where $\delta\Gamma$ is the flaw induced change in the signal transmitted from transmitter to receiver; u_i^R, τ_{ij}^R are the displacement and stress patterns that would be produced if the receiving transducer irradiated a flaw free material; u_i^T and τ_{ij}^T are the displacement and stress fields produced when the flaw is irradiated by the transmitting transducer, P is the electrical power incident on the transmitting transducer, and integration is performed over the surface of the scatterer, which has a normal n_j.

The flaw-free fields have been estimated using a scalar Gaussian approximation to the radiation of the receiving transducer [26]. The fields in the presence of the flaw have been estimated by combining a similar description of the radiation pattern of the illuminating transducer, with a spatially varying form of Equation (3) to predict the COD of the partially closed crack. This is conceptually similar to the Kirchhoff approximation, in that the plane wave response is applied to predict the local COD under the assumption that the transverse variations are small with respect to a wavelength and the characteristic dimensions of the contacts. The resulting expression for the normalized received signal Γ^N is

$$\Gamma^N = C\int_{-\infty}^{\infty} dx \left[\frac{1}{1+j\alpha(x)}\right] e^{-(x-x_1)^2/w^2} e^{-jk(x-x_1)Sin\theta} \tag{5}$$

where C is a normalization constant, w is a beam width parameter, k is the wave vector, x_1 is the position of the beam center, and θ is the angle of the receiver with respect to the transmitter axis.

Equation (5) has been found to accurately predict the signal transmitted through many cracks if one views κ as a variable to be adjusted to fit the data. This fitting constitutes an experimental procedure for the determination of the degree of contact of a fatigue crack. Figure 5 illustrates a) a set of experimental data obtained on 7075 aluminum [24] and b) the theoretical predictions based on c) a spatial variation of κ chosen for best fit. This set of data provides information of particular interest to material scientists studying fatigue crack propagation, in that a region of closure is identified at the end of the first of two temporally separate (by a few months) stages of growth.

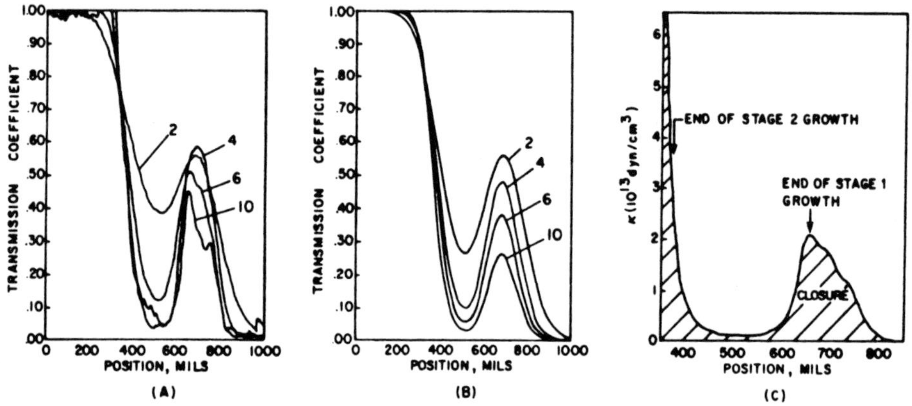

Figure 5: Experimental determination of interface stiffness for a fatigue crack grown in two stages in 7075 aluminum. (A) Experimental transmission versus position with frequency as a parameter. (B) Theoretical transmission versus position with frequency as a parameter. (C) Spatial variation of stiffness used in theoretical predictions.

The parameter θ in Equation (5) defines the angle of the receiving transducer. For through transmission experiments, $\theta = 180°$ and the exponential factor does not enter the analysis. For other angles, exponential factor oscillates sinusoidally. Physically, one may view this as a consequence of the lack of parallelism of the phase fronts of the radiation patterns of the transmitting and receiving transducers, which causes signals transmitted between the two to be weak. Comparison to experiment for such cases, with $\kappa(x)$ being independently determined from the longitudinal wave transmission data, has revealed that this spring model predicts signal strengths which can be orders of magnitude less than the experimental observations [24]. It is postulated that the explanation of this result lies in the unrealistic smoothing of the field singularities surrounding the asperities in the vicinity of the crack tip. Integration of the rapidly varying phase factor times the slowly varying $[1+j\alpha]^{-1}$ produces values that are probably much less than would have occurred had more accurate representation of the dynamic COD been used.

As a test of this hypothesis, a second model has been formulated in which the crack opening displacement is represented by the functional form shown in Figure 6. Part a) schematically shows the actual COD. This is replaced by an approximate form, which is equivalent to representing the COD by a slowly varying component, essentially as predicted by the spring model, plus a series of delta functions which represent the perturbations introduced by the discrete contacts, as sketched in part b). The resulting expression for the diffracted fields is identical to Equation (5) if one replaces

$$\left(\frac{1}{1+j\alpha}\right)\rightarrow\left(\frac{1}{1+j\alpha}\right)\left\{1-\frac{j8\pi f\rho dv(1-\nu^2)}{E(1-N\pi d^2)}\right.$$
$$\left.+\frac{1}{N^{1/2}}\sum\frac{j8(1-\nu^2)\pi f\rho dv\,\delta(x-x_i)}{E(1-N\pi d^2)}\right\} \tag{6}$$

where ν is Poisson's ratio, E is Young's modulus, d is the effective contact diameter, and N is the number of contacts per unit area [24].

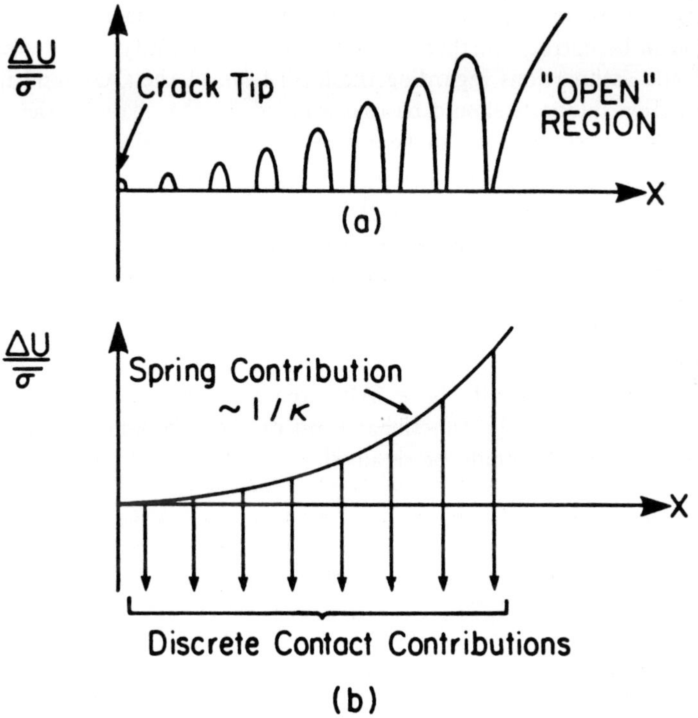

Figure 6: Dynamic crack opening displacement of a fatigue crack. (a) conceptual sketch of true COD; (b) spring plus delta function idealization.

The first term is identical to the quasi-static spring model. The third is the contribution of the delta functions representing diffraction from the contact points. The second balances the average weight of the delta functions so that the spring result is reproduced on the average. For sparse contacts, it is small with respect to the first term.

For through transmission geometries, the first term in Equation (6) is dominant. For the cases studied, evaluation of the second and third terms with reasonable choices of parameters led to insignificant changes. However, when predicting the diffracted waves, the third term is generally dominant. Its omission is believed to be the cause of the previously discussed disagreement between the strengths of diffracted fields predicted by the spring model and observed experimentally. This belief is supported by the fact that, when one fixes $\kappa(x)$ and performs computations for a set of decreasing N, the predicted diffracted signal increases rapidly [24].

Although the qualitative nature of the above predictions, and in particular the sensitivity of the diffracted signals to the number density of contacts (for constant κ), are believed to be correct, further work is needed to quantify the predictions. In particular, ad hoc assumptions regarding the local form of the COD and its contribution to diffraction integrals should be re-examined. In the analysis, the weight of the delta functions were selected to be proportional to the area of the contacts. However, this may be a significant underestimate. For example, the surface deformation of a half-space, induced by a local force, decreases as r^{-1}, where r is the distance from the force [27]. Since the element of area at constant radius from the force increases as rdr, the surface integral of deformation over a circular area is proportional to the radius of the circle. This leads to the possibility that the area per contact, not the area in contact, may be the appropriate weight, and contact contributions to diffraction may be somewhat greater than predicted on the basis of the ad hoc COD form assumed above. Although the absolute values of the diffracted signal strengths and the functional form of their dependence on N might change, the sensitivity to N should be retained.

4. Applications of the Models to the Study of Fatigue Crack Closure

In Section 2, the potential of ultrasonic scattering as a tool in studying crack closure and its effects on fatigue crack growth rates has been discussed. Buck et al. have used the models discussed in Section 3 to propose some specific experimental procedures [12]. They note that, assuming isolated, circular contacts, the contact topography can be described by the number density N and diameter d of the contacts. These parameters can be related to an experimentally determined κ through the relationship

$$\kappa(x) = \frac{d(x)}{8N(x)\pi E'} \tag{7}$$

where $E' = E/(1-\nu^2)$. κ is obtained by fitting the longitudinal through transmission data to Equation (5). Keeping κ fixed, N can then be obtained by varying its value to obtain best fit to the diffracted data. This entails the use of Equation (7) to obtain d in terms of κ and N, and substitution into Equations (6) and (5) sequentially. The average contact stress can then be predicted from the expression of Haines [14,24]

$$\sigma_o = \frac{\pi d p_m \kappa}{2k^* E} \tag{8}$$

where p_m is the flow pressure and k^* is a constant of order 2.

This contact stress distribution must finally be related to $K_{closure}$, in order to predict crack growth rates in accordance with Equation (1). A first step has been taken by Buck et al. [12], who note that the parameters N and d are closely related to those appearing in Beevers model for the effect of contacting asperities on crack growth rate [28].

5. Conclusions

A number of initial steps in the modeling of ultrasonic scattering from partially contacting interfaces have been taken. Included are exact solutions for some two-dimensional, periodic problems, numerical solutions for other contacting cracks with bounded two-dimensional geometries, and approximate results applicable to experimental configurations. The latter have been used to define elementary inversion procedures for deducing crack contact parameters from experimental data.

Future work should be conducted in two areas. The present techniques are sufficiently refined that they can provide new information to materials scientists interested in experimentally studying the closure phenomenon. For example, samples can be subjected to overloads and/or environmental changes and the resulting closure changes can be ultrasonically monitored to see if they are in agreement with theoretical expectations. Based on these closure changes, subsequent crack growth rates should be predicted and compared to direct observation.

Simultaneously, the theoretical background should be extended. An important topic is the improved understanding of the effect of discrete contacts, such as quantification of the effects of varying number densities and investigation of the sensitivity of the results to random versus periodic contact positions. The possibility of obtaining important contact information from the nonlinear response at the cracks [29,30], and effects of property changes in the plastic zone near the crack tip should also be considered. Finally, the relationship of the contact parameters to $\Delta K_{effective}$ should be examined more fully.

Although not discussed in this paper, the implications of the improved understanding of elastic wave scattering from partially closed cracks on the reliability of ultrasonic detection and sizing methodologies should also be investigated, as it has an important bearing on the field implementation of nondestructive evaluation techniques [31].

References

[1] Achenbach, J.D., Gautesen, A.K. and McMaken, M., *Ray Methods for Waves in Elastic Solids with Applications to Scattering by Cracks*, Pittman, London, 1982.

[2] Thompson, R.B., "Quantitative Ultrasonic Nondestructive Evaluation Methods", *J. Appl. Mech., 50*, 1983, p. 1191.

[3] Bowles, C.Q. and Schijve, J., *Fracture Mechanisms: Advances in Quantitative Measurement of Physical Damage*, ASTM STP 811, 1983, p. 400.

[4] Buck, O. and Tittmann, B.R., *Advances in Crack Length Measurement*, C.J. Beevers (Ed.), Engineering Materials Advisory Services (EMAS), Cradley Heath, Warley, West Midlands, U.K., 1982, p. 413.

[5] Elber, W., *Damage Tolerance in Aircraft Structures*, ASTM STP 486, 1971, p. 230.

[6] Colby, D.M., Packman, P.F. and Pearson, H.S., "The Accuracy and Precision of Ultrasonic Shear Wave Flaw Measurements as a Function of Stress on the Flaw", *Mat. Eval., 103*, 5, 1970.

[7] Birchak, J.R. and Gardner, C.G., "Comparative Response of Machined Slots and Fatigue Cracks in 7075 Aluminum", *Mat. Eval. 109*, 275, 12, 1976.

[8] Becker, F.L., Doctor, S.R., Heasler, P.G., Morris, C.J., Pitman, S.G., Selby, G.P. and Simonen, F.A., "Integration of NDE Reliability and Fracture Mechanics, Phase I Report", *NUREG/CR-1696* and *PNL-3469, Vol. 1*, U.S. Nucl. Reg. Comm., Washington, D.C., 1981.

[9] Buck, O., Frandsen, J.D. and Marcus, H.L., "Crack Tip Closure and Environmental Crack Propagation", *Eng. Fract. Mech., 7*, 1975, p. 167.

[10] Golan, S. and Arone, R., *New Procedures in Nondestructive Testing*, P. Höller (Ed.), Springer-Verlag, Berlin, 1983, p. 557.

[11] Resch, M.T., London, B.D., Yuce, H.M. and Nelson, D.V., "Use of Nondestructive Evaluation Techniques in Studies of Small Fatigue Cracks", presented at *International Symposium on Fundamental Questions and Critical Experiments on Fatigue*, ASTM, Dallas, Texas, October, 1984.

[12] Buck, O., Thompson, R.B. and Rehbein, D.K., "The Interaction of Ultrasound with Contacting Asperities: Application to Crack Closure and Fatigue Crack Growth", *J. Nondestr. Eval., 4*, 1984, p. 203.

[13] Kendall, K. and Tabor, D., "An Ultrasonic Study of the Area of Contact Between Stationary and Sliding Surfaces", *Proc. Roy. Soc. Lond., A323*, 1971, p. 321.

[14] Haines, N.F., "The Theory of Sound Transmission and Reflection at Contacting Surfaces", Report RD/B/N4711, Central Electricity Generating Board, Research Division, Berkeley Nuclear Laboratories, Berkeley, England, 1980.

[15] Thompson, R.B., Skillings, B.J., Zachary, L.W., Schmerr, L.W. and Buck, O., *Review of Progress in Quantitative Nondestructive Evaluation, Vol. 2A*, D.O. Thompson and D.E. Chimenti (Eds.), Plenum Press, New York, 1983, p. 235.

[16] Thompson, R.B. and Fiedler, C.J., *Review of Progress in Quantitative Nondestructive Evaluation, Vol. 3A*, 1984, p. 207.

[17] Baik, J.-M. and Thompson, R.B., "Ultrasonic Scattering from Imperfect Interfaces: A Quasi-Static Model", *J. Nondestr. Eval., 4*, 1984, p. 177.

[18] Angel, Y.C. and Achenbach, J.D., "Reflection and Transmission of Elastic Waves by a Periodic Array of Cracks", *J. Appl. Mech., 107*, 1985, p. 33.

[19] Angel, Y.C. and Achenbach, J.D., "Reflection and Transmission of Elastic Waves by a Periodic Array of Cracks: Oblique Incidence", *Wave Motion, 7*, 1985, p. 375.

[20] Visscher, W.M., *Review of Progress in Quantitative Nondestructive Evaluation*, D.O. Thompson and D.E. Chimenti (Eds.), Plenum Press, New York, 1984, p. 187.

[21] Visscher, W.M., "Eastic Wave Scattering by a Surface-Breaking or Subsurface Planar Crack. II. Three-Dimensional Geometry", *J. Appl. Phys., 57*, 1985, p. 1538.

[22] Punjani, M. and Bond, L.J., "Scattering of Plane Waves by Partially Closed Cracks", *Review of Progress in Quantitative Nondestructive Evaluation, Vol. 5*, D.O. Thompson and D.E. Chimenti (Eds.), Plenum Press, New York, in press.

[23] Thompson, R.B., Fiedler, C.J. and Buck, O., *Nondestructive Measurements for Materials Property Determination*, C.O. Ruud and R.E. Green (Eds.), Plenum Press, New York, 1984, p. 161.

[24] Rehbein, D.K., Thompson, R.B. and Buck, O., *Review of Progress in Quantitative Nondestructive Evaluation 4A*, D.O. Thompson and D.E. Chimenti (Eds.), Plenum Press, New York, 1985, p. 61.

[25] Auld, B.A., "General Electromechanical Reciprocity Relations Applied to the Calculation of Elastic Wave Scattering Coefficients", *Wave Motion, 1*, 1979, p. 3.

[26] Thompson, R.B. and Lopes, E.F., "The Effects of Focusing and Refraction on Gaussian Ultrasonic Beams", *J. Nondest. Eval. 4*, 1984, p. 107.

[27] Timoshenko, S. and Goodier, J.N., *Theory of Elasticity*, 2nd ed., Chpt. 13, McGraw-Hill, New York, 1951.

[28] Beevers, C.J., Bell, I., Carlson, R.L. and Starke, E.A., "A Model for Fatigue Crack Closure77, *Engr. Fract. Mechanics, 19*, 93, 1980.

[29] Buck, O., Morris, W.L. and James, M.R., "Remaining Fatigue Life Prediction in the Initiation Regime Using SAW NDE", *J. Nondestr. Eval., 1*, 3, 1980.

[30] Achenbach, J.D. and Norris, A.N., "Loss of Specular Reflection due to Nonlinear Crack Face Interactions", *J. Nondestr. Eval., 3*, 3, 229, 1982.

[31] Thompson, R.B. and Gray, T.A., "Use of Ultrasonic Models in the Design and Validation of New N.D.E. Techniques", *Phil. Trans. R. Soc. Lond. A.*, in press.

ULTRASONIC NONDESTRUCTIVE EVALUATION, MICROSTRUCTURE, AND FRACTURE TOUGHNESS INTERRELATIONS

Alex Vary
National Aeronautics & Space Administration
Lewis Research Center
Cleveland, Ohio

Abstract

Conceptual models are advanced for explaining and predicting empirical correlations found between ultrasonic measurements and fracture toughness of polycrystalline solids. The models lead to insights concerning microstructural factors governing fracture processes and associated stress wave interactions. Analysis of the empirical correlations indicates that, in addition to grain size and shape, grain boundary reflections, elastic anisotropy, and dislocation damping are factors that underly both fracture toughness and ultrasonic attenuation.

1. Introduction

Ultrasonic methodology for determining the fracture toughness of structural materials is of high interest. A major incentive is the need for rapid, inexpensive, and nondestructive methods for verifying fracture toughness and related mechanical properties prior to placing critical parts in service and after the parts have been exposed to service. The viability of ultrasonics for verifying fracture toughness of materials and components is being investigated. Thus far, correlations of ultrasonic measurements with toughness have been demonstrated only on laboratory samples of polycrystalline solids. Further work is needed to establish underlying principles and appropriate approaches for applications to a variety of materials and hardware configurations.

Fracture toughness is an extrinsic mechanical property. It is a measure of a material's fracture resistance that quantifies the stress intensity at which a particular size crack becomes unstable and grows catastrophically (Brown and Scrawley, [2]; Hahn, et al, [11]). It is known that fracture toughness is governed by microstructure and morphology in polycrystalline solids. Because the attenuation of ultrasonic waves is also governed by similar factors, one should expect correlations between toughness and ultrasonic properties of polycrystallines.

Prior works have presented empirical evidence of correlations between ultrasonic attenuation measurements and fracture toughness in polycrystalline solids (Vary [20]; Vary and Hull [27,28]). A theoretical basis has been suggested for the correlations found between ultrasonic attenuation and fracture toughness (Vary [21]). This paper describes some conceptual foundations that can help explain and predict the empirical correlations.

2. Preliminary Considerations

2.1. *Microstructural Factors*

The role of the microstructure in governing the mechanical properties and behavior of engineering solids is well established (Green, [10]; MacCrone, [14]; Fores, et al, [5]). Ultrasonic evaluation of mechanical properties depends on the characterization of microstructure. Figures 1 and 2 illustrate examples of the dependence of mechanical properties and also ultrasonic factors on microstructure. Inferences of material properties are often based on metallographic and other destructive methods that reveal material composition, microstructure, and morphology. Ultrasonic methods are alternatives to the conventional approaches for characterizing microstructure and morphology.

By examining the micrographs in Figure 1, one might infer that decreasing grain size corresponds to increasing toughness. The micrographs in Figure 2 illustrate an opposite case where increasing grain (tungsten carbide crystal) size corresponds to increasing toughness. But, this apparent correlation with grain size is deceptive because it is the volume of cobalt "cement" (between the carbide crystals) with its high dislocation density that actually governs the fracture toughness.

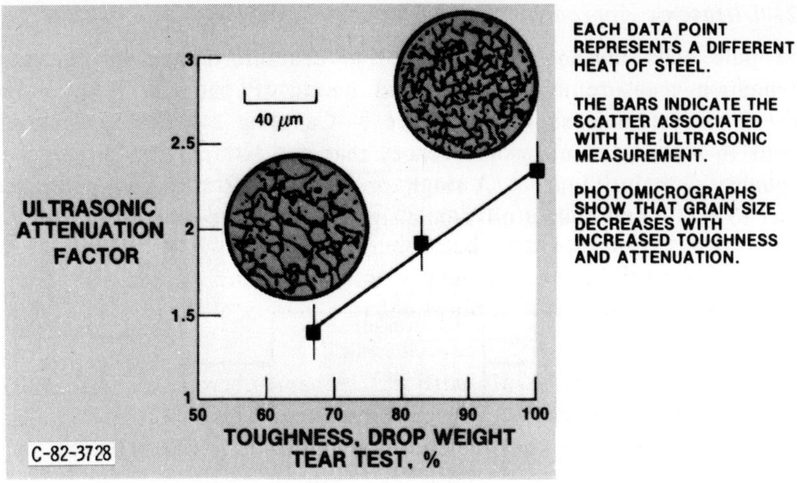

Figure 1: Ultrasonic attenuation factor as function of toughness as measured by drop weight test for low carbon steel (Vary [27]).

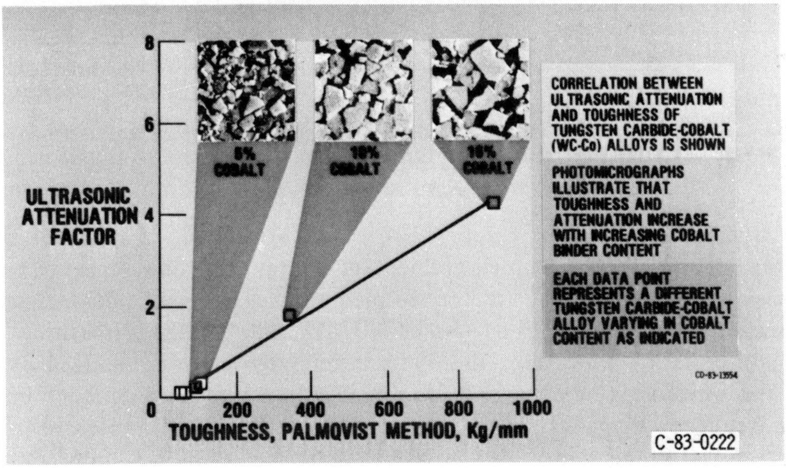

Figure 2: Ultrasonic attenuation factor as function of toughness as measured by Palmqvist method for cobalt cemented tungsten carbide (Vary and Hull [28]).

Clearly, there is more to consdier than grain size (or even grain shape and aspect ratio) in identifying factors that govern toughness. It will be seen that ultrasonic interrogation supplements the information appearing in photomicrographs and, indeed, can add to our understanding of factors that govern dynamic fracture behavior.

2.2. *Ultrasonic Approach*

The pulse-echo method for quantitative characterization of microstructure via attenuation measurements is assumed herein (Papadakis, [16]; Vary [23]). A pulse-echo system is shown in Figure 3. Coupling a probe to a material sample results in a series of ultrasonic echoes that can be analyzed either in the time or frequency domain, Figure 4. A single probe serves as a sending and receiving transducer to excite and collect ultrasonic signals in a pulse-echo mode.

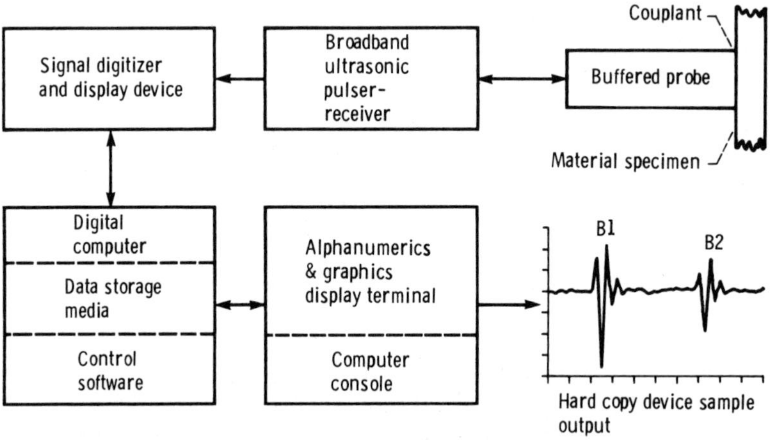

Figure 3: Block diagram of computer system for ultrasonic signal acquisition and processing for pulse-echo velocity and attenuation measurements.

Specific signals (e.g., the first two echoes B_1 and B_2 in Figure 4) are selected for examination and the measurement of energy loss due to various attenuation mechanisms. The signals are transformed to the frequency domain using digital Fourier transform algorithms (Bracewell, [1]; Vary, [22, 24]; Fitting and Adler, [4]). Information concerning the material microstructure is obtained by deconvolving the waveforms of B_1 and B_2 to obtain attenuation as a function of frequency. The frequency domain approach, associated concepts, and some salient results will be discussed in this paper.

3. Conceptual Foundations

3.1. *General*

Quantitative ultrasonic characterization of mechanical properties is a relatively new and essentially unexplored area. Lack of established theoretical foundations makes it necesasry to postulate interrelations that connect ultrasonic wave propagation properties with microstructure and mechanical properties. We begin with a consideration of hypothetical stress wave interactions with microstructural factors

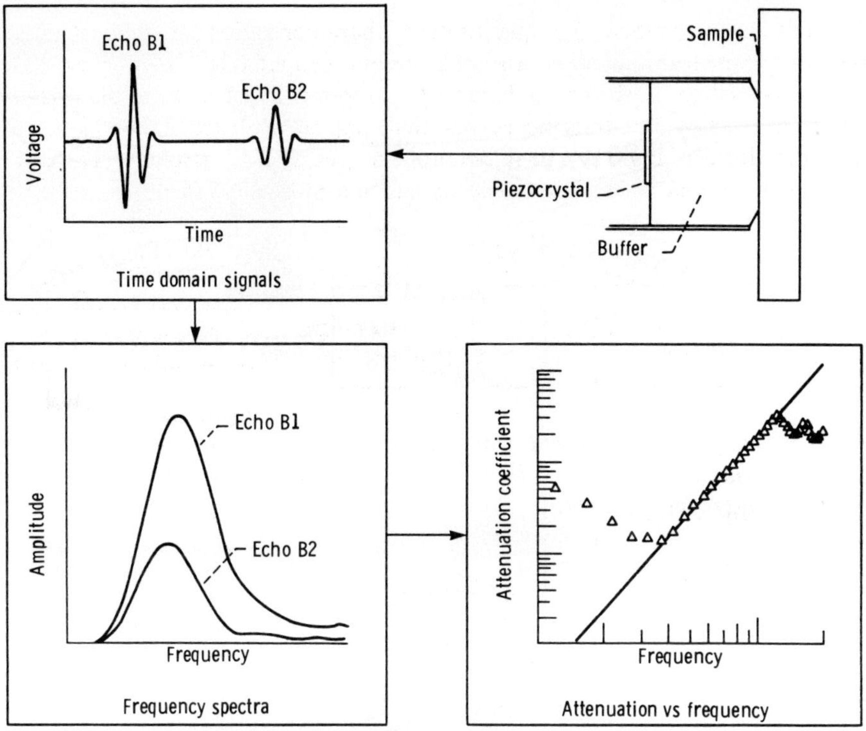

Figure 4: Illustration of ultrasonic signal processing procedure for attenuation measurements using pulse-echo method.

during crack nucleation events at the onset of fracture.

Spontaneous stress waves that arise during dislocation movements, microcrack nucleation, plastic deformation, and fracture are ultrasonic in nature. It is reasonable, therefore, to expect that ultrasonic probing can interrogate the microstructure to infer its effect on the stress waves that interact with it during the aforementioned processes. This idea, illustrated in Figure 5, can be pursued by considering three complementary concepts: (1) a stress wave interaction, (2) a microstructure transfer function, and (3) a microcrack nucleation mechanics concept. Models based on these concepts are shown schematically in Figures 6 to 8. They may be used to derive expressions for explaining and predicting empirical correlations that have been found among ultrasonic attenuation measurements, microstructue, and fracture toughness.

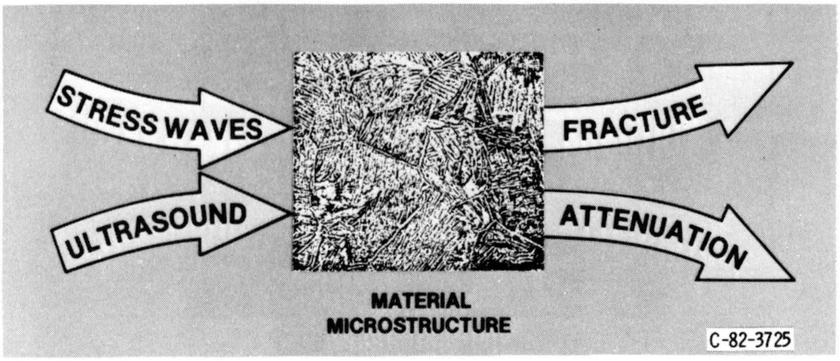

Figure 5: Diagram of concept wherein ultrasonic attenuation measures microstructural factors that govern stress wave propagation during microfailure events.

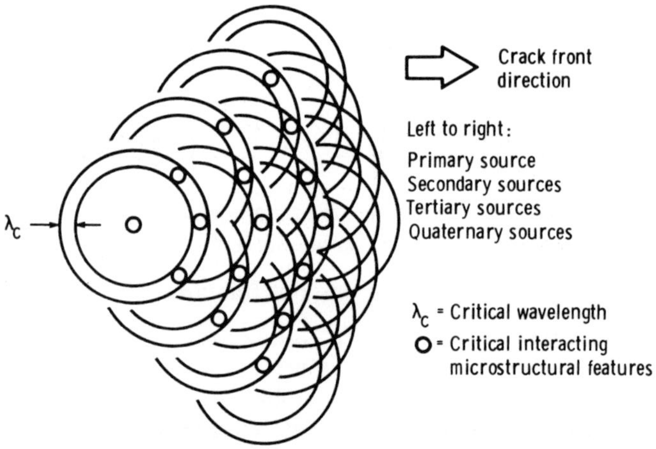

Figure 6: Diagram illustrating stress wave interaction (SWI) model showing cascade effect during interactions with critical microstructural features.

3.2. *Stress Wave Interaction (SWI) Concept*

The stress wave interaction (SWI) concept is needed to explain the correlations found between ultrasonic attenuation and fracture toughness. Experimental evidence indicates that the stress wave attenuation properties of polycrystalline metals determine toughness (Vary, [20]; Vary and Hull, [27]).

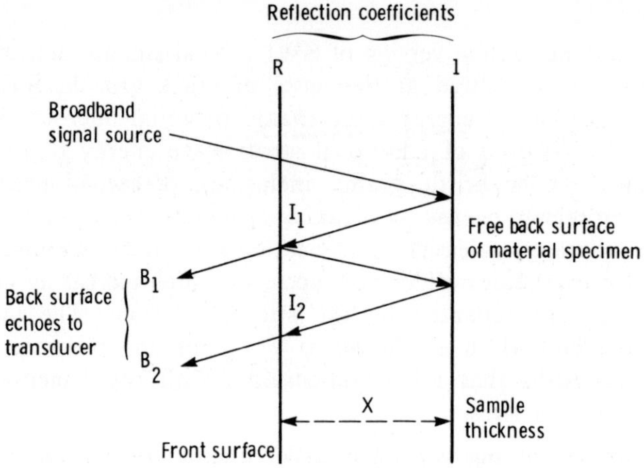

Figure 7: Diagram of echo system for defining the microstructure transfer function (MTF).

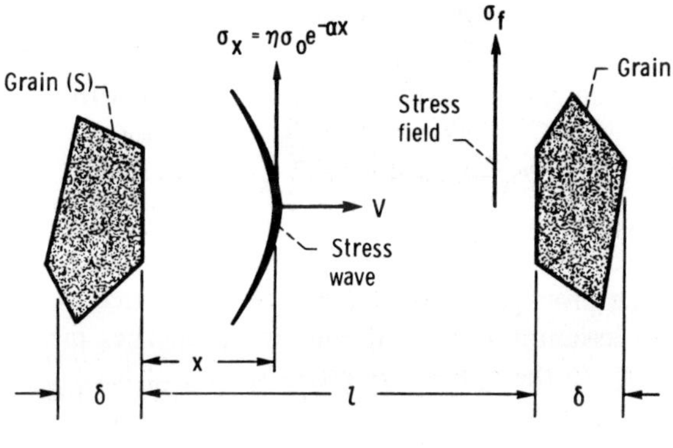

Figure 8: Diagram of microcrack nucleation mechanics (MNM) model wherein crack nucleation in grain S produces crack nucleation in grain R via stress wave interaction.

According to the SWI concept spontaneous ultrasonic stress waves produced at the onset of crack initiation interact with microstructural features ahead of a crack front (Vary, [21,24]). One corollary hypothesis is that stress wave interactions promote further cracking (Kolsky, [12,13]). Another corollary is that the stress wave energy simply dissipates in dislocation motions during the formation of a crack blunting zone (Hahn et al, [11]). In the former case stress wave interactions enhance crack nucleation (e.g., grain cleavage), coalescence, and growth. In the latter case cracking is inhibited by localized plastic deformation. Some combination of both cases undoubtedly applys in the dynamics that underlie the fracture behavior and the fracture toughness exhibited by many engineering solids (Curran, et al, [3]; Fu [7,8]).

In the crack nucleation version of SWI critical sits are activated by the spontaneous stress waves emitted at the onset of crack growth. Stress wave energy added to the local strain energy field around potential microcrack nucleation sits may result in the release of additional stress wave energy that acts on adjacent sites. The sites can be brittle grains, inclusions, or second phase particles that absorb and then release energy by breaking apart. If the advancing stress wavefront is reinforced by energy released by nucleation sites then, as indicated in Figure 6, an avalanch (or cascading) effect may occur in which increasing numbers of crack nucleations sites are activated by the stress wavefront. The uptake of energy depends on the bandwidth of the stress wave and the presence of critical (i.e., resonant) wavelengths that are commensurate with the dimensions of potential microcrack nucleation sites.

In the crack blunting version of SWI microcrack formation is inhibited by localized containment and dissipation of stress wave energy. Solids containing grains (i.e., crystallites) with high dislocation densities exhibit higher attenuation and absorption of ultrasonic stress wave energy via dislocation motions reduces the relative amount available for crack nucleation. Therefore, a greater initial stress intensity is required to induce dislocation pile-ups and conseqent crack growth. The relative energy absorption by dislocation motions (damping) depends on the stress wave bandwidth (i.e., content of wavelengths that interact with dislocations).

3.3. *Microstructure Transfer Function (MTF) Concept*

Underlying the microstructure transfer function (MTF) concept is the hypothesis that the propagation of probe ultrasound is governed by the same microstructural factors that govern the propagation of stress waves generated during fracture initiation. There is a tacit assumption that attenuation properties measured with benign probe ultrasound apply to the aggressive stress waves, at least at the onset of fracture.

Specification of the magnitudes of attenuation and energy transmission during stress wave interactions requires the definition of an appropriate MTF model. In particular the MTF model must specify how stress wave energy varies with ultrasonic wavelengths (i.e., with frequency). Accordingly, considering material microstructures as mechanical "filters" that have a transfer function definable in terms of frequency dependent ultrasonic attenuation mechanisms proves to be a useful concept (Vary, [24]; Vary and Kautz, [30]). A polycrystalline solid is assumed for the purposes of the ensuing discussion.

The conditions under which a MTF can be defined are restricted: The sample should have flat, parallel opposing surface and satisfy the conditions necessary to obtain two back surface echoes as shown in Figures 4 and 7. These constraints are for mathematical convenience and also for ease of signal acquisition (Truell et al, [19]). Signal acquisition and processing may be accomplished as described by Vary

[22]) and Generazio [9]).

The ultrasonic probe acting as sender and receiver, Figure 4, produces a broadband pulse and a series of back surface echoes in the material specimen. The first two back echoes, B_1 and B_2, Figure 7, are defined in terms of the internal signals E_1 and E_2 and R, the reflection coefficient at the specimen-transducer interface (Papadakis, [16]). The reflection coefficient is unity, 1, at the free surface.

$$B_1 = (1+R)E_1 \quad (1)$$

$$B_2 = TR(1+R)E_1 \,. \quad (2)$$

The quantities B_1, B_2, E_1, E_2, T, and R, in Figure 7, are Fourier transforms of the corresponding time domain quantities (Bracewell [1]). Combining Equations (1) and (2) yields,

$$T = B_2/RB_1 \quad (3)$$

The quantity T is the MTF and incorporates signal modulation (i.e., attenuation) and associated factors typical to polycrystalline microstructures (e.g., grain scattering, absorption) (Serabian and Williams [17]; Serabian, [18]). Papadakis [16] has demonstrated that the attenuation coefficient α can be found by frequency spectrum analysis and that it may be written as,

$$\alpha = (1/2x) ln(RB_1/B_2) \quad (4)$$

Combining Equations (3) and (4) we have,

$$\alpha = (1/2x) ln(1/T) \quad (5)$$

where, x is the specimen thickness.

Equation (5) forms the basis for identifying microstructural factors that govern fracture toughness. An expression for α for polycrystalline solids based on the MTF approach is (Vary and Kautz [30]),

$$\alpha = -\left(\frac{1}{a}\right) \ln\left(1 - G\left\{\frac{(\pi u f)^2}{1+(\pi u f)^2}\right\}\right) + \left(\frac{h}{v}\right) f + \left(\frac{16K}{3a}\right)\left\{\frac{(\pi u f)^4}{[1+(\pi u f)^2]^2}\right\} \quad (6)$$

where, a is mean grain size, f is ultrasonic frequency, G is grain boundary reflection coefficient, h is damping constant, K is elastic anisotropy, u is grain time constant, v is ultrasonic velocity, and $u = a/v$.

Use of analytial expressions of the form of Equation (6) requires a priori knowledge of the quantities a, G, h, and K. An alternative expression for the attenuation coefficient that does not require knowing explicit values for these quantities is desirable. One such expression that is useful for fitting experimental data is (Vary [21,23]),

$$\alpha = cf^m \tag{7}$$

Equation (7) has been found appropriate within the frequency range where attenuation is due primarily to Rayleigh scattering (Vary and Kautz [30]). The quantities c and m are empirical constants that substitute for the explicit quantities in Equation (6) (i.e., a, G, h, K, v). It will be seen later that Equation (7) forms a basis for demonstrating empirical correlations between fracture toughness and ultrasonic attenuation and that Equation (6) provides a means for analyzing the correlations.

3.4. *Microcrack Nucleation Mechanics (MNM) Concept*

The microcrack nucleation mechanics (MNM) concept combines the SWI and MTF concepts in the derivation of a relation among the attenuation properties of a microstructure, stress wave sources, and potential crack nucleation sites. The MNM concept is used the model energy transmission mechanisms that incorporate the principal factors involved in interactions between stress wave sources and potential microcrack nucleation sites (Vary [21]; Fu [6,7,8]).

The MNM model depicted in Figure 8 assumes that stress wave interactions promote microcrack nucleation in accordance with the SWI concept. A stress wavefront is shown traveling from grain S to grain R. These "grains" represent critical, interacting microstructural features such that, when grain S releases energy by fracturing, grain R will absorb some of this energy. Given an existing static stress field around R, it is only necessary for the impinging stress wave to impart enough energy to take R above a fracture threshold. The energy imparted to R depends on the ultrasonic stress wave attenuating properties of the ligament between grains S and R. The MNM model depicted in Figure 8 was applied specifically to plane strain fracture toughness (Brown and Srawley [2] to derive an expression connecting ultrasonic factors and the fracture toughness quantity (K_{Ic}/σ_y) (Vary [21]).

$$(K_{Ic}/\sigma_y)^2 = M\sqrt{v_\ell \beta_\delta / m} \tag{8}$$

where, K_{Ic} is plane strain fracture toughness and σ_y is yield strength. The right hand side of Equation (8) consists of ultrasonically determined quantities, where v_ℓ is velocity, M is an experimental constant for the materials being evaluated, and m is the exponent on frequency in Equation (7). The quantity β_δ in Equation (8) is the derivative

$$\beta_\delta = d\alpha/df\,|_\delta \tag{9}$$

where, $d\alpha/df$ is evaluated at a fequency that coresponds to a "critical" ultrasonic wavelength, λ_c, in the material. This wavelength is defined by the critical dimension, δ, in the microsructure (e.g., mean grain size or other feature that dominates during crack nucleation). This dimension is taken as the average for all microcrack nucleation sites that interact with the stress waves in accordance with the SWI concept. The critical dimension links the material transfer function, T, to specific microstructural features that govern fracture toughness.

The quantity on the left side of Equation (8), $(K_{Ic}/\sigma_y)^2$, is termed the "characteristic length". This characteristic length quantity is also a measure of the fracture toughness (Hahn et al, [11]). It is proportional to the size of the microcrack blunting zone that develops at the crack tip due to dislocation motions during the onset of crack growth in materials with plastic yield.

Equation (8) relates factors that govern toughness with ultrasonic propagation properties of the microstructure and predicts that the characteristic length or crack blunting zone size will be determined primarily by the attenuation properties associated with dislocation interactions. Materials that develop larger crack bunting zones confine and absorb more stress wave energy locally and exhibit greater fracture toughness.

All the quantities in Equation (8) are functions of material microstructure. Although the parameter M may remain essentially constant for a given alloy, the characteristic length and ultrasonic quantities (e.g., c and m in Equation (7) and a, G, h, K, v in Equation (6)) will change with heat treatment, phase composition, and other factors that affect toughness.

4. Verification and Analysis

4.1. *Experimental Correlations*

The correlation predicted by Equation (8) has been experimentally verified (Vary [20,21]) and Figure 9 shows the predicted correlation for three metals (two maraging steels and a titanium alloy). In these metals the critical microstructural dimension is the average grain size (or the subgrain "lath" spacing in the 200 grade maraging steel).

In the case of the titanium alloy described by Vary and Hull [27] there are three levels of microstructure: (prior) grains, colonies (within grains), and (within colonies) alternating alpha/beta phase platelets. The question regarding which of these features exerts the greatest influence on fracture toughness has been answered on the basis of the peviously described concepts. The results shown in Figure 10 indicate that the best empirical correlation and also the best agreement with theory (i.e., with Equation (8)) occur with data based on the beta phase thickness (Vary and Hull [27]). The alpha phase thickness was found to be somewhat less significant than the beta phase thickness (e.g., correlation coefficients were 0.977, and 0.998, respectively). The colony size was weakly influential, while grain size influence was indeterminate.

For the titanium alloy both the alpha and beta phases appear to be comparably critical microstructural features. Fracto-graphic studies identify the alpha phase aspect ratio as a critical factor. The alpha platelets appear to act as obstacles to microcrack extension and dissipate energy by making the crack path tortuous via frequent, abrupt changes in direction (Froes et al. [5]). However, the best correlation coefficient for Equation (8) is obtained with the beta phase. From an ultrasonic viewpoint the beta phase has the greatest influence on toughness. This is

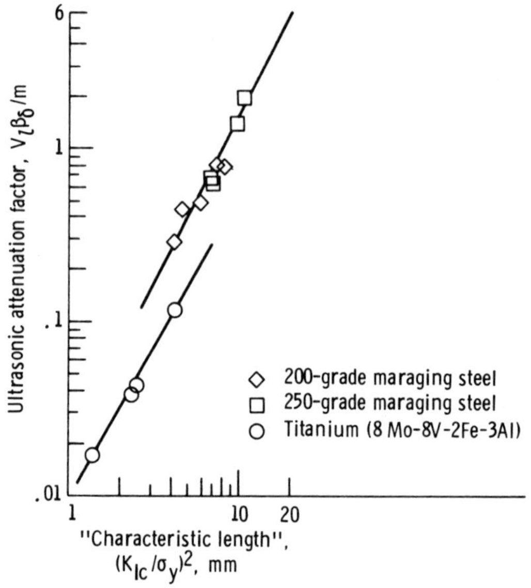

Figure 9: Experimental results showing predicted correlation of ultrasonic attenuation factor and fracture toughness (characteristic length) factor (Vary [20,21]).

undoubtedly due to its greater dislocation density (about tenfold greater than the alpha phase) and concomittantly greater attenuation and greater absorption and dissipation of stress wave energy.

4.2. *Analysis*

Equation (6) provides a basis for analyzing the correlations in Figures 9 and 10. First, note that the quantity $v_\ell\beta_\delta/m$ in Equation (8) is equivalent to the product of the size of the critical microstructural feature, a, and the attenuation coefficient, α_a, evaluated at the frequency corresponding to the critical wavelength δ, that is,

$$a\alpha_a = (v/m)\beta_\delta = v(/m)d(cf^m)/df\,|_\delta \tag{10}$$

where, δ is set equal to a and v is taken as longitudinal velocity v_ℓ. The quantity $a\alpha_a$ is the specific attenuation coefficient for the critical microstructural feature. Recalling that $u=a/v$ and $f_\delta = v/\delta = v/a$, Equation (6) is rearranged:

$$a\alpha_a = -ln(1-GF) + h + (16/3)KF^2 \tag{11}$$

where, $F=\pi^2/(1+\pi^2)$, is a dimensionless geometrical factor. Combining Equations (8), (10) and (11), we have,

$$(K_{Ic}/\sigma_y)^2 = M\sqrt{-ln(1-GF)+h+(16/3)KF^2} \tag{12}$$

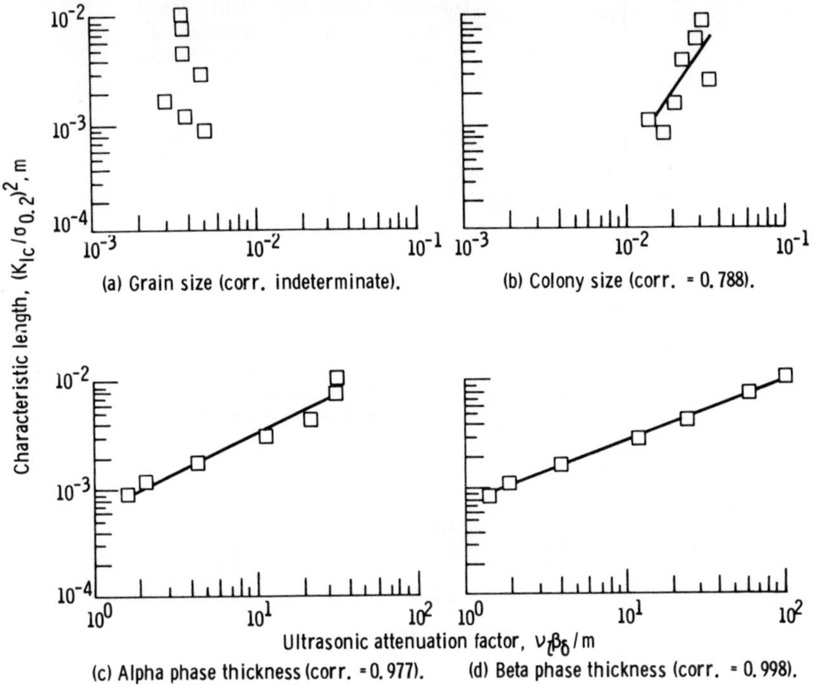

Figure 10: Comparison of toughness (characteristic length) and attenuation factors for four microstructural features in a two-phase titanium alloy (Vary and Hull [27]). ($\sigma_{0.2}=\sigma_y$ yield strength at 0.2% offset).

As might be expected, characteristic length (toughness) is a function of G, h and K the boundary reflection, damping, and elastic anisotropy factors, respectively. In Equation (12) fracture toughness is independent of explicit grain size, velocity, and frequency because it is defined in terms of a generalized geometric factor, F, and the elastic-anelastic attributes of a critical microstructural feature. In particular, $(K_{Ic}/\sigma_y)^2$ is directly proportional to the damping constant, an intrinsic property of crystallites (grains) that is related to their dislocation densities (Nowick and Berry [15]).

Insight into the quantitative effects of variations in G, h, and K can be gained by examining interrelations with the empirical parameters c and m in Equation (7). It has been found that c and m are interdependent for variously conditioned samples of a polycrystalline material. As an example, experimentally determined values for c and m are plotted and tabulated in Figure 11. Apparently, the parameters c and m will be interdependent for samples of a material that have undergone heat treatment or other thermomechanical processing that preserves global microstructural patterns while altering mechanical properties like toughness.

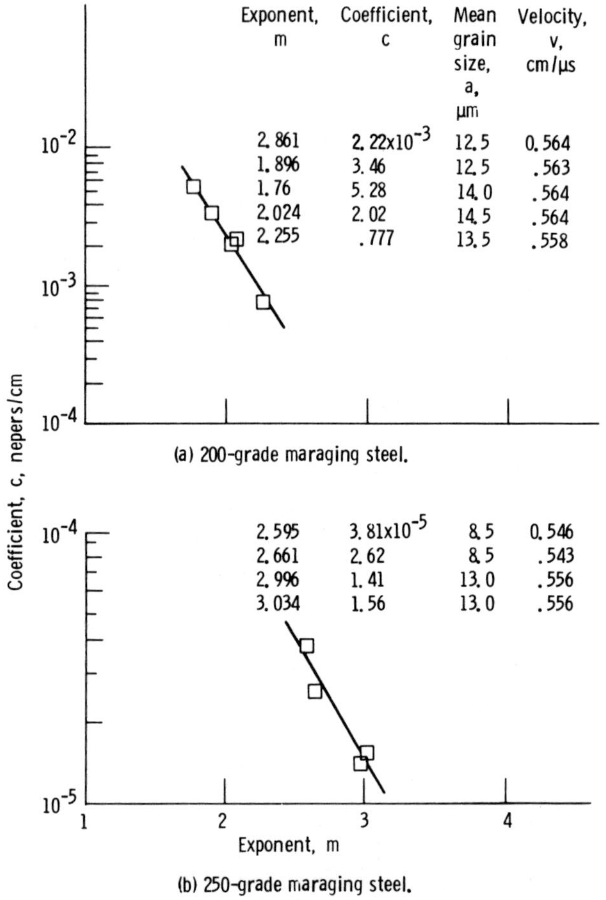

Figure 11: Interdependence of the experimentally determined attenuation parameters c and m for two maraging steels, based on Equation (7) (Vary and Kautz [30]).

A graphical method for evaluating the quantities G, h and K in terms of c and m was devised (Vary and Kautz [30]). By varying a, G, h, K and v in Equation (6) over a range of representative values, the c and m parameters for a bounded frequency range (e.g., the Rayleigh frequencies) may be computed with Equation (7). The parametric fields in Figure 12 were generated by assuming a range of values for a, G, h and K and $v = 0.55\ cm/\mu s$ corresponding to a 250 grade maraging steel (250MS). Co-plotted in Figure 12 are data for the 250MS from Figure 11. In Figure 12(a) the 250MS data fall near the curve for mean grain size equal to 10 μm, since $8.5 \leq a \leq 13 \mu m$. In Figures 12(b) and 12(c) $a = 10 \mu m$ is assumed and in Figure 12(b) the 250MS data lie on the $G \simeq 0.007$ curve while in Figure 12(c) they lie near the $h = 2 \times 10^{-5}$ curve.

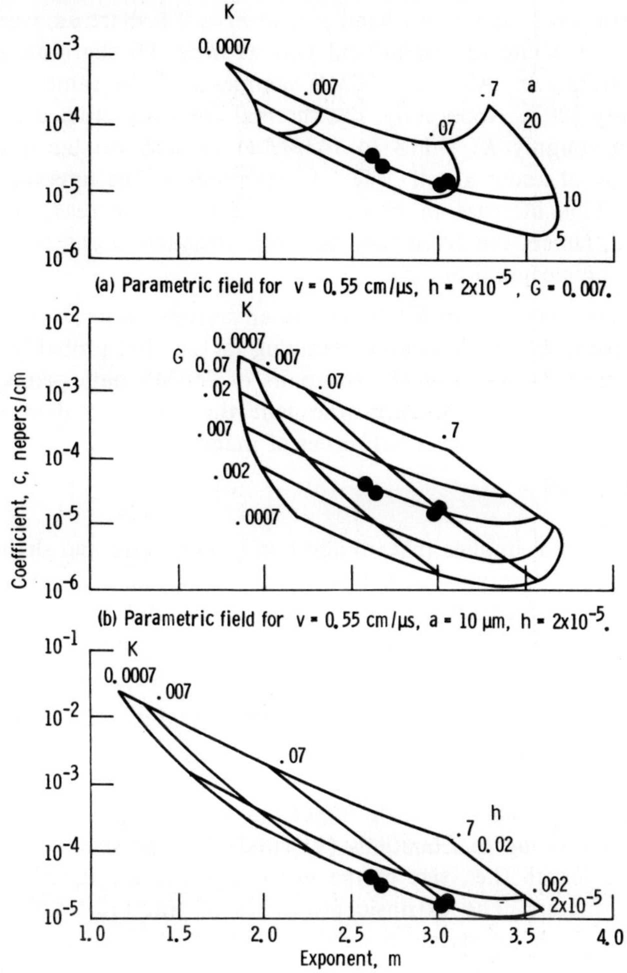

(c) Parametric field for v = 0.55 cm/μs, a = 10 μm, G = 0.007.

Figure 12: Parametric fields of c versus m based on Equations (6) and (7) for assigned values of mean grain size, a, boundary reflection coefficient, G, damping constant, h, elastic anisotropy, K, and velocity, v. Data for 250 grade maraging steel from Figure 11 are co-plotted within the fields (Vary and Kautz [30]).

All the fields in Figure 12 indicate that the elastic anisotropy, K, varies by roughly a half order of magnitude between the two pairs of 250MS data points. This is significant because the left-hand pair at $m \simeq 2.6$ share an average toughness $K_{Ic} \simeq 118 MPa\sqrt{m}$ while the right-hand pair at $m \simeq 3.0$ share an average toughness $K_{Ic} \simeq 143 MPa\sqrt{m}$. All four 250MS samples had the same yield strength of 1400 MPa (Vary [20]). Apparently, the thermal treatment that increased fracture toughness from roughly $K_{Ic} = 118$ to $143 MPa\sqrt{m}$ also produced an increase in elastic anisotropy of about a half order of magnitude in the interval between K = 0.007 to 0.07. This increase in K can account for an increase in (stress wave) attenuation and, hence, the correlation between attenuation and fracture toughness predicted by the conceptual models.

The prominent increase in K, the elastic anisotropy factor, overshadows a comcomitant increase in h, the dislocation damping factor, that probably contributed to the increased toughness between the two pairs of 250MS data points in Figure 12. The scale of Figure 12(c) obscures the fact that the two data pairs lie near h-curves that differ by perhaps a half order of magnitude.

5. Discussion

The preceding analysis implies that, in addition to grain size and shape, grain boundary reflections (and transmissions), elastic anisotropy, and dislocation damping are among factors that determine the toughness of polycrystalline solids. This interpretation is predicated on the empirical correlations and applicability of the SWI, MTF, and MNM concepts described herein.

Of course, it is not surprising that in polycrystallines dislocations and local plastic deformations determine toughness via crack blunting. The new information inferrable from Figures 9 to 12 and from Equations (8) to (12) is that metals that are more attenuating tend to develop larger crack blunting zones in the vicinity of active cracks. According to Equations (11) and (12) the specific attenuation and the charactristic length (i.e., size of the blunting zone) are functions of the same microstructural factors. The extrinsic properties of fracture toughness and stress wave attenuation both ultimately depend on dislocation densities. Tougher metals are apparently those in which more stress wave energy transmission is impeded at grain boundaries by reflections and elastic scattering and then dissipated in localized plastic deformation zones by dislocation motions.

The preceding observations should be contrasted with experimental findings for fiber reinforced composites. In fiber reinforced plastics greater strength and toughness correspond to less attenuation (Vary and Bowles [26]; Vary and Lark [29]). Many composites have brittle matrices that cannot sustain plastic deformation. In this case it is better to have prompt, efficient transmission of stress wave energy away from microcrack nucleation sites. Therefore, unlike metals it is desirable for brittle matrix composites to have lower attenuation and less local concentration of stress wave energy.

6. Conclusion

Three conceptual models for interrelating ultrasonic attenuation, microstructure, and fracture toughness were described: (1) a stress wave interaction model, (2) a microstructure transfer function model, and (3) a microcrack nucleation mechanics model. It was shown that the conceptual models may be combined to explain and predict the experimental correlations found between ultrasonic attenuation and fracture toughness and also to identify microstructural factors that underly the stress wave attenuation and associated toughness properties of polycrystalline solids.

7. References

[1] Bracewell, R.N., *The Fourier Transform and Its Applications*, McGraw-Hill, New York, N.Y., 1978.

[2] Brown, W.F., Jr. and Srawley, J.E., "Plane Strain Crack Toughness Testing of High Strength Metallic Matrials," *ASTM STP 410*, American Society for Testing and Materials, Philadlephia, Pennsylvania, 1966.

[3] Curran, D.R., Seaman, L. and Shockey, D.A., *Phys. Today 30:1*, 1977, pp. 202-210.

[4] Fitting, D.W. and Adler, L., *Ultrasonic Spectral Analysis for Nondestructive Evaluation*, Plenum Press, New York, New York, 1981.

[5] Froes, F.H., Chesnutt, J.C., Rhodes, C.G. and Williams, J.C., in *Toughness and Fracture Behavior of Titanium*, R.G. broadwell and C.F. Hickey, Jr. (Eds.), ASTM STP. American Society for Testing and Materials, Philadelphia, Pennsylvania, 1978.

[6] Fu, L.S., *Applied Mechanics Reviews, 35:8*, 1982, pp. 1047-1057.

[7] Fu, L.S., *Engineering Fracture Mechanics, 18:1*, 1983, pp. 59-67.

[8] Fu, L.S., in *Developments in Mechanics*, Vol. 12, E.J. Haug and K. Rim, (Eds.), University of Iowa Press, Iowa City, Iowa, 1983, pp. 263-265.

[9] Generazio, E.R., "The Role of the Reflection Coefficient in Precision Measurement of Ultrasonic Attenuation", *TM-83788 National Aeronautics and Space Administration*, Washington, DC., 1984.

[10] Green, R.E., Jr., "Ultrasonic Investigation of Mechanical Properties", *Treatise on Materials Science and Engineering*, Vol. 3, Academic Press, New York, 1973.

[11] Hahn, G.T., Kanninen, M.F. and Rosenfeld, A.R., in *Annual Reviews of Materials Science*, Vol. 2, R.A. Huggins (Ed.), Annual Reviews, Inc., Palo Alto, California, 1972, pp. 381-404.

[12] Kolsky, H., *Stress Waves in Solids*, Dover Publishers, New York, New York, 1963.

[13] Kolsky, H., in *International Conference on Dynamic Crack Propagation*, G.C. Sih, (Ed.), Noordhoff International Publishing Co., Leyden, The Netherlands, 1973.

[14] MaCrone, R.K., (Ed.), in *Treatise on Materials Science and Technology*, Vol. 11, Academic Press, New York, New York, 1977.

[15] Nowick, A.S. and Berry, B.S., *Anelastic Relaxation in Crystalline Solids*, Academic Press, New York, New York, 1972.

[16] Papadakis, E.P., in *Physical Acoustics – Principles and Methods*, Vol. 12, W.P. Mason and R.N. Thurston (Eds.), Academic Press, New York, 1976, pp. 277-374.

[17] Serabian, S. and Williams, R.S., *Matl. Eval., 36:8*, 1978, pp. 55-62.

[18] Serabian, S., *Brit. J. of Nondestruc. Test, 22:2*, 1980, pp. 69-77.

[19] Truell, R., Elbaum, C. and Chick, B.B., *Ultrasonic Methods in Solid State Physics*, Academic Press, New York, 1969.

[20] Vary, A., *Matl. Eval., 36:7*, 1979, pp. 55-64.

[21] Vary, A., in *Fracture Mechanics*, C.W. Smith (Ed.), ASTM STP677, American Society for Testing and Materials, Philadlephia, Pennsylvania, 1979.

[22] Vary, A., in *Proceedings of the Twelfth Symposium on Nondestructive Testing*, Columbus, Ohio and Southwest Research Institute, San Antonia, Texas, 1979, pp. 33-46.

[23] Vary, A., in *Research Techniques in Nondestructive Testing*, Vol. 4, R.S. Sharpe, (Ed.), Academic Press, London, England, 1980, pp. 159-204.

[24] Vary, A., in *Mechanics of Nondestructive Testing*, W.W. Stinchcomb, (Ed.), Plenum Publishing Co., New York, 1980, pp. 123-141.

[25] Vary, A., *Advanced Materials Technology*, CP-2251, NASA, Washington, DC, 1982, pp. 411-412.

[26] Vary, A. and Bowles, K.J., in *Proceedings of the Eleventh Symposium on Nondestructive Evaluation*, American Society for Nondestructive Testing, Columbus, Ohio and Southwestern Research Institute, San Antonia, Texas, 1977, pp. 242-258.

[27] Vary, A. and Hull, D.R., *Matl. Eval., 41:3*, 1982, pp. 309-314.

[28] Vary, A. and Hull, D.R., "Ultrasonic Ranking of Toughness of Tungsten Carbide", *TM-83358, NASA*, Washington, DC, 1983.

[29] Vary, A. and Lark, R.F., *J. of Test. and Eval., 7*, 1979, pp. 185-191.

[30] Vary, A. and Kautz, H.E., in *Analytical Ultrasonics in Research and Testing*, NASA CP-2383, 1986, pp. 257-298.

ACOUSTIC MICROSCOPY

ACOUSTIC MICROSCOPY OF SURFACE CRACKS: THEORY AND PRACTICE

G.A.D. Briggs
Department of Metallurgy & Science of Materials
University of Oxford, England
M.G. Somekh
Department of Electronic & Electrical Engineering
University College, London

Abstract

Recent developments in acoustic microscopy of surface cracks are briefly reviewed. Particular attention is devoted to the dominant role which is played by the excitation of Rayleigh waves on the surface of the specimen.

1. Introduction: Rayleigh Wave Imaging

The development of acoustic microscopy with a resolution equal to or better than a light microscope has been a formidable achievement [1]. By focussing acoustic waves at frequencies typically in the GHz range in water on the axis of a sapphire lens, and then scanning the lens, it is possible to obtain images with submicron resolution. But the advantage of scanning acoustic microscopy does not lie in its resolution alone. The significance of the technique is that enables one to image, with this resolution, how the acoustic waves interact with the elastic properties of the specimen. In many applications in materials science it is possible to be more specific than this: for a specimen with flat topography, and under suitable operating conditions, a dominant role in the reflected signal is played by the excitation of Rayleigh waves in the surface of the specimen. The instrument may for these purposes be regarded as a Rayleigh wave microscope, and the images will be sensitive to factors that affect the propagation of Rayleigh waves in the specimen [2]. Such features may include variations of bulk elastic properties, surface coatings or implanted layers, elastic anisotropy such as grain structure, or attenuation due to damping or scattering.

The contrast theory for the acoustic microscope is now well developed for a specimen whose properties are uniform across the surface. If the surface of specimen is at the focal plane of the lens, then the returned signal from the transducer due to the reflected acoustic wave is the sum over its area of the acoustic wave incident on it. In two dimensions this is:

$$V(o)=\int_{-k}^{k} P(k_x)R(k_x)dk_x \qquad (1)$$

where k is the wavenumber in the coupling fluid and k_x is the component of the wavevector parallel to the specimen surface. $P(k_x)$ is the two-way pupil function of the lens for waves passing through the focus and $R(k_x)$ is the reflectance function of the specimen surface. The expression can easily be extended to three dimensions by integrating over k_y as well, but throughout this paper the theory will be confined to two dimensions because the solutions for the contrast from cracks have not yet been extended to three dimensions. If the specimen is moved towards the lens so that the surface is a distance $-z$ from the focal plane, then a phase shift $exp(-2ik_z z)$ is introduced for each wave, where k_z is the component of the wavevector normal to the specimen surface. The signal on the transducer, expressed as a function of the defocus, z, is then:

$$V(z)=\int_{-k}^{k} P(k_x)R(k_x)exp(-2ik_z z)dk_x \,. \qquad (2)$$

By suitable changes of variable this can be expressed as a Fourier transform, with z and $2k_z$ as the transform pair. This means that $V(z)$ can be regarded as a Fourier transform of the pupil function of the lens and the reflectance function of the specimen. The pupil function may be regarded as a constant for an instrument with a given lens, and if it is well behaved the interest is in the reflectance function of the specimen. This is a complex function, because there are phase changes associated with it, and for an isotropic half-space the dominant phase change occurs around $k_x=k_R$, the Rayleigh wavevector. In the Fourier transform this would be expected to lead to oscillations of periodicity:

$$\Delta z=\frac{2\pi}{(2k_{z1}-2k_{z2})} \qquad (3)$$

where k_{z1} and k_{z2} correspond to $k_x=0$ and k_R respectively. If θ is the angle of incidence, and θ_R is the angle that corresponds to $k_x=k_R$, then this may be written as:

$$\Delta z = \frac{\lambda}{2(1-\cos\theta_R)} \tag{4}$$

where $\lambda = 2\pi/k$ is the wavelength in the fluid. This formula, which was first derived from ray considerations [3] and also by an angular spectrum approach [4], has now been extensively tested experimentally [5], and it is well established that oscillations do indeed occur in $V(z)$ with periodicity that is accurately described by this equation. This provides very strong evidence for the dominant role played in the contrast by Rayleigh waves.

Further evidence for the role that Rayleigh waves play is provided by the interference fringes that are often seen alongside cracks or surface steps in specimens [6]. The explanation for this is that as the lens of the microscope is scanned across the surface of the specimen, waves that are geometrically reflected from the surface will travel a constant path length. However, waves incident at the Rayleigh angle, θ_R will excite Rayleigh waves in the surface of the specimen. If these Rayleigh waves are partially reflected by a feature such as a crack, then their total path length back to the transducer would lead to interference fringes half a wavelength apart. It is found that such fringes do indeed have a spacing of half a Rayleigh wavelength, and this confirms that the contrast is very sensitive to factors that affect the propagation of the Rayleigh waves that are excited in the specimen surface. It also emphasises that surface cracks should be included as one of the factors that can have a very marked effect on that propagation.

2. Enhanced Contrast from Surface Cracks

The contrast from surface cracks can be very strong; much stronger than would be expected from geometrical considerations alone [7]. Figure 1a shows an optical micrograph of an Al-20%Si plain bearing alloy that had been subjected to a fatigue test. Figure 1b shows the same area at the same magnification, but viewed in a scanning acoustic microscope at 0.73 GHz. The cracks are somewhat easier to find in the acoustic image. This is because the cracks are so fine that they have little effect on waves, whether optical or acoustic, that are reflected geometrically from the surface. But the cracks can scatter strongly Rayleigh waves incident on them from within the surface, and this is why they appear with such strong contrast in the acoustic microscope.

There is a serious difficulty in accounting for the contrast from cracks in terms of the $V(z)$ theory outlined above, because it depends on having a reflection function for the surface being imaged. A reflectance function is only defined for the case where infinite plane waves incident on a surface are reflected into infinite plane waves at the same angle, i.e. with conservation of the tangential component of the wavevector. This requires that the surface is uniform in planes parallel to the surface (variations along the normal to the surface, as in multilayered structures, are more easily dealt with). But once a crack or other surface discontinuity is present, this approach is no longer valid. Instead, it is necessary to use a scattering function for the surface, to allow for the fact that the crack can scatter waves of

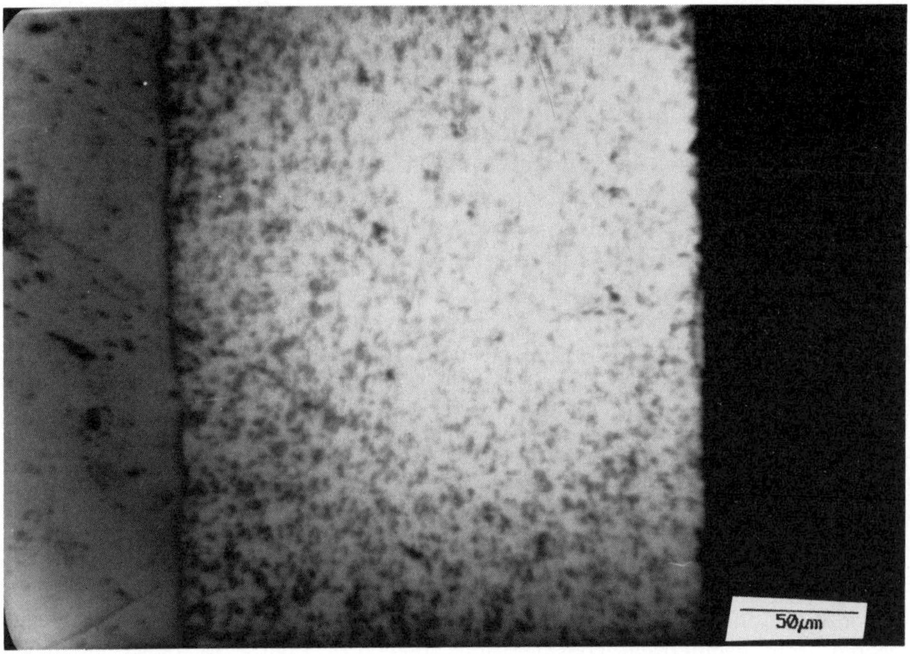

Figure 1a: An Al-20%Si plain bearing that has been subject to fatigue – optical.

any incident angle into waves of any reflected angle.

In its general form this problem requires solutions that are not available, but the previous knowledge that it is the behaviour of the Rayleigh waves that are likely to dominate the solution can be used to reduce the problem to a manageable form. First the reflectance function must be separated into a part that contains the Rayleigh wave phenomena and the remainder. This may be written in terms of k_p and k_o, the pole and the zero respectively of the reflectance function:

$$R(k_x) \equiv R_o(k_x)\left[1+\frac{k_p^2-k_o^2}{k_x^2-k_p^2}\right] \tag{5}$$

k_p is equal to the complex wavenumber for the propagation of the leaky Rayleigh waves, and for a lossless material k_o is equal to the complex conjugate of k_p. This identity is simply a definition of $R_o(k_x)$, but it has the advantage that to a very good approximation of the whole effect of the excitation of Rayleigh waves on $R(k_x)$ is described by the second term in the brackets, and that over the range of

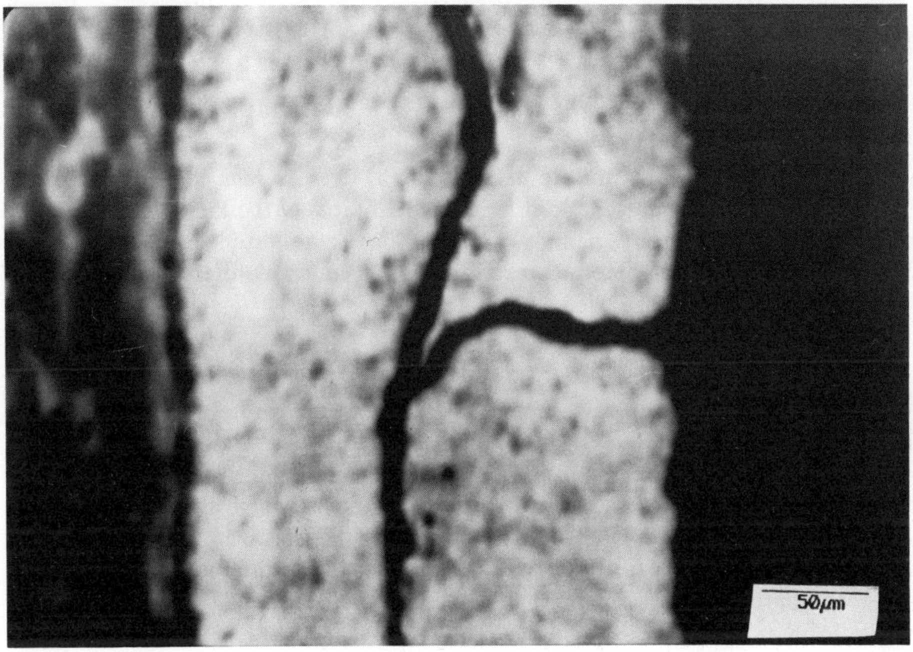

Figure 1b: An Al-20%Si plain bearing that has been subject to fatigue – acoustic (0.73 GHz).

k_x for which this makes a significant contribution (i.e. close to $\pm k_p$), $R_o(k_x) \simeq 1$. The assumption is now made that the crack being considered is much less wide than the size of the focal spot, and therefore will have negligible effect on the geometrically reflected wave, which is described by the first term in the brackets in (5). The effect of the crack is therefore accounted for by its effect on the part of the reflected wave that is associated with Rayleigh wave excitation.

The simplest situation to consider is a plane surface-breaking crack normal to the surface, and a cylindrical (or line-focus) lens directly over the crack and with its axis in the same plane as the crack. This has the advantage that, while there may be phase shifts associated with the scattering of Rayleigh waves by the crack, there are no other phase shifts to consider. This is because, by symmetry, any wave that suffers reflection by the crack travels back along a path to the transducer equal in length to the path that it would have travelled in a forward direction had it not been reflected. This means that the effect of the crack can be characterized by a scattering parameter S_R that multiplies the Rayleigh part of the reflection function in (5). S_R allows for both backward and forward scattering, i.e. for reflection and transmission by the crack, though in some cases for very deep cracks only the reflection may be important. It may be a complex parameter, to allow for changes in both amplitude and phase at the crack.

For this rather simple case the $V(z)$ theory expressed in [2] can be directly modified [7]:

$$V(z)=\int_{-k}^{k} P(k_x)R_o(k_x)\left[1+S_R\frac{k_p^2-k_o^2}{k_x^2-k_p^2}\right]\exp(-1ik_z z)dk_x. \tag{6}$$

This formula has been evaluated to calculate $V(z)$ over a crack in a material similar to that imaged in Figure 1, and it predicts qualitatively the kind of features that are observed experimentally, namely a change in the amplitude and phase of the oscillations in $V(z)$, and therefore reversals in the contrast observed in images of cracks as the defocus, z, is varied.

3. Green Function Theory

A more satisfactory theoretical account [8] proceeds along the same principles as those outlined above, but allows for any position of the lens relative to the crack. It does not assume that scattered Rayleigh waves will radiate energy back to the fluid with the same value of k_x as that which excited them, and it allows for radiation by Raleigh waves that have not been scattered by the crack as well as those that have. However, to enable a number of integrals to be evaluated analytically, the fairly good approximation is made that:

$$R(k_x)\simeq R_o(k_x)+\frac{k_p^2-k_o^2}{k_x^2-k_p^2} \tag{7}$$

Using this, together with the other good approximation that:

$$i(k_p^2-k_o^2)/(2k_p)=-2\alpha \tag{8}$$

where α is the imaginary part of the complex wavevector for the leaky Rayleigh wave propagation (and therefore describes the attenuation of the Rayleigh wave due to leakage into the fluid), the contribution due to Rayleigh wave excitation in the surface of the reflected field $p_1(x)$ in the fluid at the surface from an incident field $p_i(x')$ is:

$$P_L(x)=-2\alpha\{\int_{-\infty}^{x} p_i(x')\exp[ik_p(x-x')]dx' + \int_{x}^{\infty} p_i(x')\exp[-ik_p(x-x')]dx'\}. \tag{9}$$

This expression for the field in the fluid close to the surface at x. There are two contributions, one from all the Rayleigh waves excited to the left of x and propagating to the right, and the second from those excited to the right and propagating to the left. When a crack is present at x_o, then for $x<x_o$ first term is unaffected. The second term, however, must be integrated only from x to x_o, because

any excitation beyond x_o is affected by the coefficient of transmission of the crack. There is therefore a third term describing all the Rayleigh wave that is excited to the right of the crack, and is then transmitted past the crack. Finally, there is a fourth term describing the Rayleigh wave that is excited in the surface to the left of the crack and then reflected by it. These last two terms are both subject to an additional Green function to allow for the propagation from the crack back to x. Gathering these four terms together, and swapping the order of the third and fourth for tidiness, for $x<x_o$ the Rayleigh reflected field at the surface is:

$$\begin{aligned} p_1(x) = -2\alpha[&\int_{-\infty}^{x} p_i(x')\exp[ik_p(x-x')]dx' \\ &+\int_{x}^{x_o} p_i(x')\exp[-ik_p(x-x')]dx' \\ &+\exp[-ik_p(x-x_o)]\{R\int_{-\infty}^{x_o} p_i(x')\exp[ik_p(x_o-x')]dx' \\ &+T\int_{x_o}^{\infty} p_i(x')\exp[-ik_p(x_o-x')]dx'\}]. \end{aligned} \tag{10}$$

A similar expression can be written for the Rayleigh reflected field to the right of x_o. These expressions relate the Rayleigh field $p_1(x)$ to the incident field $p_i(x')$. In order to use these results to compute $V(z)$ it is necessary to transform them to the spatial frequency domain by a double Fourier transform [9]. This gives a scattering function:

$$\begin{aligned} S(k_x,k_x') = R_o\delta(k_x-k_x') + i2\alpha\delta(k_x'-k_x)k_p\left[\frac{1}{k_x^2-k_p^2}+\frac{1}{k_x'^2-k_p^2}\right] \\ +\frac{2\alpha}{\pi}\left[\frac{(T-1-R)k_xk_x'+(T-1+R)k_p^2}{(k_x^2-k_p^2)(k_x'^2-k_p^2)}\right]. \end{aligned} \tag{11}$$

In this expression, it can be seen that the first two terms on the right are independent of the crack. They do not contain R or T, and they are governed by Dirac δ-functions so that they are confined to specular reflection, i.e. $k_x=k_x'$. They describe the geometrical and Rayleigh reflected components respectively. The third term on the right describes the scattering by the crack. If $R=0$ and $T=1$, then this is the same as having no crack and the third term vanishes. But for any other values of R and T this term contributes and leads to scattering from a given value of k_x' to different values of k_x. Because this is a Rayleigh wave effect, the third term contributes significantly only when both k_x' and k_x are close to k_p (where α is the measure of what is meant by close).

To use $S(k_x,k_x')$ in the formalism for $V(z)$ it is necessary to consider the summations of the transmitted and the received waves separately, the two being related by $S(k_x,k_x')$. The origin of x is now taken at the crack, with the lens centered above x_o, and suitable Green functions are included in the phase of the transmitted and received waves to allow for this and for the defocus z. Also, the lens function is split up into L_1 and L_2 for the two directions of propagation through it. With these definitions, finally, the contrast is:

$$V(z)=\int_{-\infty}^{\infty}\int_{-\infty}^{\infty} L_2(k_x)\exp[i(k_z z-k_x x_o)]S(k_x,k_x')$$
$$L_1(k_x')\exp[i(k_z' z+k_x' x_o)]dk_x' dk_x \, . \tag{12}$$

4. Theoretical and Experimental Results

In the formulation that has been expounded here, the crack is characterized by two parameters, R and T. These are external to the contrast theory of the acoustic microscope. They must be calculated quite independently of the theory presented here, and the corollary of this is that where solutions exist from independent calculations, they can be used directly in the acoustic microscope contrast theory. As a first check, if values $R=0$ and $T=1$ are used, then for a given lens characteristic the form of the $V(z)$ curve is found to be the same as would be predicted using the theory for a uniform half-space (Equation (2)). For a preliminary evaluation of what the theory would predict when the specimen contains a deep crack the following values of R and T were taken [10]:

$$R=0.4\exp(i0.6)$$
$$T=0 \, . \tag{13}$$

Various qualitative features can be observed from $V(z)$ curves calculated in this way. If the lens is taken to be directly over the crack ($x_o=0$), then the form of the $V(z)$ curve is similar to that predicted by equation (6), but it is more accurate because even in this symmetrical case it is necessary to take account of Rayleigh reflection that is not affected by the crack (i.e. the first two terms in equation (10)). It is found that there is again a decrease in the amplitude of the oscillations in $V(z)$, and also a shift in their phase, corresponding to the 0.6 radian phase shift in the Rayleigh reflection coefficient in (13). If the lens is displaced to one side of the crack by an amount $1/4(n+1/2)$ Rayleigh wavelengths, where n is zero or a small integer, then the oscillations are reduced still further, because the reflections from the two sides of the crack then have a path difference of half a wavelength and therefore reach the transducer in antiphase. If the lens is moved further from the crack, then $V(z)$ is similar to the uncracked specimen until a defocus is reached where the incident beam is so wide that the crack lies within it, and then the larger deviations start to appear.

For direct comparison with images, it is more useful to plot a line scan for a given defocus; this can be called $V(x)$. Figure 2 shows line scans for various values of defocus for a cylindrical lens scanning horizontally over a crack in glass, scaled for a frequency of 215 MHz. Various qualitative features can be seen in these curves. First, even in the line scan that corresponds to no defocus (a), there are considerable oscillations in amplitude. These have a spacing of half a Rayleigh wavelength from one maximum to the next, and in an image they would appear as fringes parallel to the crack, exactly as has been observed. At a defocus of

$z = -27\mu m$ (b), this ripple is reduced in amplitude, but now another feature is apparent, namely there is a gross contrast associated with the crack of considerably greater amplitude than the Rayleigh fringes. At this particular defocus it has width about two Rayleigh wavelengths, and would make the contrast of the crack in image bright relative to the surround. Since the wavelength in water is about half the Rayleigh wavelength in glass, and since the resolution of a microscope would be about equal to the wavelength, this means that the width of a crack would be expected to be about four times the resolution of the instrument at focus. If the defocus is increased a little to $z = -34\mu m$, this contrast reverses, so that the crack would then appear dark in an image. At this defocus the effect of the crack is least when the lens is directly over it, and is greatest when the lens is about 1.54 Rayleigh wavelengths to one side. Thus in an image at this defocus the crack would be dark, and it would appear as a double dark line with a bright centre.

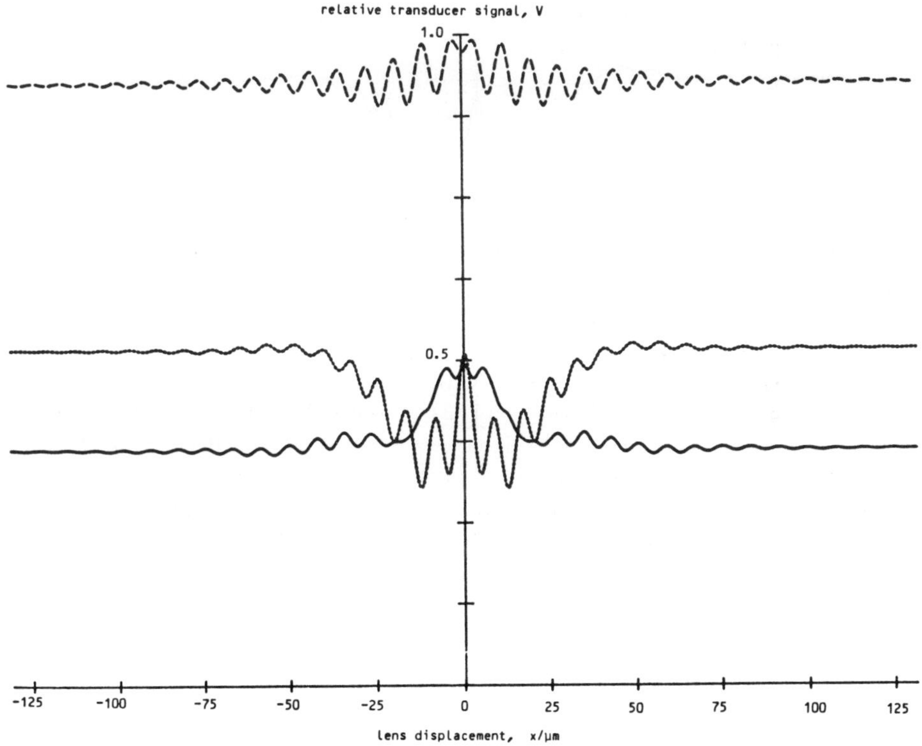

Figure 2: Theoretical line scans for a cylindrical (line-focus) lens scanning over a vertical surface crack in glass, scaled for 215 MHz: (a) at focus -----; (b) z = -27 μm – ; (c) z = -34 μm ····.

These calculations have been tested by direct experiment with a 215 MHz cylindrical lens over a crack in glass [11]; this corresponds directly to the two dimensional geometry considered in the theory above. Experimental curves are shown in Figure 3 corresponding to the theoretical curves in Figure 2. The greatest uncertainties in relating the two quantitatively concern the exact value of z (the results are very sensitive to this), the characterization of the lens, and the values of the crack parameters R and T. Nevertheless it can be seen that many of the qualitative features predicted by the theory are indeed found experimentally.

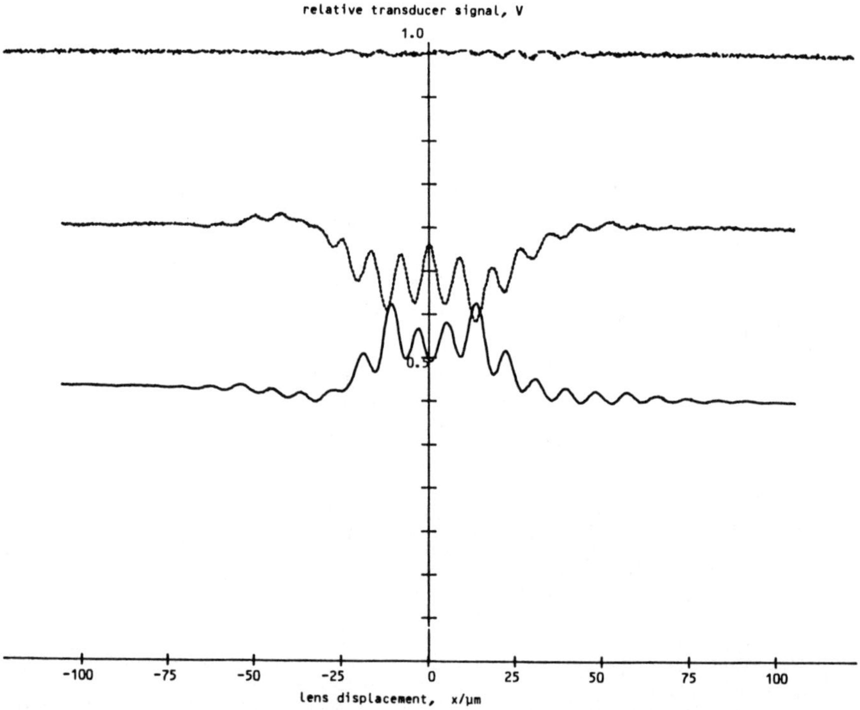

Figure 3: Experimental line scans for the same parameters as Figure 2.

Although this theory is at present restricted to a two-dimensional form, it seems to give a useful account of many of the features that are found in images obtained with a full spherical lens in the scanning acoustic microscope. Figure 4 shows an acoustic micrograph of thermal stress relief cracks in the TiN coating on a hard metal cutting tool. As the tool cools down from the coating process the metal contracts more than the ceramic, and so the cracks are tightly closed in compression. In Figure 4a relative little contrast is seen from the cracks. In Figure 4b, which was taken with some defocus, the cracks appear bright against a dark background. The width of the cracks in the image corresponds to about 10 μm, which at 0.73 GHz is about two Rayleigh wavelengths in a ceramic. At the slightly greater defocus of Figure 4 the contrast reverses and the cracks appear dark, but in some places a fringe pattern or a bright line in the centre of the crack can be discerned. The most important point of all, however, is that there is strong contrast from these cracks even when they are much narrower than the resolution spot size. In measurements by scanning electron microscopy it was found that the width of such cracks was less than 0.1 μm. At 0.73 GHz the focussed spot size is about 2 μm in diameter, so that the effect of the crack on the geometrically reflected waves would indeed be very small, and so the enhanced contrast from the cracks must indeed be explained in terms of their effect on the Rayleigh waves.

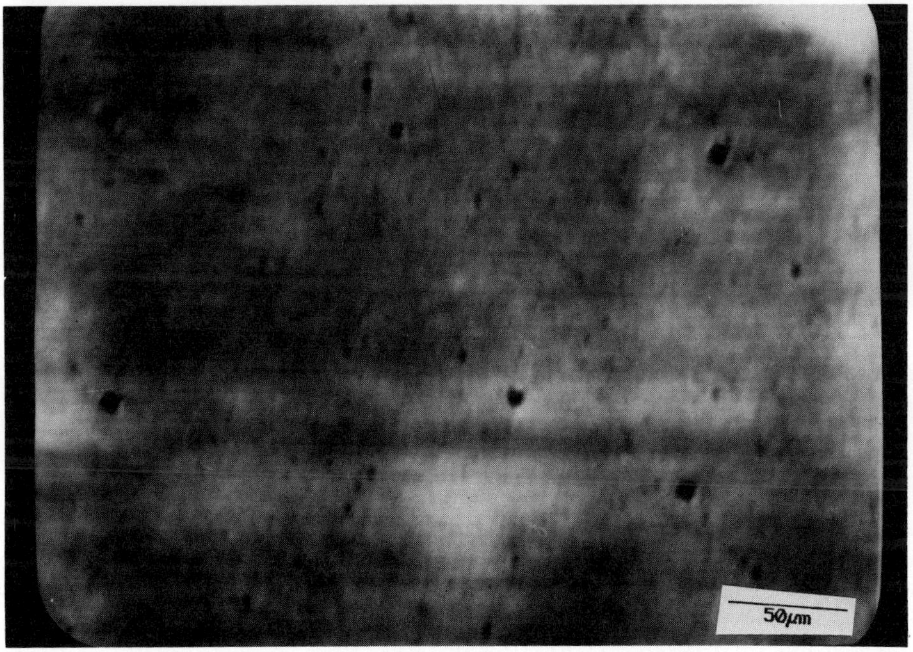

Figure 4a: Acoustic micrographs (0.73 GHz) of a TiN coating (approximately 10 µm thick) on a hard metal cutting tool – at focus. The cracks formed because of thermal stresses on cooling from the chemical vapour deposition, the microhardness indents that can be seen were made later for location purposes.

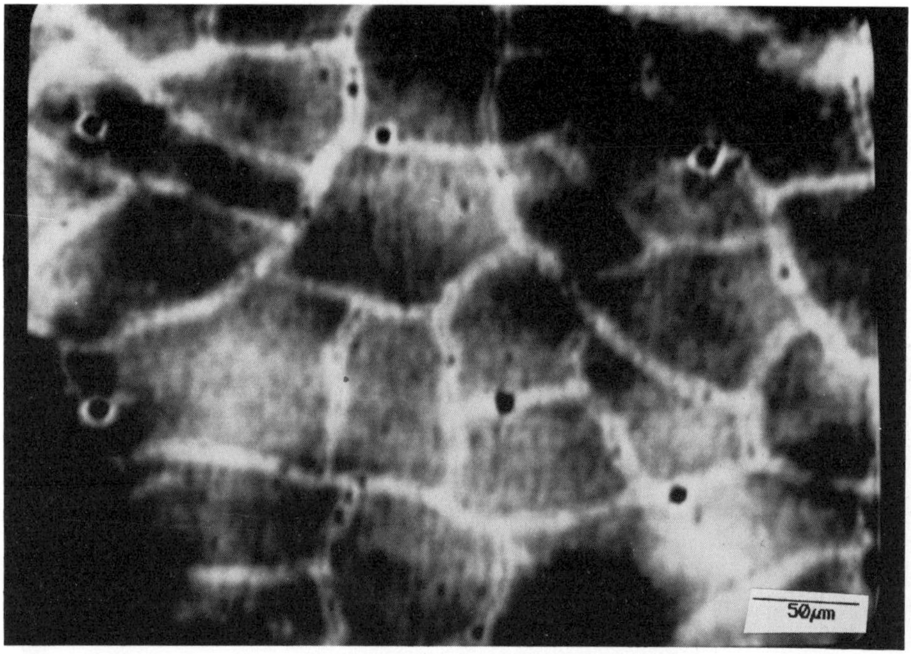

Figure 4b: Acoustic micrographs (0.73 GHz) of a TiN coating (approximately 10 μm thick) on a hard metal cutting tool – z = -7 μm.

5. Prospects for Crack Characterization

In order to develop practical methods of quantitative crack characterization, further progress is needed in various areas. On the experimental side it is desirable to be able to make the kind of measurements that will enable R and T to be determined. R is probably best determined from a scan of $V(x)$, because this is much more sensitive to R than to T. $V(z)$, on the other hand, depends equally on R and T when the lens is directly over the crack (as can be seen by setting $x_o=0$ in (10) and in the complementary equation for the Rayleigh reflected field to the right of the crack, and then considering the symmetry). The values of R and T will be particularly sensitive to the crack depth when it is of order one Rayleigh wavelength deep, and for a given crack they will then vary with the acoustic frequency [12]. Therefore it will be especially valuable to develop methods of spectroscopy that will enable the frequency dependence of R and T for a given crack to be determined.

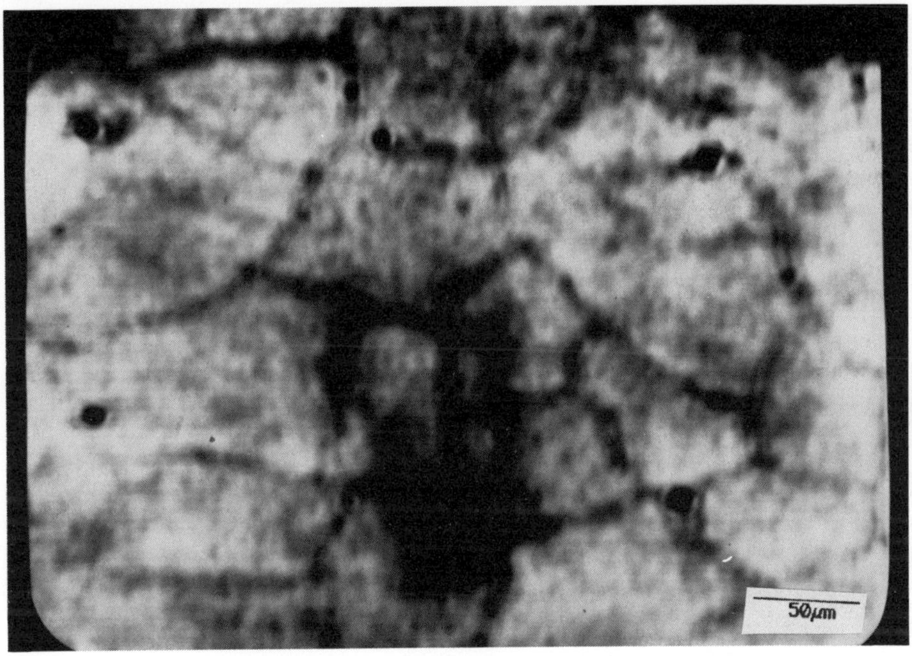

Figure 4c: Acoutic micrographs (0.73 GHz) of a TiN coating (approximately 10 µm thick) on a hard metal cutting tool – z = -16 µm.

On the theoretial side, the first requirement is to extend the theory for the contrast from a crack to three dimensions, in order to describe the case of a spherical lens. Attention often needs to be turned to the calculated values of R and T. At present most of the solutions that are available are for a Rayleigh wave excited at infinity scattered by a crack in an unloaded half-space. In the acoustic microscope there is always a coupling fluid (usually water) on the surface of the specimen, and although this is relatively light loading, so that the unloaded solutions are useful approximations, it should be taken into account for greater accuracy. This inevitably means that the Rayleigh waves cannot propagate to the crack from infinity, and indeed in the integrals of equation (10) excitation is included right up to the crack itself, and so the calculations of the scattering by the crack need to be extended to include near-field excitation. However, it must be appreciated that if the scattering is a function of the distance of the points of excitation and observation from the crack, many of the transforms that have been solved analytically in the present theory will need to be tackled numerically.

A great advantage of the present formulation of the theory is that it is possible to take calculations that are already available for different crack geometries and use them to calculate the contrast in the scanning acoustic microscope for that geometry. Thus solutions for different kinds of cracks, with different orientations, perhaps with varying amounts of crack closure or asperity contact across the crack faces, and taking into account the possibility of fluid in the crack, can all be incorporated directly. For example, solutions already exist for the scattering of surface waves by a surface [14] or a sub-surface crack [15], and, for the three dimensional case where all angles of incidence must be considered, for oblique incidence on a surface breaking crack [16]. Calculations also exist for the resonances in the scattering of Rayleigh waves by rectangular [17] and half-penny shaped cracks [18], and for the effect of roughness on the crack faces [19]. If such results can be combined with comparable progress in the development of experimental techniques to enable the relevent parameters to be measured for cracks that are found, then quantitative characterization of the cracks may indeed become possible.

References

[1] Lemons, R.A. and Quate, C.F., "Acoustic microscopy", in *Physical Acoustics*, W.P. Mason and R.N. Thurston (Eds.), Vol. 14, 1979, pp. 1-92.

[2] Briggs, G.A.D., *An Introduction to Scanning Acoustic Microscopy*, Oxford University Press and The Royal Microscopical Society, 1985.

[3] Parmon, W. and Bertoni, H.L., "Ray interpretation of the materials signature in the acoustic microscope", *Electron Lett.*, 15, 1979, pp. 684-6.

[4] Atalar, A., "A physical model for acoustic signatures", *J. Appl. Phys.*, 50, 1979, pp. 8237-9.

[5] Kushibiki, J. and Chubachi, N., "Material charactrization by line-focus-beam acoustic microscope", *IEEE*, SU-32, 1985, pp 189-212.

[6] Yamanaka, K. and Enomoto, Y., "Observation of surface cracks with scanning acoustic microscope", *J. Appl. Phys.*, 53, 1982, pp. 846-50.

[7] Ilett, C., Somekh, M.G. and Briggs, G.A.D., "Acoustic microscopy of elastic discontinuities", *Proc. R. Soc. Lond.*, A 393, 1984, pp. 171-83.

[8] Somekh, M.G., Bertoni, H.L., Briggs, G.A.D. and Burton, N.J., "A two-dimensional imaging theory of surface discontinuities with the scanning acoustic microscope", *Proc. R. Soc. Lond.*, A 401, 1985, pp. 29-51.

[9] Saad, A., Bertoni, H.L. and Tamir, T., *Proc. IEEE 62*, 1974, pp. 1552-61.

[10] Achenbach, J.D., Gautesen, A.K. and Mendelsohn, D.A., "Ray analysis of surface-wave interaction with an edge crack", *IEEE Trans.*, SU-27, 1980, pp. 124-9.

[11] Rowe, J.M., Kushibiki, J., Somekh, M.G. and Briggs, G.A.D., "Acoustic microscopy of surface cracks", *Phil. Trans. R. Soc. Lond., A320*, 1986, (in press).

[12] Clark, L.R., Chan, C-H. and Khuri-Yakub, B.T., "Acoustic evaluation of grinding damage in ceramic materials", *Proc. IEEE Ultrasonics Symposium*, 1982, pp. 979-982.

[13] Khuri-Yakub, B.T., *Proc. IEEE Ultrasonics Symposium*, 1985, (in press).

[14] Mendelsohn, D.A., Achenbach, J.D. and Keer, L.M., "Scattering of elastic waves by a surface-breaking crack", *Wave Motion*, 2, 1980, pp. 277-92.

[15] Achenbach, J.D. and Brind, R.J., "Scattering of a surface wave by a sub surface crack", *J. Sound and Vibration*, 76, 1981, pp. 43-56.

[16] Angel, Y.C. and Achenbach, J.D., "Reflection and transmission of obliquely incident Rayleigh waves by a surface-breaking crack", *J. Acoust. Soc. Am.*, 75, 1984, pp. 313-18.

[17] Domarkas, V., Khuri-Yakub, B.T. and Kino, G.S., "Length and depth resonances of surface cracks and their use for the crack size estimation," *Appl. Phys. Lett.*, 33, 1978, pp. 557-9.

[18] Ayter, S. and Auld, B.A., "On the resonances of surface breaking cracks", (submitted).

[19] Ayter, S. and Auld, B.A., "Inverse scattering from rough cracks", *IEEE Ultrasonics Symposium Proc.*, 1982, pp. 988-93.

QNDE USING LOW-FREQUENCY ACOUSTIC MICROSCOPY

B.T. Khuri-Yakub
P.A. Reinholdtsen
Edward L. Ginzton Laboratory
W. W. Hansen Laboratories of Physics
Stanford University
Stanford, California 94305

Abstract

An acoustic microscope operating at frequencies of 3 and 10 MHz has been built for the purpose of detecting subsurface defects in structural materials. Samples of fused quartz were examined with vertically- and horizontally-oriented seeded subsurface defects. By defocusing the transducer, it was possible to enhance detection of subsurface defects and to estimate their depth below the surface.

The acoustic microscope was also used to detect subsurface defects in composite materials such as carbon-carbon, carbon-epoxy, and Kevlar-epoxy. Imaging can also yield information about the layering as well as the texture and orientation of fibers in undamaged as well as damaged composites. The depth and orientation of cracks was determined by defocusing. The images were compared with those produced by C-scan and dye-enhanced x-ray radiography. We believe this is the only technique where defects can be detected and tracked to find how they propagate below the surface.

Our results indicate that low-frequency acoustic microscopy is an excellent tool for investigating metals and composite materials for surface and internal damage.

1. Introduction to Acoustic Microscopy

A schematic diagram of the acoustic microscope's operation is shown in Figure 1. An efficient, focused transducer was excited with a tone burst in order to interrogate the sample of interest. It was also used as the receiving transducer in order to detect the echo from the surface. The transducer receives the reflected signal which is rectified and filtered to give a signal proportional to the amplitude of the returned signal. The transducer is mechanically scanned in a raster style in order to create a two-dimensional acoustic image. This image is then displayed on a high-resolution color monitor using various pseudo-color and grey-level color schemes.

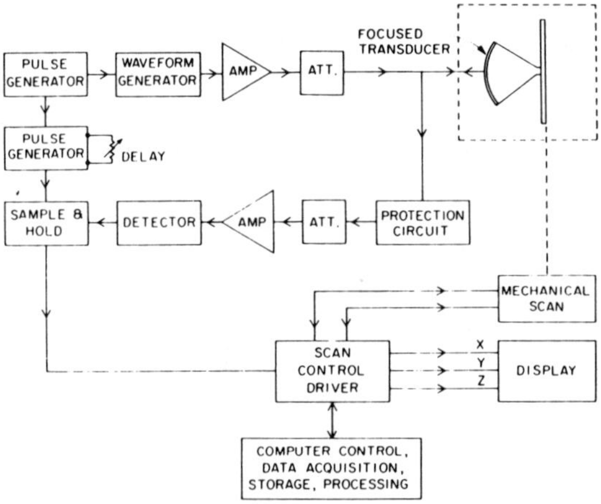

Figure 1: Schematic diagram of the operation of the acoustic microscope.

Variations in the echo come from two effects. First, changes in the surface topography lead to differences in the amount of energy reflected back to the transducer. Secondly, the return signal depends on the impedance of the surface of the sample. This impedance not only depends on the material properties of the surface material, but also on the presence of any features that may be below the surface. We were interested in being able to detect subsurface features, in spite of variations in the surface topography (i.e., surface roughness). This is analogous to finding hidden caves behind a rock wall by "yelling" at it and listening to the echo, even if the wall has many cracks and ledges on its surface.

Because of the low impedance of air, water is usually used as the coupling medium between the transducer and the sample. More energy can be coupled into water than into air from the transducer, as well as from water into the samples because of the better impedance match with water. By using a higher impedance coupler, smaller changes in the sample's surface impedance are detectable. Thus, the above analogy is really only accurate if the wall and caves are submerged under water.

The focus of the transducer is normally located at the surface of the sample being scanned. It was found that increased sensitivity to subsurface features could be obtained by defocusing, that is by moving the transducer closer to the sample so that the focus would be located within the sample, assuming the sample did not affect the acoustic fields.

There are primarily three types of waves transmitted into the sample: longitudinal bulk waves, shear bulk waves, and pseudo-Rayleigh surface waves. The longitudinal and shear waves have different velocities, which leads to their being transmitted into the sample at differing angles because of refraction. Thus, as the transducer is defocused, these components of the transmitted waves are focused at different depths below the surface, as shown in Figure 2. Likewise, the surface waves are "focused" at the intersection of the axis of the transducer and the sample's surface. Therefore, it is not accurate to say that, as the transducer is defocused, the focus has moved inside the sample. There are now several foci.

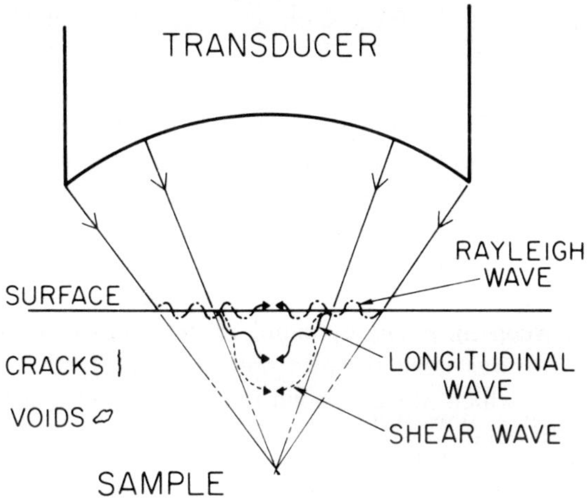

Figure 2: Three types of acoustic waves transmitted into sample.

Defocusing the transducer, however, increases the concentrations of acoustic energy below the surface. If there is a defect below the surface, it will lead to larger changes in the acoustic field, and therefore to larger changes in the received echo. Only defects within an acoustic wavelength of the surface affect the pseudo-Rayleigh waves because those waves decay exponentially into the material. Defects further into the sample affect the bulk waves more or less, depending on where the corresponding focus is located. If a defect is near one of the foci of the bulk waves, the component of the bulk waves will be affected a large amount, leading to a larger change in the received signal.

2. The Transducer

The transducer we used was very efficient with only a 3 dB two-way insertion loss. It had an aperture diameter of 16 mm and a focal length of 16 mm. Operating at its fundamental center frequency of 3 MHz, this gave a focal spot size of 500 μm (at 3 MHz, the acoustic wavelength in water is 500 μm). The transducer could also be operated at its third harmonic, which actually occurred at 10 MHz, giving a focal spot size of 150 μm.

3. The Samples of Interest

3.1. *NDE of Fused Quartz Samples*

We considered two different types of samples in our work. One set of samples was made of fused quartz with seeded defects. Another set of samples was made of different types of composite materials.

The quartz samples we looked at were designed to simulate machined metal with various defects below the surface, such as voids and vertical and horizontal cracks. In one type of sample, shown schematically in Figure 3 (optically-bonded sample schematic), a block of fused quartz had its top surface polished optically flat and defects machined into the surface. Another optically flat block of quartz was then brought into close contact, and because of intermolecular forces between the two surfaces, it bonded directly to the first block. The resulting bond was both optically and acoustically invisible. In other samples, shown schematically in Figure 4 (vertical crack schematic), defects were machined into vertical faces of blocks of fused quartz, the faces were brought close together, and a thin layer of paraffin was drawn between the faces by capillary action to form a paraffin bond only 0.3 μm thick. This was done because optical bonds tend to separate when exposed directly to water, which makes a bond that is even more noticeable acoustically than a thin paraffin bond.

4. Experimental Results

4.1. *Samples with Polished Surfaces*

In earlier work we looked at subsurface defects when the surface of the sample was polished smooth. In Figure 5 (vertical crack sampl, on focus), note that the smaller defects are barely visible when the image is taken with the transducer on focus. As the transducer is defocused, the smaller defects show with increased contrast.

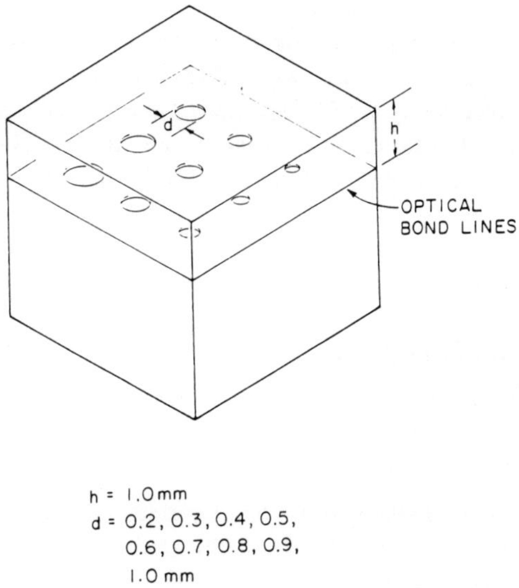

Figure 3: *Schematic of optically-bonded fused quartz sample with horizontal defects.*

4.2. *Samples with Roughened Surfaces*

We then looked at the effect of surface roughness on the ability of the microscope to detect the subsurface defects. The surfaces of the sample were roughened with a coarse polish and the samples were reimaged to determine if any degradation of the images had occurred. The surfaces of the samples were roughened with diamond powder varying in size from 5 to 60 μm, which gave corresponding surface roughnesses ranging from 2-30 μm. These roughnesses are of the same size as those obtained by most machining techniques, as can be seen in Table 1.

At a frequency of 3 MHz (focal spot size of 500 μm), the images were unaffected for all roughnesses studied. This is not surprising since the dimension of the roughness is less than 1/25 of a wavelength.

At 10 MHz (focal spot size of 150 μm), the images were virtually unaffected up to surface roughnesses of 15 μm. With 30 μm roughness and the transducer on focus, the image was visibly degraded. In Figure 6 (picture of horizontal cracks, .25 mm below surface, 30 μm surface roughness, on focus), notice that the two smaller defects are hard to distinguish from the random spots produced by the roughening. Notice, however, that when the transducer is defocused, not only do the smaller defects show increased contrast, the spots from the roughening disappear.

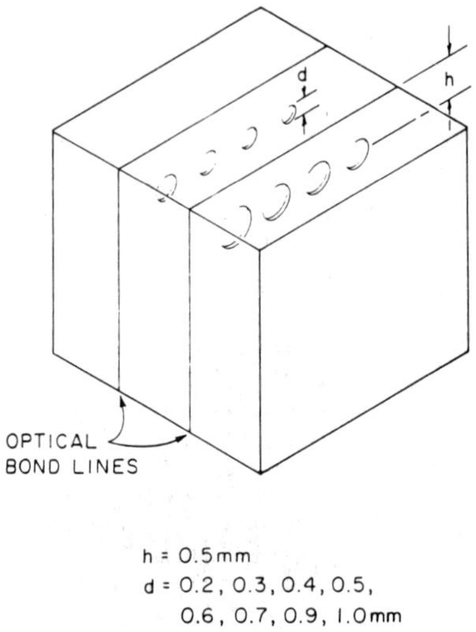

Figure 4: *Schematic of paraffin-bonded fused quartz sample with vertical defects.*

The decreased sensitivity to surface roughness is to be expected because, when the transducer is defocused, the incident beam is spread out over the surface of the sample, as seen in Figure 7. This tends to average out the effects of the random roughness. We use the microscope to evaluate different types of composite materials such as Kevlar-epoxy, carbon-epoxy, and carbon-carbon samples, and we found in all cases that defects such as delamination, fiber breakage, and porosity are easy to locate, not only in the plane of the image, but also in depth below the surface.

4.3. *NDE of Composite Materials*

The major difference between composite materials and other structural materials is their lower acoustic impedance and acoustic wave velocities. Consequently, considerable amounts of acoustic energy penetrate the sample in the form of focused longitudinal and shear waves. Also, surface waves are not converted on the surface of the sample (the surface wave velocity is comparable to the longitudinal wave velocity in the water and would not be excited by a focused transducer with an acceptance angle of less than 90 degrees) and would play no role in the subsurface defect detection or material characterization, as is the case in most acoustic microscopy work that is in the literature [2]. Only the specular reflection, the focused longitudinal waves, and the focused shear waves are the contributors to the image formed of the sample.

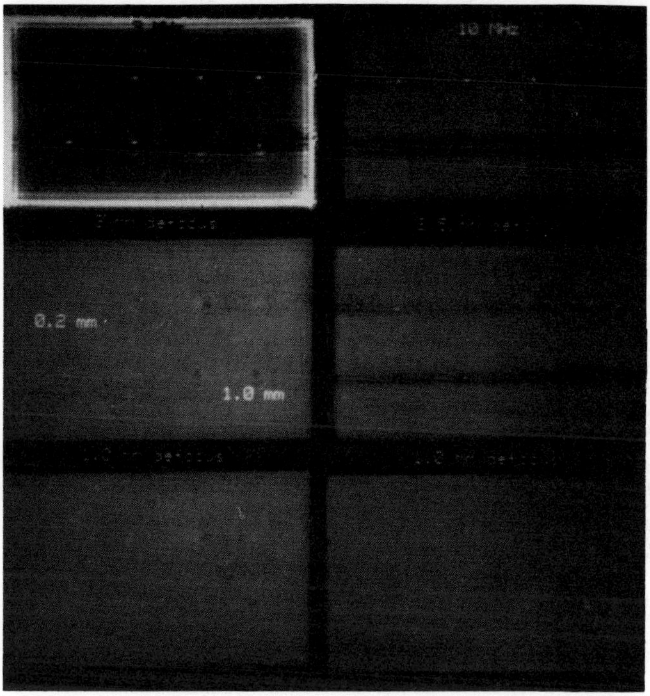

Figure 5: *Acoustic images of vertical crack sample with varying defocus levels. Note increased contrast as transducer is defocused.*

Figures 8a, 8b, 8c and 8d show four scans of a Kevlar-epoxy sample with a seeded defect. Figure 8a is a scan of the sample on focus, and shows clearly the kind of weaving of the fiber in the material. Also present in Figure 8a is a dim indication of the presence of a defect. Figures 8b, 8c and 8d are taken for a defocus of 1, 2 and 3 mm, respectively. It is seen that the seeded defect comes more clearly into focus, and seems to be located 1.0 mm below the surface (2 mm defocus corresponds to focusing at a depth of 1 mm below the surface. This assumes a velocity of 3000 m/s for the Kevlar-epoxy). More importantly, we are able to detect a 1.5 mm below the surface, the rectangular shape of an inclusion that seems to have been used to generate the defect. It is clear in these results that the instrument is capable of not only locating the defect in the plane of the scan, but also at depths below the surface.

The next samples we evaluated were made of carbon laminates with twenty-four plies at 0, 45, 90 and 135 degrees. The samples were impact damaged by dropping an indentor from various heights corresponding to damaging the sample

Table 1

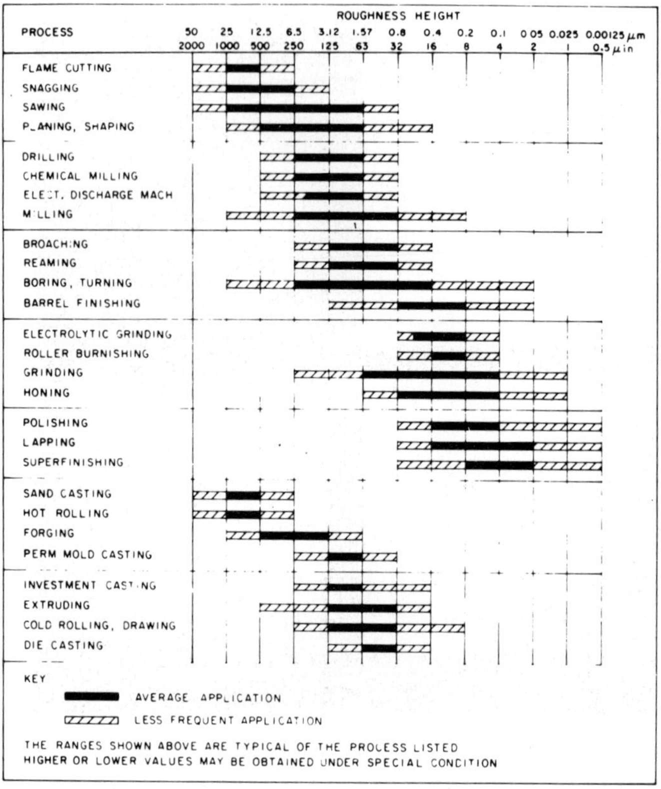

with different amounts of energy. The samples were evaluated by three methods: (1) with our acoustic microscope, operating at a frequency of 3 MHz and interrogating the undamaged side of the sample (samples were 3 mm thick); (2) with a transmission focused C-scan system operating at a frequency of 5 MHz; and (3) with dye-enhanced x-radiography. Figure 9 shows the result of the transmission C-scan at a frequency of 5 MHz. It is seen in Figure 9 that an indication of damage is present; however, the data is typical of C-scans and contains no information as to the direction of the damage or its location in depth. Figure 10 is the result of the dye-enhanced radiograph; it shows, with excellent clarity, the extension of the damage in the plies and its variation with direction. Figures 11a, 11b and 11c are results of acoustic microscopy scans at different values of defocusing. We see that the microscope results show the same extent of damage as the radiograph, and indicate more clearly the depth below the surface where the damage is located. In some cases, it is seen that the microscope shows larger damage regions than the radiograph, which could be attributed to the fact that dyes do not penetrate into very tight spaces, while very tight disbonds show a very large acoustic impedance mismatch. Overall, the acoustic microscope results show the most detail, and are

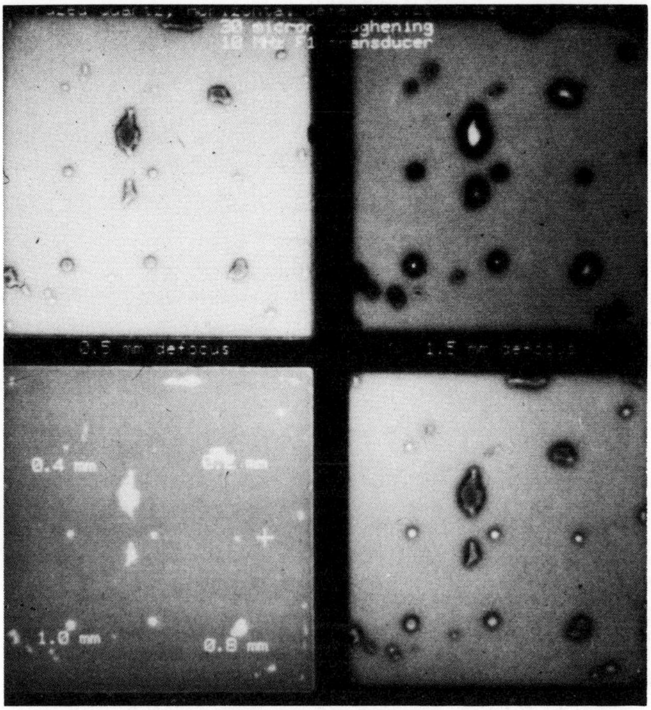

Figure 6: *Acoustic image of sample with horizontal defects 0.25 μm below surface. The sample's surface roughness was 30 μm. The images were taken at 10 MHz with varying defocus levels. Note the increased contrast with defocusing. The effects of the surface roughness also disappear.*

the only ones where the damage depth below the surface of the sample is obtained.

We have seen that the ability of a low-frequency acoustic microscope to detect subsurface defects is largely unhindered by small random surface roughness, especially when defocusing is employed. This applies to roughnesses corresponding to those produced by many machining techniques and due to the usual rough finish of composite materials. Also, the microscope is an excellent tool for evaluating composite materials. Although there are no surface waves excited on the surface of the samples, the good impedance match between water and the samples allows for good penetration and defect detection in the samples. Therefore, low-frequency acoustic microscopy appears to be an excellent technique for examining machined parts for subsurface flaws, in spite of surface roughness created by the machining technique.

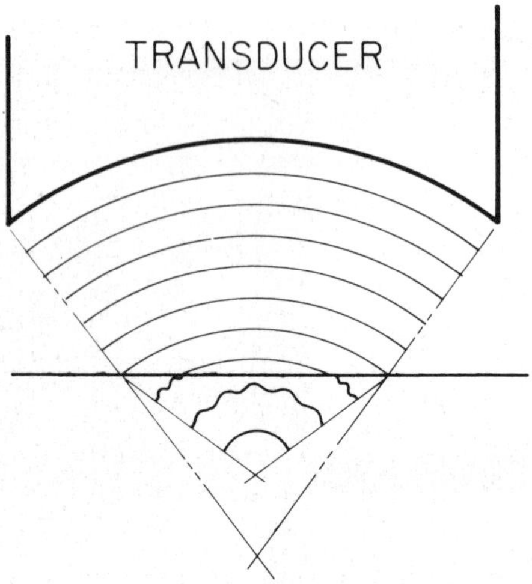

Figure 7: Acoustic waves transmitted through rough surface.

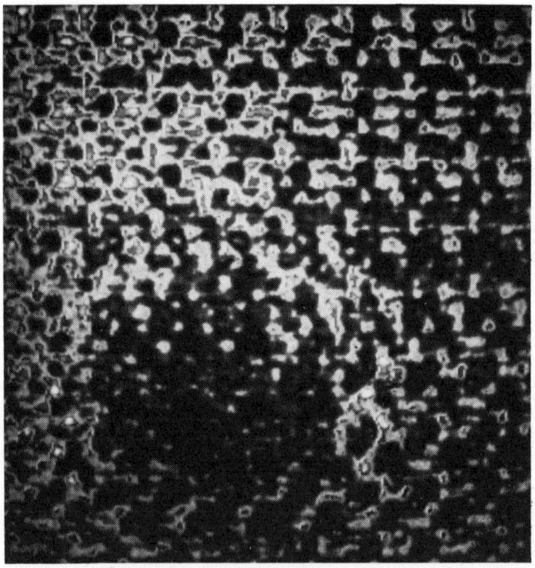

Figure 8a: Acoustic microscope image of a Kevlar-epoxy composite with a seeded defect. Focus on the surface.

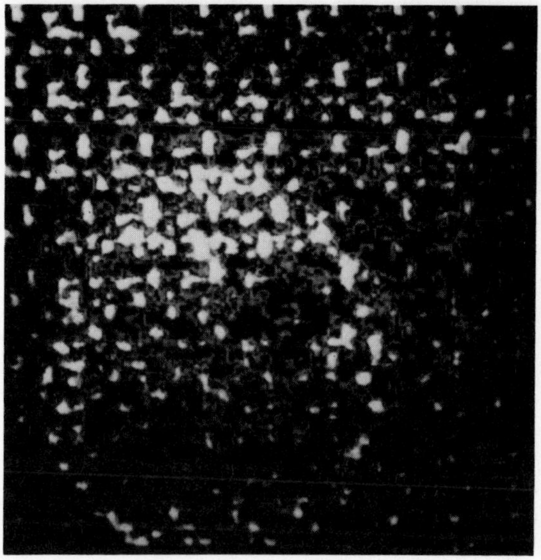

Figure 8b: Acoustic microscope image of a Kevlar-epoxy composite with a seeded defect. 1 mm defocus.

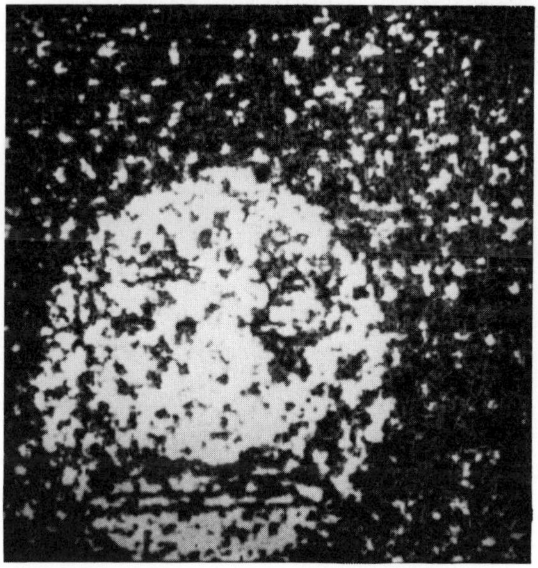

Figure 8c: Acoustic microscope image of a Kevlar-epoxy composite with a seeded defect. 2 mm defocus.

Figure 8d: *Acoustic microscope image of a Kevlar-epoxy composite with a seeded defect. 3 mm defocus.*

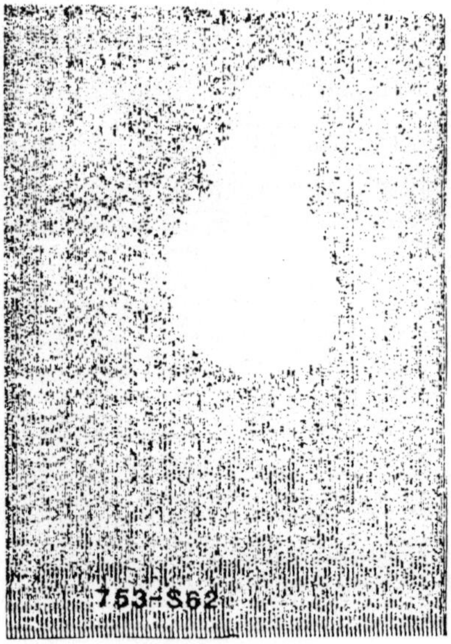

Figure 9: *Focused C-scan image at 5 MHz of a carbon laminate with impact damage.*

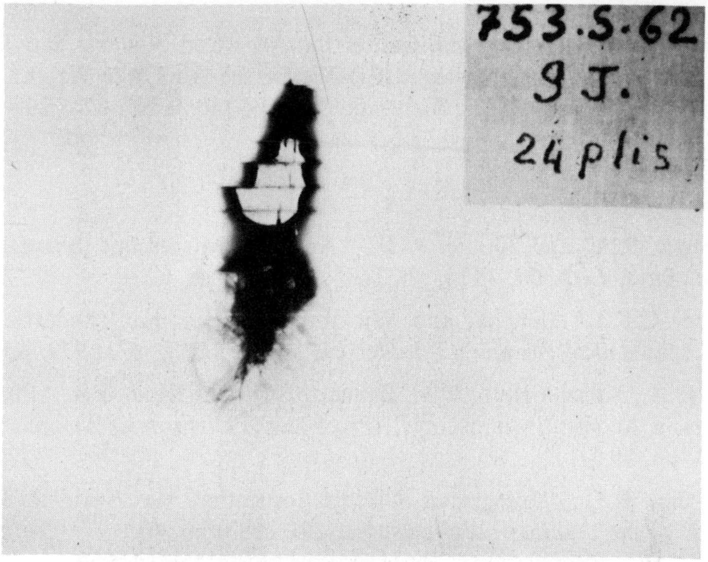

Figure 10: Dye-enhanced x-radiograph of the carbon laminate of Figure 9.

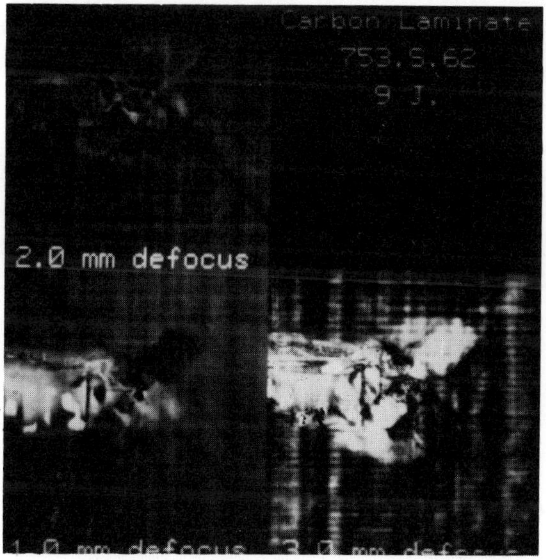

Figure 11: Acoustic microscope image of the carbon laminate of Figures 9 and 10. (a) 1 mm defocus. (b) 2 mm defocus. (c) 3 mm defocus.

Acknowledgment

This work was sponsored by the Center for Advanced Nondestructive Evaluation, operated by the Ames Laboratory, USDOE, for the Air Force Wright Aeronautical Laboratories/Materials under Contract No. W-7405-ENG-82 (Subcontract No. SC-86-102).

References

[1] Lemons, R.A. and Quate, C.F., "Acoustic Microscope Scanning Version", *Appl. Phys. Lett., 24*, 1974, pp. 163.

[2] Quate, C.F., Atalar, A. and Wickramasinghe, H.K., "Acoustic Microscopy with Mechanical Scanning - A Review", *Proc. IEEE, 67*, 1979, pp. 1052.

[3] Liang, K., Khuri-Yakub, B.T., Bennett, S.D. and Kino, G.S., "Phase Measurements in Acoustic Microscopy", *Proc. IEEE Ultrasonics Symp.* 83CH1547-2, 1983, pp. 594.

[4] Weglein, R.D., "Integrated Circuit Inspection via Acoustic Microscopy", *IEEE Trans. Sonics and Ultrasonics, 30*, 1983, pp. 40.

[5] Sinclair, D.A. and Smith, I.R., "Tissue Characterization Using Acoustic Microscopy", *Acoustic Imaging, 12*, Eds.: E.A. Ash and C.R. Hill, Plenum Press, 1982.

[6] Wickramasinghe, H.K., "Contrast and Imaging Performance in the Scanning Acoustic Microscope", *J. Appl. Phys., 50*, February, 1979, pp. 664-668.

[7] Atalar, A., "An Angular Spectrum Approach to Contrast in Reflection or Acoustic Microscopy", *J. Appl. Phys, 45*, October, 1979, pp. 5130-5139.

[8] Reinholdtsen, P., Hipkiss, W.W., and Khuri-Yakub, B.T., "Low-Frequency Acoustic Microscopy", presented at the DARPA/AF Review of Progress in Quantitative NDE, San Diego, CA, July, 1984.

[9] See *IEEE Trans. on Sonics and Ultrasonics*, Special Issue on Acoustic Microscopy, *SU-32* (2), March, 1985.

QNDE OF COMPOSITE MATERIALS I

ULTRASONIC NDE OF COMPOSITES – A REVIEW

Yoseph Bar-Cohen
Douglas Aircraft Company
McDonnell Douglas Corporation
Long Beach, California

Abstract

The application of fiber-reinforced composites for structural components requires reliable NDE methods to determine their integrity and serviceability at the various production stages and in service. The detection of flaws in composites is difficult because of their anisotropic, nonhomogeneous, and layered nature. Further, the failure mechanism of composites is more complicated than the failure mechanism of metals. New ultrasonic techniques have been developed in recent years for the detection and characterization of defects and the evaluation of the material properties. Several of these techniques will be reviewed and discussed.

1. Introduction

The anisotropic, nonhomogeneous, layered nature of composites limits the application of traditional nondestructive evaluation (NDE) techniques. Moreover, the failure mechanism of composites is more complex than in metals and is a result of the degradation effect of multiple defects rather than the growth of a single crack. These facts indicate the need for an adequate approach to deal with the various types of possible critical defects, rather than developing methods to characterize a single discontinuity type.

The term "composite" covers a wide variety of materials. This paper concentrates on the fiber-reinforced plastic systems that are widely used by the aircraft industry today. The fibers include graphite, glass, Kevlar, and boron, whereas the matrix is mainly epoxy resin. Boron/epoxy was the first material to be applied to primary structures, but it has been replaced by graphite/epoxy for economical reasons. Glass/epoxy and Kevlar/epoxy are mainly used for secondary structures, where quality is less critical. In recent years, efforts have been made to combine graphite/epoxy and either Kevlar/epoxy or glass/epoxy to a hybrid composite for improved toughness. The diameter of different types of fiber varies from 1 to 10 micrometers for graphite, 5 to 15 micrometers for glass and Kevlar, and 100 to 150 micrometers for boron.

2. Composites as Anisotropic Homogeneous Media

As a first approximation of handling composites, one can assume the use of low frequency, where the wave number-length is much smaller than 1. This range covers the frequencies below about 20 MHz for graphite/epoxy, glass/epoxy, and Kevlar/epoxy. For this condition, the wave equation in homogeneous, anisotropic (orthotropic) media is applicable.

$$\sigma_{ij,j} = \rho U_i \; . \tag{1}$$

Where σ_{ij} is the stress tensor component, ρ the material density, and U_i the displacement vector. Assuming the media is elastic, Hook's law can be applied.

$$\sigma_{ij} = C_{ijkl} U_{kl} \tag{2}$$

where C_{ijkl} is the elastic constant. Equations (1) and (2) are leading to the wave equation.

$$C_{ijkl} U_{k,ij} - \rho \ddot{U}_i = 0 \; . \tag{3}$$

Assuming a plane wave solution, one obtains a determinant with the normals n_i [1].

$$Det \mid C_{ijkl} n_j n_i - \rho V^2 \delta_{ik} \mid = 0 \; . \tag{4}$$

This set of equations can be simplified once the orthotropic symmetry of a unidirectional composite is assumed. From the solutions of these equations, three wave velocities are determined. One major difficulty in predicting the acoustic velocities of an arbitrary composite material is the unavailability of its elastic constants. These constants can be predicted from the constituent's properties using various models, such as Hashin [2]. However, this requires knowledge of the fiber elastic constants, which are difficult to measure. In Figure 1, the velocities are plotted as a function of angle of wave propagation for two typical composite laminates.

Knowledge of the acoustic velocities of a defect-free composite is essential for characterization of defects that are changing its elastic properties. Examples of such defects are poor curing, porosity, and weathering damage. This approach is limited to unidirectional composites. The general case of the multiorientation laminate has not yet been solved rigorously.

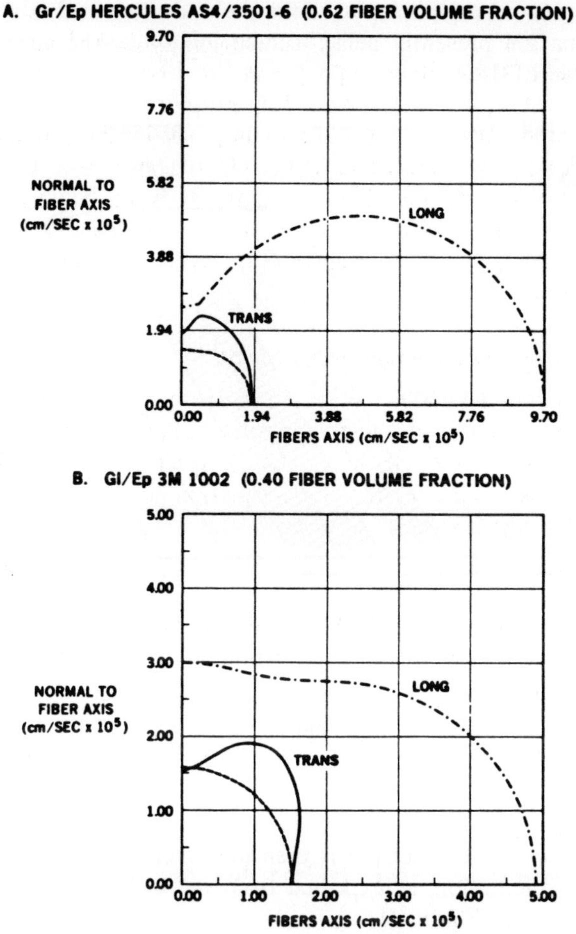

Figure 1: Velocities of the three modes at various angles of propagation, with the fibers, for graphite/epoxy and glass/epoxy

3. Composites as Anisotropic, Homogeneous, Layered Media

Composite laminates are mostly utilized in the thickness range (0.0125 to 1 cm) that supports the generation of Lamb waves at 1 to 10 MHz. If a laminate is loaded with fluid, the Lamb waves leak energy into the fluid at an angle that is established by Snell's law from the Lamb wave phase velocity. The Lamb wave can be induced by insonifying a plate in an angle and by proper selection of transmission frequency, through a mode conversion. The specular component of the wave and the leaky wave interfere to generate two components with a null between them. A schematic description of the field behavior is shown in Figure 2. To visualize the wave behavior, a pulsed schlieren system has been used. An example of a wave burst before impinging on a glass/epoxy laminate and afterward is shown in Figure 3. These phenomena are presently being studied jointly by the author and his colleague, D.E. Chimenti [3].

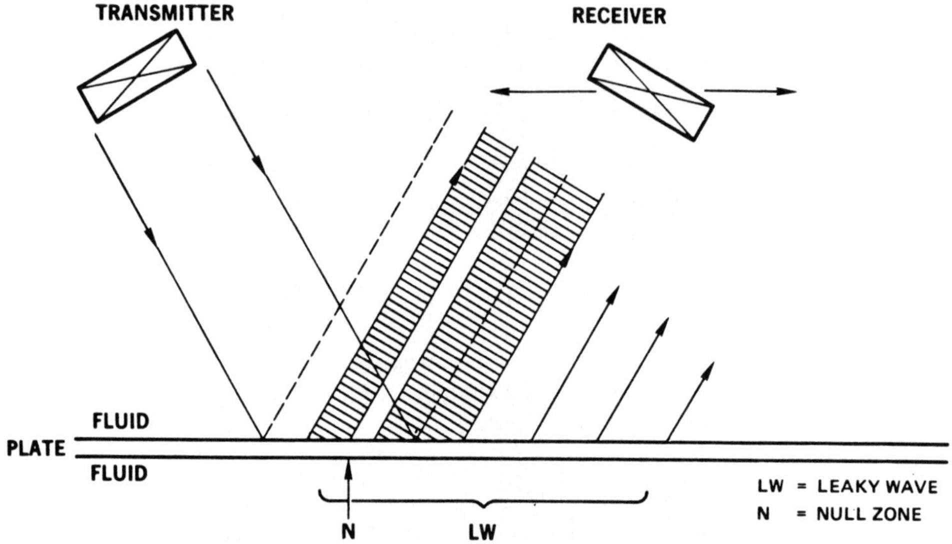

Figure 2: Schematic description of leaky Lamb wave acoustic field

Experimentally, the frequency of the leaky Lamb wave (LLW) modes is determined by sweeping through a given frequency range and searching for minima in the reflected amplitude, at the null zone. As an alternative approach, pulses can be used where the reflected signal is processed by fast-Fourier transform (FFT), and the transducer characteristic spectrum is then deconvolved. The dependence of the modes on the fiber orientation, as expected from the anisotropic nature of composite laminates, is shown in Figure 4.

To limit the number of variables in the present study, efforts were concentrated on unidirectional laminates, and tests were conducted in the plane of the fibers' orientation and the normal to the composite laminate surface. The dispersion curve for graphite/epoxy AS4/3501-6 has been determined and is shown in Figure 5.

The leaky Lamb wave phenomena have been found to provide a very sensitive tool for NDE of defects. Various defects can be identified by their characteristic modes. For example, delaminations, which are expressed by a decoupling of two or more layers, modify the boundary conditions of the top part of the laminate to a thin layer loaded with fluid on one side and with a stress-free surface on the other side. As a result, fewer modes are observed (see Figure 6), as expected from the dispersion curve. Porosity can also be identified from the LLW spectral response – a typical response is shown in Figure 7. At low frequencies, the modes are unchanged, whereas at high frequencies the porosity layer behaves as a reflector

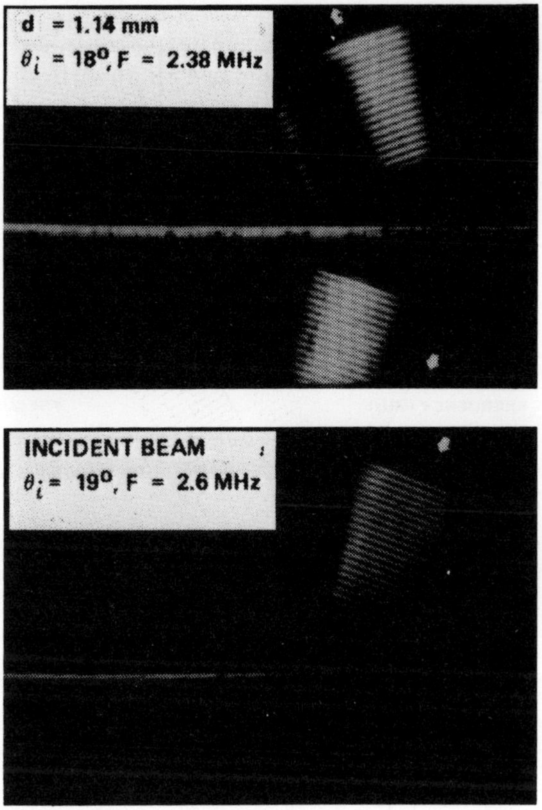

Figure 3: Pulsed Schlieren visualization of tone-burst signal before and after impinging on a glass/epoxy sample for a LLW mode

and excites modes of a thinner layer. At the intermediate frequency region, no modes are formed.

In the various tests of LLW modes, a relatively small coefficient of variation (= 3 percent) was measured over the area of the laminate. This small variance allows a composite to be characterized in a more reproducible measure than attenuation, which can vary more than 150 percent in many cases.

This new NDE technique demonstrates a superior capability over conventional ultrasonics when it is combined with a C-scan setup, as shown in Figure 8. This figure presents a comparison of an image obtained from porosity, imbedded between layers of three different depths, using LLW and an image obtained by a conventional pulse-echo technique.

The results of the present research for unidirectional laminates are expected to be applicable to the general multiorientation laminates. Preliminary tests conducted on graphite/epoxy $[0,90]_{2S}$ and $[0,\pm 45,90]_S$ samples showed a modal behavior

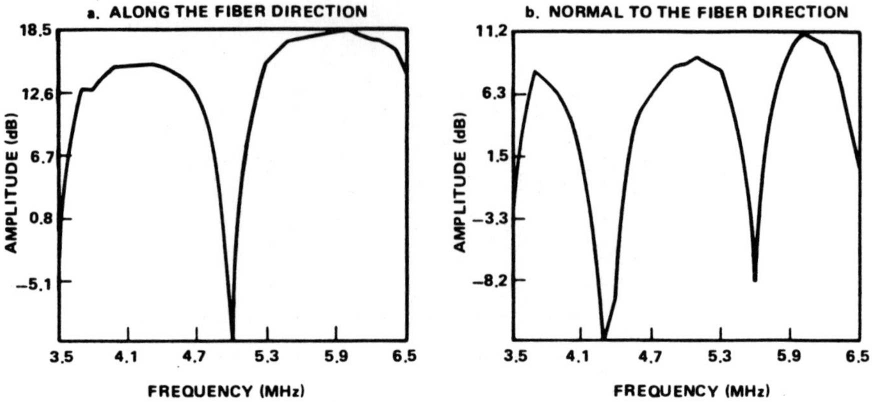

Figure 4: Deconvolved characteristic response of leaky Lamb wave for AS4/3501-6 Graphite/Epoxy

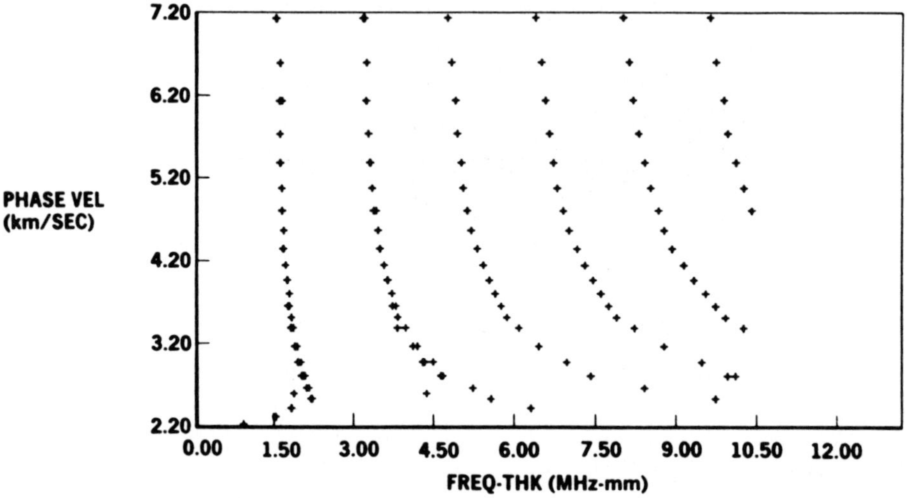

Figure 5: Dispersion curve of LLW for graphite/epoxy T300/CG914 unidirectional laminate tested along the fiber direction

similar to the one in unidirectional laminates. However, more experiments need to be conducted before any conclusion can be drawn.

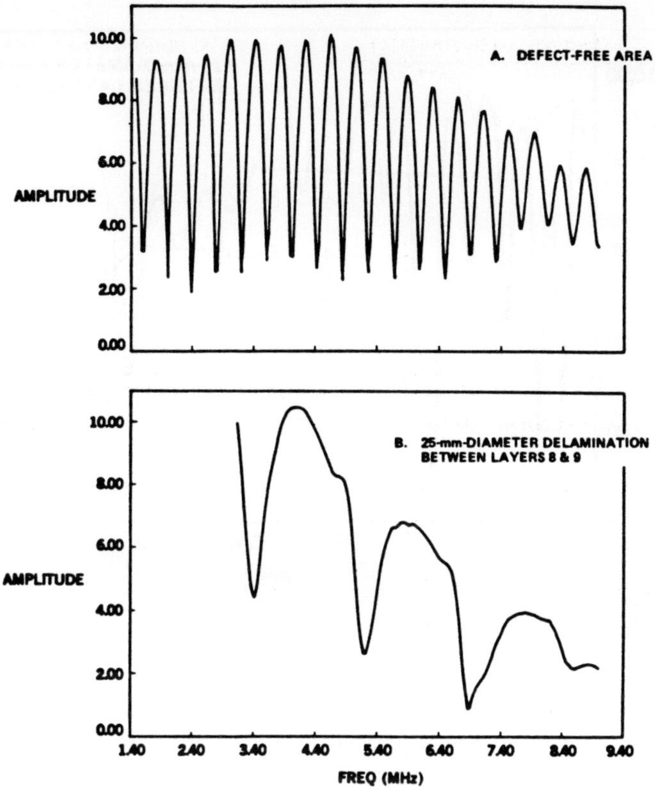

Figure 6: Reflection spectra for LLW in graphite/epoxy AS4/3501-6 $[0]_{32}$ tested at 22 degrees along the fibers

4. Composites as Anisotropic Nonhomogeneous Layered Media

With the increase in frequency or wave number, composites can be modeled as multiple cylindrical inclusions embedded in an elastic media stacked in layers of various orientations, where the fibers in each layer are straight (not covering woven composites) and have a volume fraction of less than 65 percent.

Schlieren observations of the specular reflection from a unidirectional laminate indicate the generation of a diffraction pattern, when impinging in a plane normal to the fiber's axes. On the other hand, the specular reflection along a plane parallel to the fibers is unperturbed (see Figure 9). Since the acoustic field resembles a diffraction from an array, a simple model of a two-dimensional array of fibers can be assumed. The acoustic field of such an array has been well documented (see, for example, reference [4]), and it is characterized by a: $\sin(x)/x$ or $J_1(x)/x$ behavior, where $x = ka\sin\theta$, k = wave number, a-unit spacing, θ-angle of observation, and $J_1(x)$ is the Bessel function of the first order.

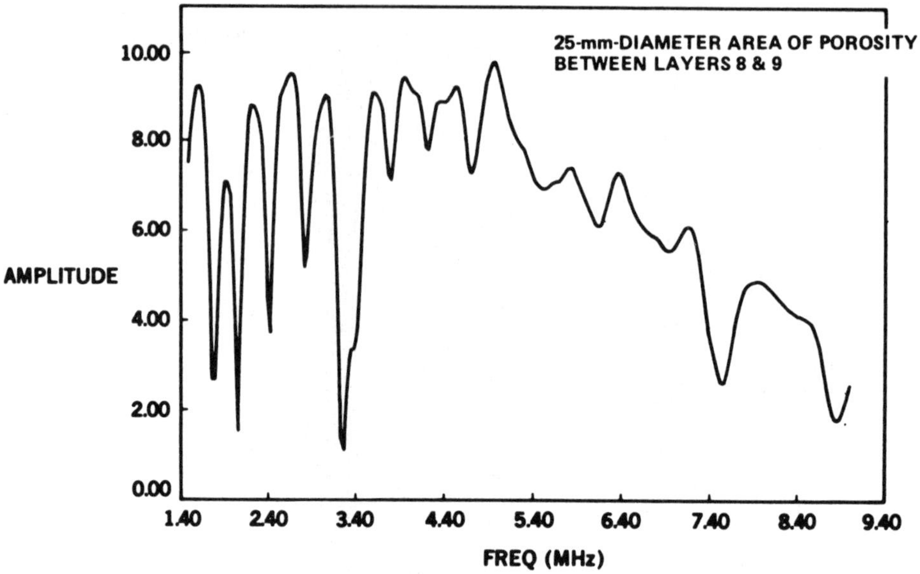

Figure 7: Reflection spectrum for leaky Lamb waves in graphite/epoxy tested along the fibers at 22 degrees

To study the acoustic scattering field, the setup shown in Figure 10 was prepared with a transmitter set in a fixed location normal to the sample and rotating with it on a turntable. The sample was prepared so that the fibers were parallel to the axis of rotation and its front surface matched this axis. The receiver was placed in a stationary position outside the turntable to allow the measurement of amplitude as a function of the angle. In this experiment, frequencies up to 12 MHz were employed in a tone-burst mode.

Using this experimental setup, no consistent trend of scattering field could be recorded to indicate the above functional field behavior. To eliminate the transducer side lobes, transducers with a Gaussian directivity were used. Even though lobes were observed in the acoustic field, in most cases the recorded lobes were random in characteristics and had no apparent symmetry. It seems that the local variations and the fact that fibers are irregularly ordered within the laminate create a strong perturbation to the fibers' order which prevents the generation of a simple or consistent diffraction field response.

While it was difficult to obtain a typical characteristic scattering field, backscattering measurements using pulses and spatial averaging have been highly successful. With the experimental setup shown in Figure 11, it was observed that for laminates tested with a single transmitter/receiver probe, fibers scatter backward only when the angle of insonification is normal to the fibers' surface. As shown in Figure 12, with a $[0, \pm 45, 90]_S$ glass/epoxy laminate at an incidence angle of 30 degrees, maximum backscattering is observed for angles normal to the fibers. The

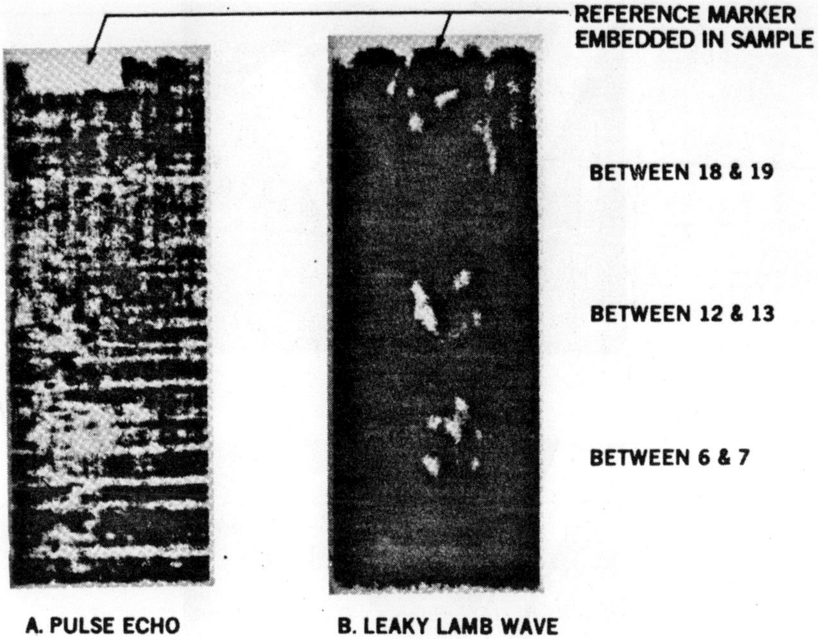

Figure 8: C-scan tests of porosity in a graphite/epoxy $[0]_{24}$

finite width of the angular spectrum is determined by the transducer directivity. This phenomenon of backscattering has been studied by the author and his colleague, R.L. Crane [5]. To reduce the effects of local variations and small discontinuities, the data, while acquired from different locations, have been averaged. Tests were conducted in the frequency range of 1 to 25 MHz, using broadband pulses with the aid of a box-car averager. In these tests, the peak amplitudes were determined. Backscattering at the plane normal to the fibers measured, as a function of incidence angle, showed two maxima at the critical angles (see Figure 13). The mechanism of generating a maximum at the critical angle is not clear and cannot be explained by any of the documented theories. Moreover, it is important to note that the wave number-length ka is relatively small and it reaches only about 0.2 at 25 MHz for graphite/epoxy.

Several multiorientation laminates were tested using this technique. For simple fibers layup, scattering amplitude decreased with the increase in depth, thus revealing the stacking order of the laminates. However, in a laminate with a complex orientation of $[0,-15,-30,+45,-60,+75,90]_S$, an unexplained trend has been observed. As shown in Figure 14, the layer of the 75-degree fibers generated scattering higher than the first ply of 0 degree. Also, the fibers with an angle higher than 90 degrees showed a different trend than the orientations below 90 degrees, and the amplitude did not correlate with depth in a simple manner.

ALONG THE FIBERS'
DIRECTION

NORMAL TO THE FIBERS'
DIRECTION

Figure 9: Pulsed Schlieren visualization of the reflection from a glass/epoxy unidirectional laminate

Matrix cracking is common in composites, and it serves as an embryo for the development of delaminations. The cracks extend throughout the total thickness of a lamina (a stack of layers of a given orientation within a laminate), and in most cases they are more than an order of magnitude larger than the fiber diameter. For example, the thickness of a graphite ply is about 125 micrometers, whereas the diameter of a fiber is about 5 micrometers. The backscattering from the transverse cracks is more than 30 dB higher than the scattering from the 90-degree fibers in a $[0,90]_{2S}$ laminate. This fact allows the detection and imaging of transverse cracks using a C-scan system, by discriminating the scattering amplitude of the cracks from that of the fibers. An example of images of transverse cracks in graphite/epoxy samples is shown in Figure 15.

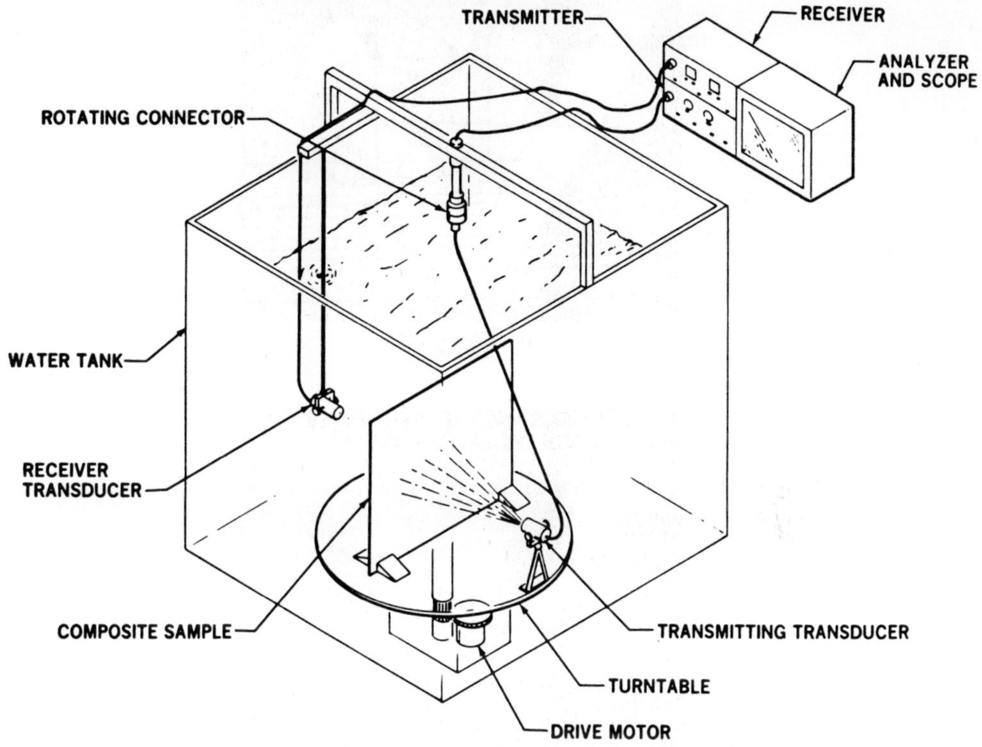

Figure 10: Turntable setup for measurement of the scattered field from a composite laminate

Porosity is also detectable by backscattering. Generally, porosity tends to accumulate between the various layers of the composite laminate. To simulate porosity, a $[0,90]_{2S}$ laminate has been prepared using microballoons 40 micrometers in diameter and a 2-micrometer shell thickness, spread in a thin coat between the first and the second layer of a $[0,90]_{2S}$ laminate. Since porosity, as randomly spread spheres, does not have any preferred orientation, it is therefore expected to generate backscattering of equal amplitude for all angles of insonification. This behavior is shown in Figure 16, where porosity causes an increase in scattering for angles that are not normal to the fiber's axes as compared to the response from a defects-free laminate. This behavior of fibers generates a spatial window in the backscattering field in which defects of different scattering directivity can be characterized.

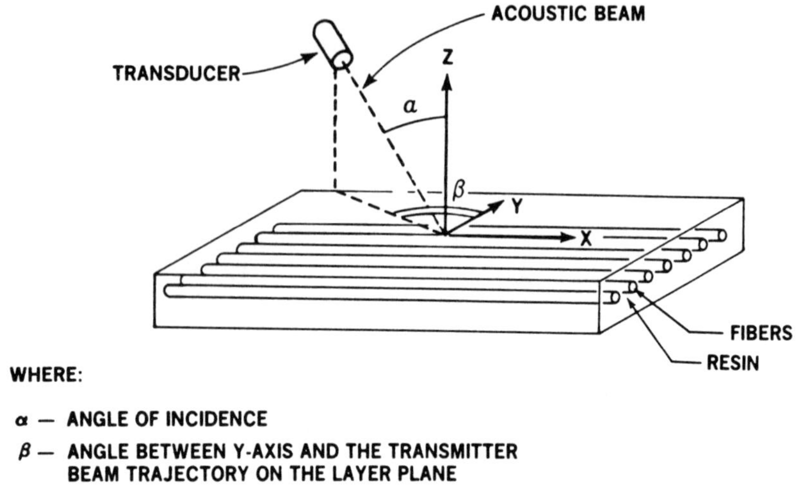

Figure 11: Schematic representation of experimental setup for backscattering measurements

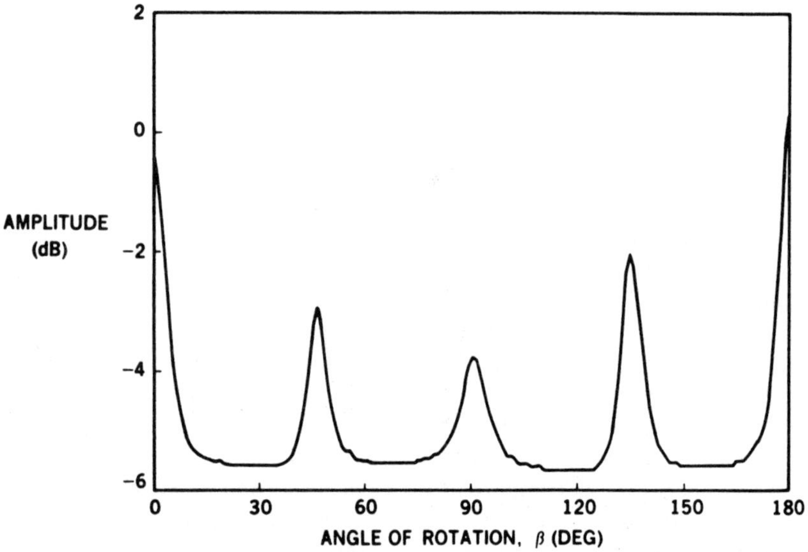

Figure 12: Acoustic backscattering as a function of rotation angle, for a 30-degree incidence angle, from a $[0,\pm45,90]_S$ graphite/epoxy laminate

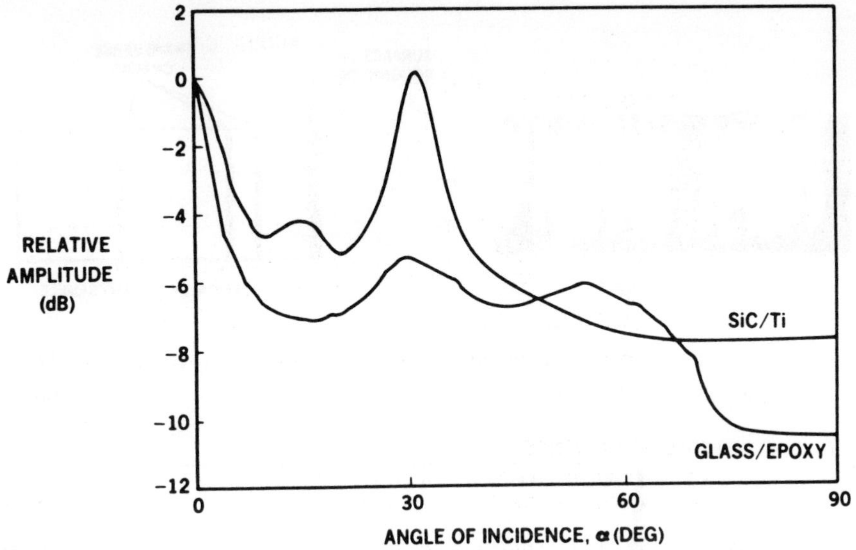

Figure 13: Backscattering as a function of angle of incidence, in a plane normal to the fiber direction

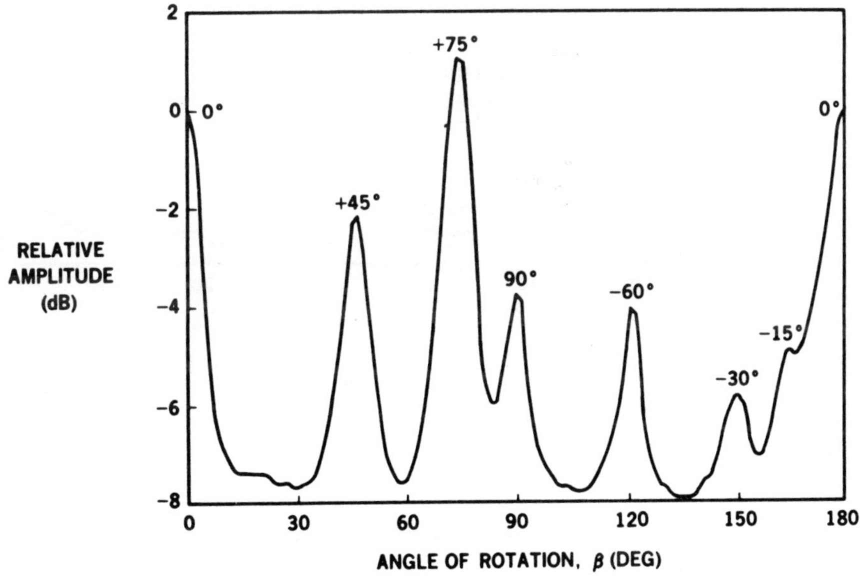

Figure 14: Acoustic backscattering from a $[0,-15,-30,45,-60,75,90]_S$ graphite/epoxy laminate tested at a 40-degree angle of incidence

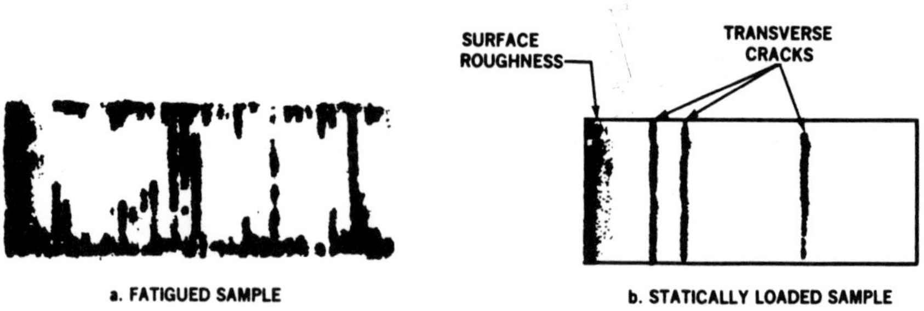

Figure 15: Acoustic backscattering images of transverse cracks at 90-degree orientations

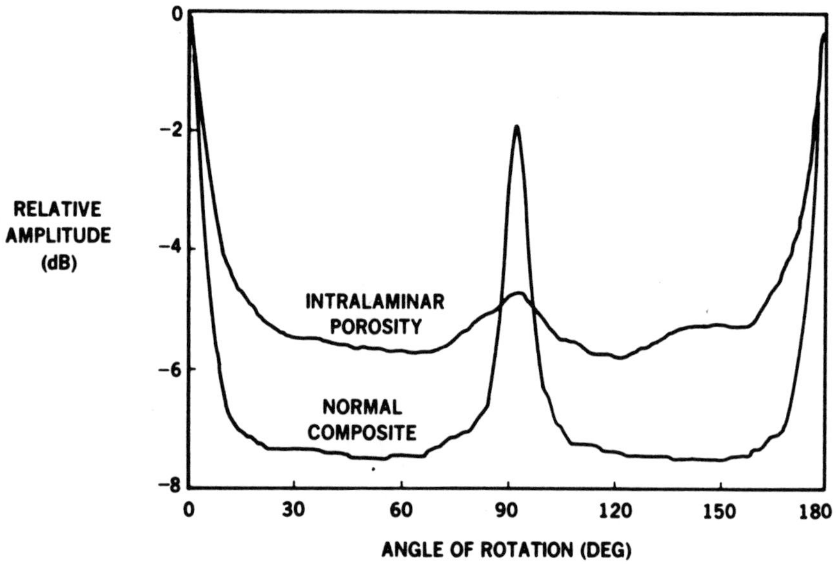

Figure 16: Backscattering from $[0,90]_{2S}$ glass/epoxy composite both with and without porosity.

5. Conclusions

The anisotropic, nonhomogeneous, layered nature of composites affects elastic wave behavior significantly. The understanding of this behavior assists in developing adequate NDE methods. Present empirical results, using backscattering and leaky Lamb wave measurements, contributed substantially to the nondestructive characterization of composites. To obtain further capability, there is a need for analytical tools to handle the complex nature of multiorientation composites, which will also explain the anomalies observed in waves' behavior in composites.

References

[1] Musgrave, M.J.P., "On the Propagation of Elastic Waves in Aeleotropic Media, II Media of Hexagonal symmetry," *Proc. Roy. Soc. 226*, 1954, p. 356.

[2] Hashin, Z., "The Elastic Moduli of Heterogeneous Materials", *J. Appl. Mech.*, March, 1962, pp. 143-180.

[3] Bar-Cohen, Y. and Chimenti, D.E., "Detection of Porosity in Composites Using Leaky Lamb Waves", *Proceedings of the 11th World Conference on Nondestructive Testing, Vol. III*, Sponsored by ASNT. Taylor Publishing, Dallas, 1985, pp. 1661-1668.

[4] Hunter, L.P., Lee, P.K. and Waag, R.C., "Forward Sound Scattering by Small Nylon Cylinders in Water", *J. Acoust. Amer. 68*, 1980, pp. 314-319.

[5] Bar-Cohen, Y. and Crane, R.L., "Acoustic-Backscattering Imaging of Subcritical Flaws in Composites", *Mater. Eval. 40(9)*, 1982, pp. 970-975.

NONDESTRUCTIVE CHARACTERIZATION OF IMPACT DAMAGE IN GRAPHITE/EPOXY LAMINATES

I.M. Daniel and S.C. Wooh
The Technological Institute
Northwestern University
Evanston, Illinois

Abstract

Transient deformation and damage were studied in graphite/epoxy plates subjected to falling weight impact. The plates were 8-ply and 16-ply quasi-isotropic laminates clamped around a 12.7 cm (5 in.) diameter circumference. They were instrumented with surface and embedded strain gages and loaded by a 202 gr (0.445 lb) mass dropped from a height of 1.83 m (72 in.). The dynamic load history and imparted energy were obtained by means of an accelerometer attached to the impactor. Transient strains at various locations through the thickness and at various distances from the loading point were obtained. The characteristic features of the strain records are associated with specific failure modes in the laminate. The deformation history was also correlated with the induced damage as detected by X-radiography and ultrasonics. Impact damage is more extensive in the thicker laminate. Damage takes the form of delaminated strips parallel to the fiber direction in each ply and increasing in length from top to bottom. The length variation of consecutive delaminated strips is more pronounced in the thin laminate. Transient strain records corroborate the ultrasonic indications regarding the location of delaminations.

1. Introduction

The ever expanding structural applications of composite materials subject them to a variety of environmental and loading conditions. One problem receiving considerable attention is the behavior of composites under localized low velocity impact. Impact velocities can range from low (3-6 ms^{-1}) for the case of tools or other foreign objects falling on the composite, to intermediate (15-60 ms^{-1}) velocities encountered by aircraft in flight, to high velocity ballistic impacts. This study deals with the low to intermediate range of velocities for which the damage caused in the composite is usually invisible to the unaided eye.

Impact studies of composites, both analytical and experimental, have been conducted by several investigators [1,7]. The general objectives of current research in the area are to study the progressive damage due to impact, assess the energy absorbing capabilities of various compsoites, and determine the residual properties of impacted laminates.

The objective of the present study was to investigate the transient deformation in composite plates during impact and correlate it with the induced damage. Load history, imparted energy, and dynamic load-deflection curves were obtained. Transient strain records obtained from embedded strain gages at various locations offer a unique means of studying damage progression in real time. The final damage state was characterized by means of X-radiography and ultrasonics.

2. Experimental Procedure

The material investigated was AS4/3501-6 graphite/epoxy. It was obtained in prepreg form and fabricated into laminates. The unidirectional material was fully characterized by determining the mechanical properties of the lamina.

The specimens were quasi-isotropic laminates clamped between two aluminum frames with 12.7 cm (5 in.) diameter cutouts as shown in Figure 1. The laminates were 8-ply and 16-ply thick of $[0/\text{-}45/90/45]_s$ and $[0/\text{-}45/90/45]_{2s}$ layups. The specimens were instrumented with surface and embedded strain gages according to the gage layouts of Figure 2.

The following procedure was developed for embedding the strain gages. Thin Nickel-clad copper ribbon leads were attached to the gages, then the gages with the leads were sandwiched between two strips of adhesive-impregnated glass scrim cloth (Figure 3). The assembly was placed between two mylar sheets and cured under pressure at a temperature of 60°C (140°F) for two hours. Bondable terminals were applied to the ends of the ribbon leads and reinforced with Kapton tape. The gage/scrim cloth sandwich was then trimmed and checked for continuity and insulation. The cured and trimmed strain gage sandwiches were then embedded between the graphite/epoxy prepreg plies during lamination.

Eight-ply and sixteen-ply coposite laminates, 15.2 cm (6 in.) square, were prepared with embedded strain gages at various locations, and cured following the standard autoclave molding process. Ribbon leads from the embedded gages

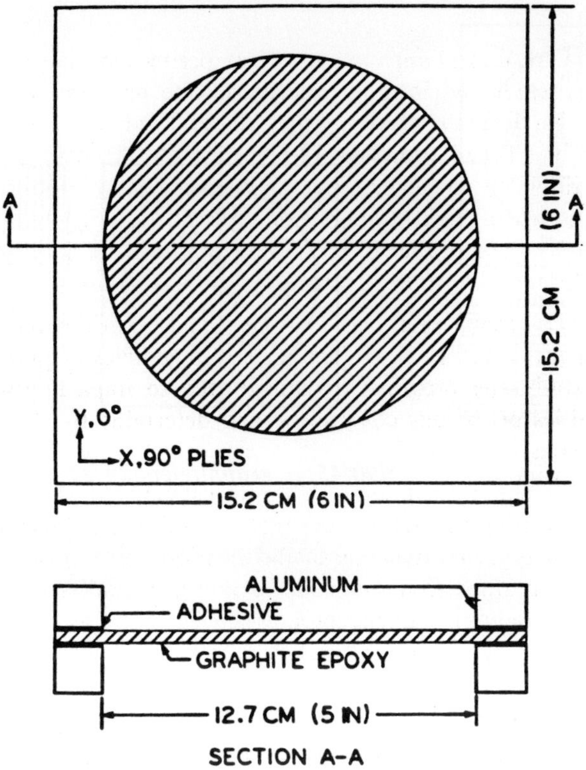

Figure 1: Specimen geometry

extended through the edges of the composite laminates. The laminate was then bonded between two square aluminum frames with a 12.7 cm (5 in.) diameter cutout (Figure 1). Additional strain gages were mounted on the outer surfaces of the composite plate.

The specimens were loaded by a 202 gr (0.445 lb.) mass impactor falling from a height of 1.83 m (72 in.). An accelerometer (Kistler 8642A5) was attached to the falling impactor to measure the dynamic load history. The impactor and experimental setup are shown in Figure 4.

The accelerometer and strain gage data were recorded on a Data Acquisition and Analysis System consisting of the following:

1. Apple IIe microcomputer
2. Digital Processing Oscilloscope (Norland 3001A Waveform and Data Analysis System)
3. Strain gage conditioning unit

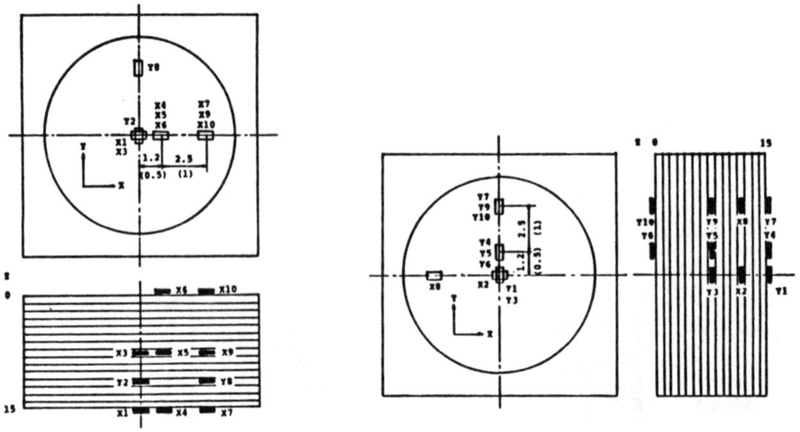

Figure 2: Gage layout for $[0/-45/90/45]_{2s}$ graphite/epoxy specimens (dimensions marked are in cm(in.))

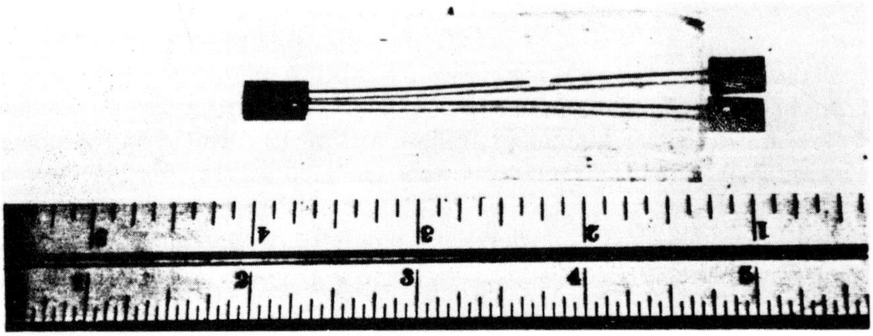

Figure 3: Gage with ribbon leads sandwiched in glass scrim cloth

4. Four-channel differential amplifier
5. Dot matrix printer (Epson FX-80)
6. Plotter (HP 7470).

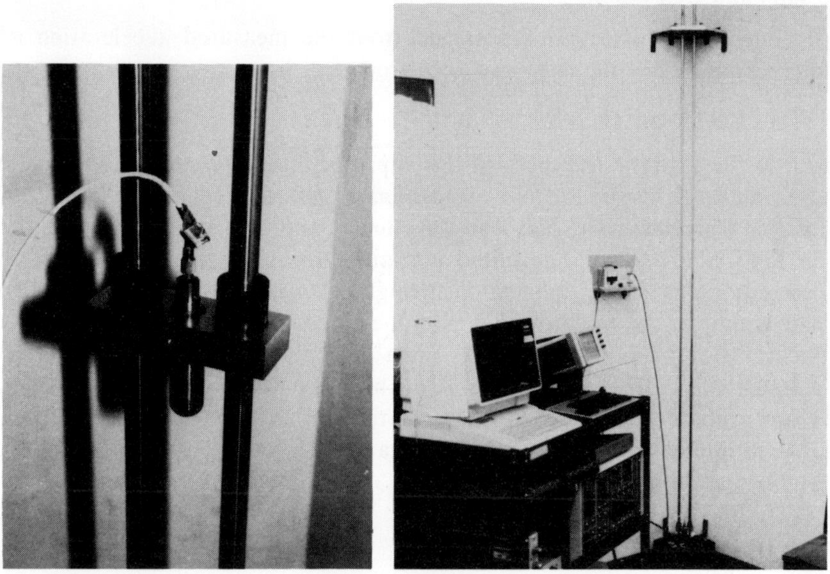

Figure 4: Impactor and experimental setup for impact testing of composite laminates

The accelerometer which has a capacity of 5000 g's was coupled to a coupler (Kistler 5116) and the output recorded on the digital oscilloscope. Strain gage signals were conditioned by Wheatstone bridge circuits, amplified, and then recorded by the digital oscilloscope. Signals can be recorded simultaneously on the four channels of this oscilloscope with each channel capable of storing 4096 points at rates as fast as one point per one-tenth of a microsecond. In this case, a sampling interval of 10 μs was selected. The oscilloscope is equipped with a microprocessor and a keyboard for preliminary data processing, such as smoothing and differentiation, by means of simple keystrokes. Further processing and analysis and storage can be performed on the microcomputer. Final results were plotted on the plotter to the desired scales.

Graphite/epoxy specimens of the same layups as the instrumented specimens were also fabricated without any embedded gages for the purpose of subsequent nondestructive inspection. This was done by means of X-radiography enhanced with a penetrant solution of zinc iodide which is opaque to X-rays and by ultrasonic scanning.

3. Results and Discussion

The dynamic load history was obtained from the measured acceleration and the combined mass of the impactor and accelerometer

$$P(t)=ma(t) \tag{1}$$

where P is the load, m the mass of the impactor assembly, and a the acceleration. The dynamic load history for the two laminates tested is shown in Figure 5. In the case of the 8-ply laminate, the load duration is approximately 1 ms and the peak load is 3547 N (797 lb.). The initial portion of the pulse shows noise-like fluctuations corresponding to the damage occurring during impact. It appears that most of the load is due to the buildup of membrane forces in the laminate. In the case of the thicker 16-ply laminate, the load has a longer duration of approximately 2 ms and a lower peak of 1850 N (416 lb). The pulse shows fluctuations throughout which are probably due to flexural oscillations of the plate. The initial noise-like features are related to damage in the laminate.

The imparted energy was determined as follows:

$$E=\int_0^t Pvdt \tag{2}$$

where

$$v(t)=v(0)-\int_0^t a(t)dt \tag{3}$$

$$v(0)=\sqrt{2gh} = \text{initial impact velocity} \tag{4}$$

h height of impactor drop and E the energy. The imparted energy for the two laminates is plotted in Figure 6. The peak energy is approximately the same for both laminates.

Transient strain records at various locations through the thickness of the thick laminate are shown in Figure 7. The peak strains under the point of impact (gages X1, X2 and X3) seem to vary nearly linearly with depth indicating a neutral plane at one ply below the midsurface. This would correspond to a major delamination at a depth of two plies from the top. The negative strain pulse at the midsurface (gage X3) is longer than the corresponding positive strain pulses (X1 and X3). At a distance of 1.27 cm (0.5 in.) from the center, the strain variation through the thickness (gages X4, X5 and X6) is nearly linear with depth corresponding to an effective neutral plane at a depth of four plies from the top. This type of deformation is compatible with pure bending of the top half of the laminate, which could be attributed to a major delamination at the midplane.

Transient strain records were obtained at many locations for both laminates. The strain data, however, are difficult to analyze as they are related to many factors, such as damage and accompanying relief of lamination residual stresses, flexural oscillations, membrane deformation and fluctuations in the load history. In any case, the transient strain records obtained corroborate indications regarding the location and distribution of delaminations obtained by X-radiography and ultrasonic

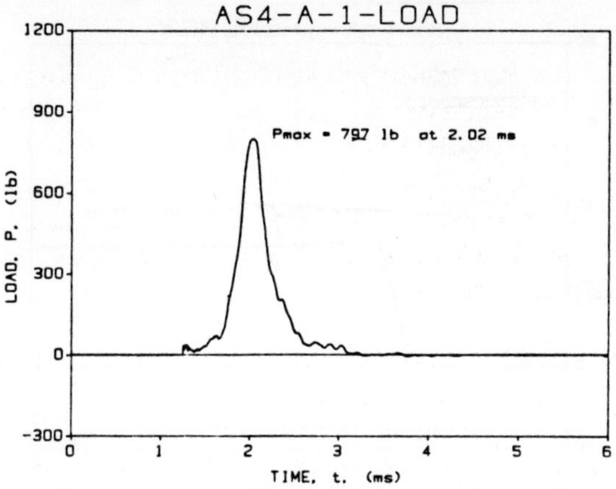

(a)

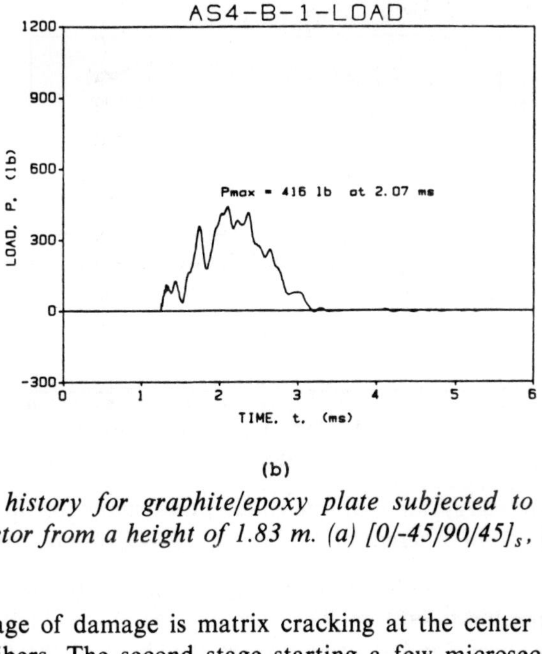

(b)

Figure 5: Load history for graphite/epoxy plate subjected to impact of 202 gr impactor from a height of 1.83 m. (a) $[0/\text{-}45/90/45]_s$, *(b)* $[0/\text{-}45/90/45]_{2s}$

inspection.

The first stage of damage is matrix cracking at the center of the bottom ply, parallel to the fibers. The second stage starting a few microseconds after impact, consists of delaminations starting from the bottom and decreasing in size towards the top. This quasi-conical distribution of delaminations was verified nondestructively. X-radiographs show no major damage directly under the impact point, except for matrix cracking concentrated at the bottom ply (Figures 8 and 9). Delaminated strips are parallel to the fibers of each ply increasing in length from top to bottom. The length variation of consecutive delaminated strips is more pronounced in the thin laminate. The overall damage area of this laminate is oblong

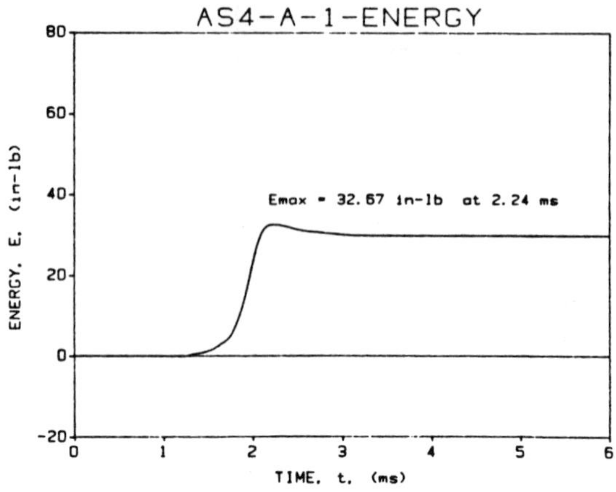

(a)

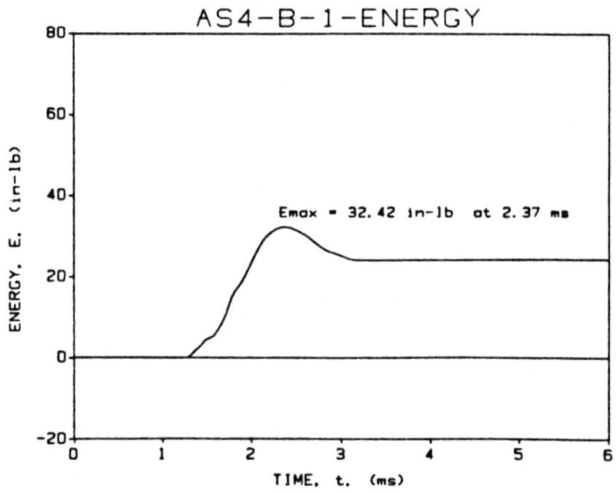

(b)

Figure 6: Imparted energy for graphite/epoxy plate subjected to impact of 202 gr impactor from a height of 1.83 m. (a) $[0/-45/90/45]_s$, (b) $[0/-45/90/45]_{2s}$

and antisymmetric in shape, reflecting the ply stacking sequence and delamination length variation from top to bottom. The damage area of the thick laminate is more elliptic or quasi-circular. Similar differences in the shape of the damage area have been observed in the same laminate by varying the height of impactor drop.

Impacted specimens were inspected ultrasonically using the pulse echo technique. A 5 MHz focused transducer with a 5.08 m (2 in.) focal length was used. The ultrasonic system used is shown in Figure 10. It includes the immersion tank and associated instrumentation for scanning the transducer, pulsing it and recording the response. The system has the capability for obtaining A-, B- and C-scans as well as three-dimensional imaging of flaws by combining both B- and C-scans. An

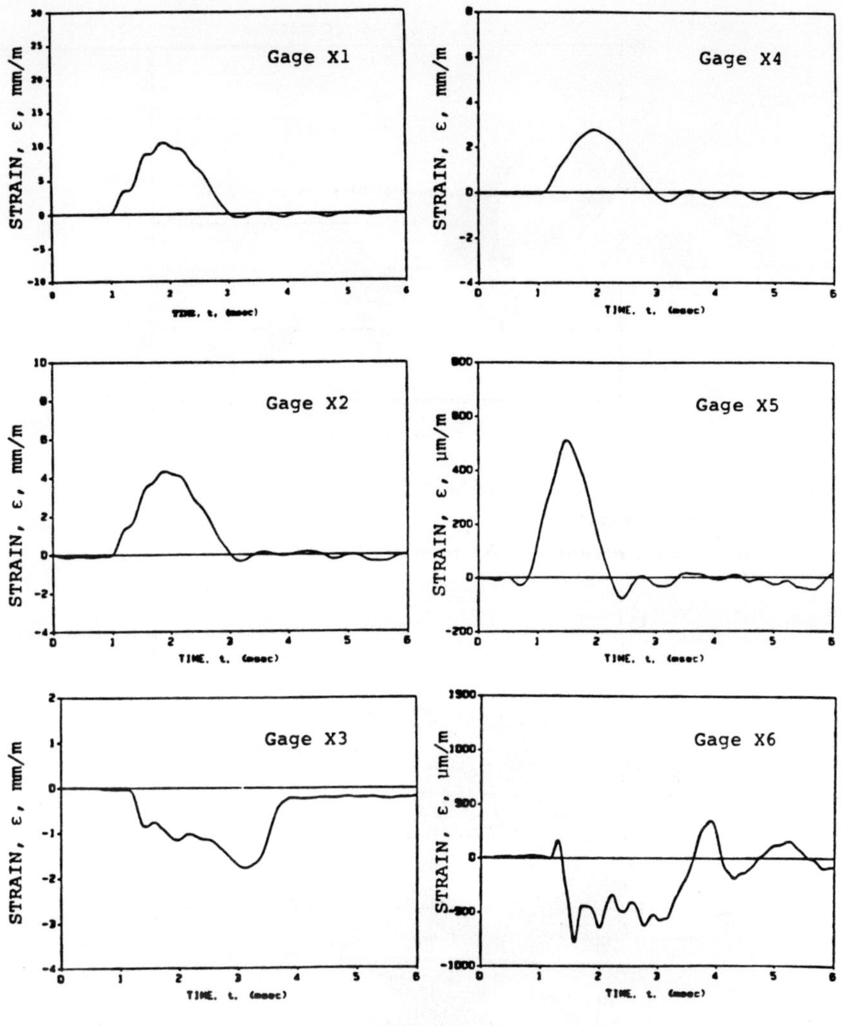

Figure 7: Transient strains at various locations through the thickness for [0/-45/90/45]$_{2s}$ graphite/epoxy plate under impact of 202 gr impactor from a height of 1.83 m. (Gage numbers refer to gage layout of Figure 2)

ultrasonic C-scan and several A-scans at various points in the damage area are shown in Figure 11 for the 16-ply laminate. The A-scans confirm that delaminations at the periphery of the damage area are located near the bottom of the plate, and those near the center of the area are located closer to the impacted side (top) of the plate. A C-scan and corresponding B-scans along three sections in Figure 12 confirm the quasi-conical distribution of delaminations through the thickness.

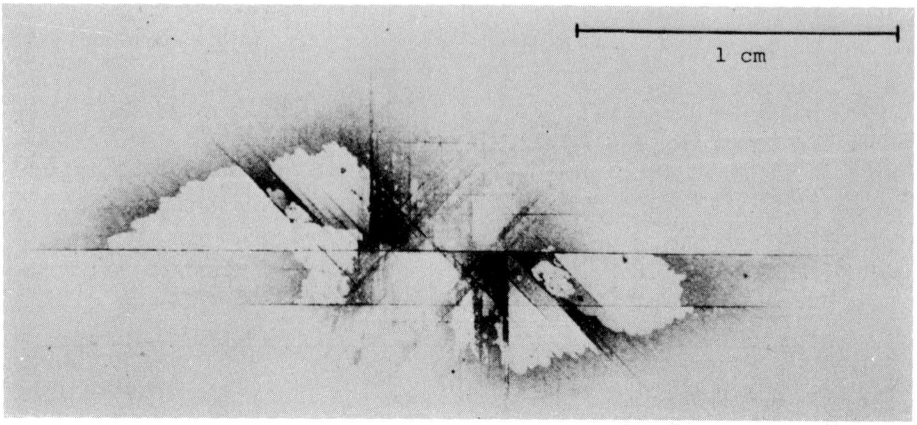

Figure 8: X-radiograph of [0/-45/90/45]$_s$ 12.7 cm (5 in.) diameter clamped graphite/epoxy plate subjected to impact of a 202 gr (0.445 lb) impactor from a height of 1.83 m (72 in.)

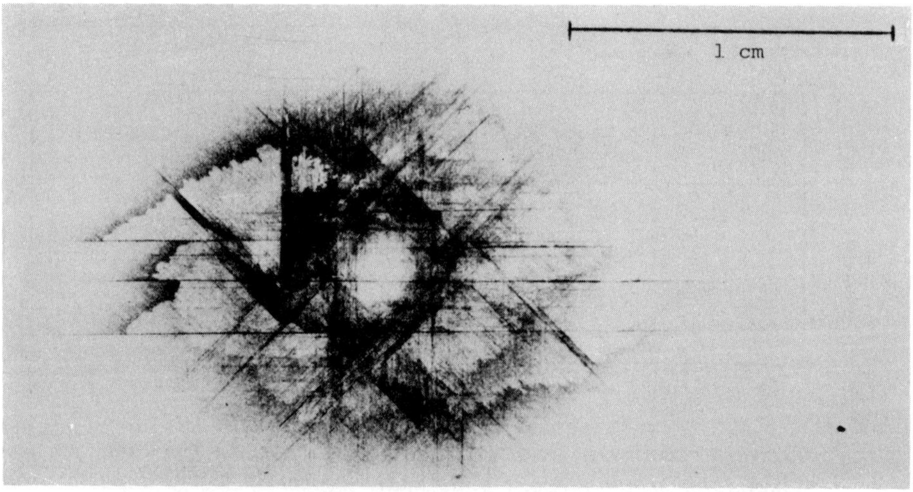

Figure 9: X-radiograph of [0/-45/90/45]$_{2s}$ 12.7 cm (5 in.) diameter clamped graphite/epoxy plate subjected to impact of a 202 gr (0.445 lb) impactor from a height of 1.83 m (72 in.)

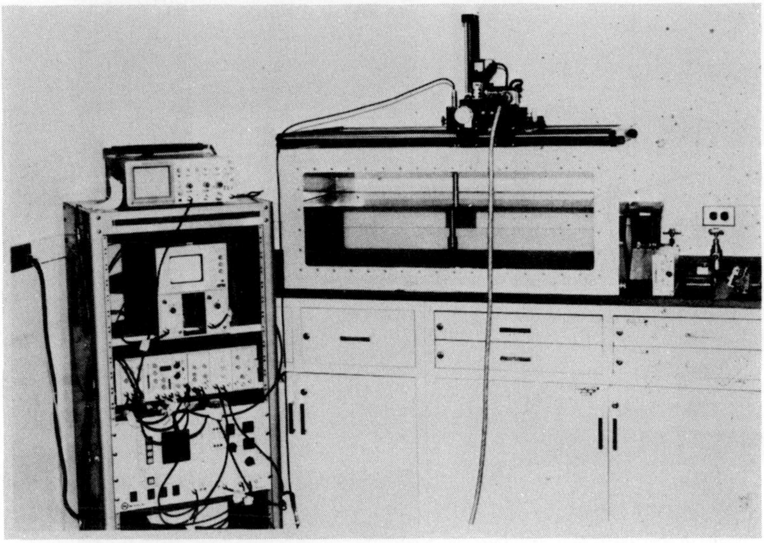

Figure 10: Ultrasonic scanning system

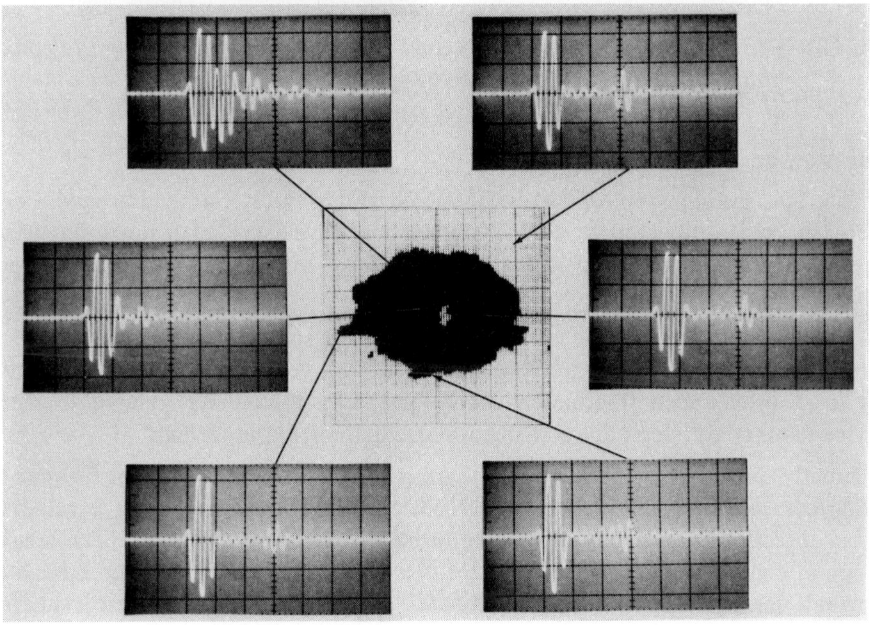

Figure 11: Ultrasonic C-scan and A-scans for $[0/-45/90/45]_{2s}$ graphite/epoxy plate subjected to impact of 202 gr (0.445 lb) impactor from a height of 1.83 m (72 in.) (AS4/3501-6 graphite/epoxy)

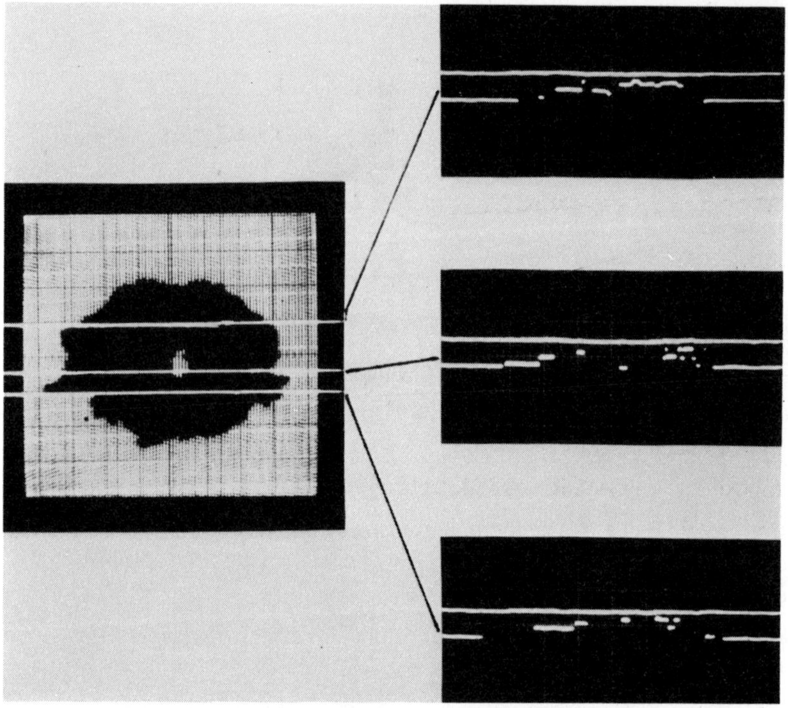

Figure 12: Ultrasonic C-scan and B-scans for $[0/-45/90/45]_{2s}$ graphite/epoxy plate subjected to impact of 202 gr (0.445 lb) impactor from a height of 1.83 m (72 in.) (AS4/3501-6 graphite/epoxy)

4. Summary and Conclusions

Quasi-isotropic graphite/epoxy plates, 8-ply and 16-ply thick, clamped around a circumference, were subjected to central impact loading. The load history, imparted energy, and transient strains at various locations through the thickness were obtained.

The dynamic load history for the thin laminate shows a higher peak load of shorter duration than for the thick laminate. The load record for the thick laminate shows more pronounced fluctuations than for the thin laminate, related to more extensive damage development and flexural oscillations in the former.

Directly under the point of impact there is no major damage, except for matrix cracking concentrated at the bottom ply. Delaminated areas are strips parallel to the fiber direction in each ply increasing in length from top to bottom. The length variation of consecutive delaminated strips is more pronounced in the thin laminate. The overall damage area of the thin laminate is oblong and antisymmetric in shape, reflecting the ply stacking sequence and delamination length variation from top to bottom. The damage area of the thick laminate is elliptical or quasi-circular.

Transient strain records through the thickness of the laminate corroborte the nondestructive indications regarding the locations of the delaminations.

Acknowledgment

The work described here was sponsored by the NASA-Langley Research Center, Hampton, Virginia. We are grateful to Dr. Walter Illg of NASA-Langley for his encouragement and cooperation.

References

[1] *Foreign Object Impact Damage to Composites*, ASTM STP 568, American Society for Testing and Materials, 1975.

[2] Ross, C.A. and Sierakowski, R.L., "Studies on the Impact Resistance of Composite Plates", *Composites*, Vol. 4, 1973, pp. 157-161.

[3] Sun, C.T. and Yang, S.H., "Contact Law and Impact Responses of Laminated Composites, NASA CR-159884, February, 1980.

[4] Humphreys, E.A., "Development of an Engineering Analysis of Progressive Damage in Composites During Low Velocity Impact," NASA CR-165778, July, 1981.

[5] Elber, W., "Failure Mechanics in Low-Velocity Impacts on Thin Composite Plates", NASA Technical Paper 2152, May, 1983.

[6] Shivakumar, K.N., Elber, W. and Illg, W., "Analysis of Progressive Damage in Thin Circular Laminates Due to Static Equivalent Impact Loads," presented at AIAA/ASME/ASCE/AHS Structures, Structural Dynamics and Materials Conference, Lake Tahoe, NV., May, 1983.

[7] Chamis, C.C. and Sinclair, J.H., "Impact Resistance of Fiber Composites: Energy-Absorbing Mechanisms and Environmental Effects", *Recent Advances in Composites in the United States and Japan*, ASTM STP 864, J.R. Vinson and M. Taya (Eds.), American Society for Testing and Materials, Philadelphia, 1985, pp. 326-345.

CHARACTERIZING THE DAMAGE STATE OF COMPOSITE LAMINATES VIA THE ACOUSTO-ULTRASONIC TECHNIQUE

Edmund G. Henneke, II, John C. Duke, Jr. and Richard C. Stiffler
Materials Response Group
Engineering Science and Mechanics Department
Virginia Polytechnic Institute and State University
Blacksburg, Virginia

Abstract

The acousto-ultrasonic, or stress wave factor, technique was developed by Vary and co-workers. This ultrasonic technique provides a quantitative ultrasonic parameter which has been found to correlate with several different mechanical properties in both homogeneous and composite materials. We have found the technique to be quite sensitive to both in-situ damage in a virgin specimen and to damage developing during mechanical loading. The technique differs from standard transmit-receive ultrasonic techniques only in that the two transducers are not in a direct line of sight (including reflection by Snell's Law). That is, one cannot follow the ray path from the sender to the receiver, even when reflections off boundary surfaces are considered. The receiver signal is composed of surface waves, body waves, and plate waves, established through scattering and diffraction processes. In this work we will present the basic technique, examples of experimental results relating the ultrasonic signals to mechanical properties of composites, and experimental results from studies of the basic nature of the waves involved in this technique.

1. Introduction

Damage in composite materials evidences itself as a change in the condition of the material properties. When this change results in a reduction of stiffness, strength, or life, as is often the case, it is obviously necessary for the material user to be aware of the damage and the consequent material condition. Numerous nondestructive techniques and methodologies exist which have as their objective the detection and physical description of material imperfections, matrix cracks, delaminations, porosity, fiber or matrix rich regions, etc. In such instances, it is necessary to analytically establish the individual and/or collective significance of the imperfections as regards service performance. Finally, the success of this analysis must be validated. Alternatively, it might be possible to directly evaluate the material condition nondestructively. A technique which possesses this capability is the ultrasonic stress wave factor, or acousto-ultrasonic, technique [1]. This technique involves the measurement of the response of the material to an ultrasonic stress wave pulse in directions which lie in the laminate plane. Because advanced composite materials such as graphite epoxy are manufactured typically in plate form for structural applications in which they are loaded along in-plane directions, the in-plane properties are of major importance.

The acousto-ultrasonic technique is briefly reviewed in the following section, followed by several examples of experimental results showing correlation of acousto-ultrasonic parameters with material failure site location, stiffness and strength. Finally, a short discussion is presented on the possible nature of the waves involved in using the acousto-ultrasonic technique.

2. Technique Description

In general, fiber reinforced laminated composite matrials are fabricated in forms that limit access for nondestructive examination. Most often the material is examined in a direction perpendicular to that in which the load is applied, using either ultrasonic or X-ray radiographic methods. If access is further limited to only one of these surfaces only pulse-echo ultrasonic inspection is generally considered. A new method of utilizing ultrasound for nondestructively examining composite material subject to such constraints has been developed by Vary [1]. The technique employs both a sending and a receiving ultrasonic transducer that are in contact with the same surface of the object being examined, Figure 1. Ultrasonic pulses are repeatedly (repetition rate - $1/P$) introduced into the specimen and the oscillations of the receiving transducer caused by these pulses are counted above a selected threshold (number of oscillations – C) for a period of time (time – T). As a means of quantifying this measurement the stress wave factor (SWF) has been defined $SWF \equiv C \cdot T \cdot 1/P$. The unique feature of this method is its sensitivity to the structural configuration of the object. That is to say that the material's internal structure, lamina interfaces, lamina orientation, and boundaries cause the stress wave to propagate out into the material to a point where the receiver is located. This feature makes it possible to interrogate the material in the same directions in which stresses resulting from applied load would act. Interrogation in these directions, as such, would be expected to find regions of the material responding in a fashion which would be peculiar to their condition, as related to mechanical performance. For example, the region which has a poor performance regarding stress wave energy propagation would be expected to test as a region of low SWF. If this number is the lowest in the object, its performance would be considered to be the poorest. Consequently, if all the regions were subject to an identical state of stress, the deformation in this region might be expected to be the most severe, and the location at which failure initiates.

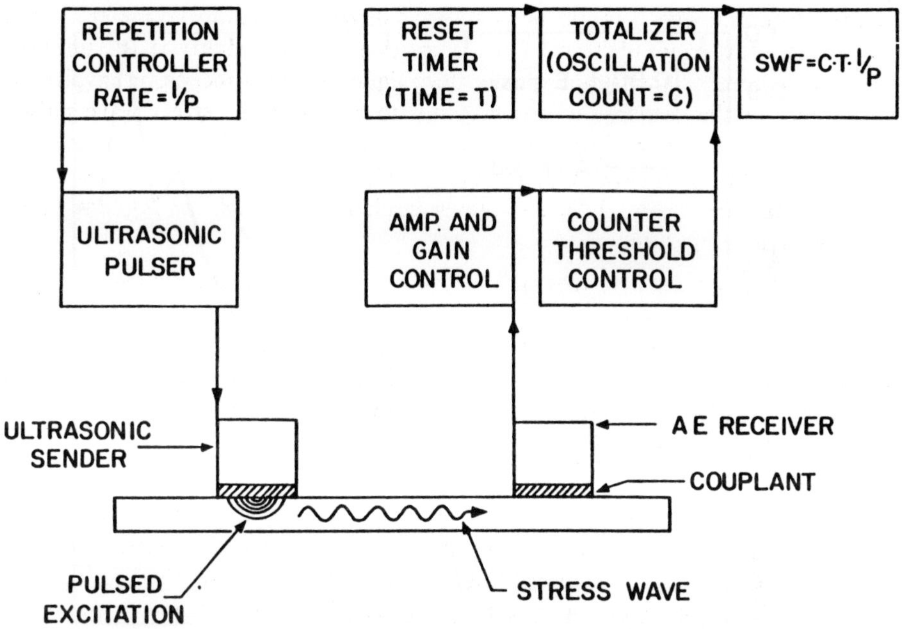

Figure 1: Schematic diagram of experimental set-up for acousto-ultrasonic technique

3. Determination of Failure Location

By repeating the measurement at different locations it is possible to determine a relative evaluation of the material. Considerable care must however be taken to reproducibly perform this examination, otherwise the comparison is pointless. Special attention has been given to adequately characterize the parameters associated with the performance of this technique. Of particular significance is the type and amount of couplant material, the transducer attachment method, the threshold level and period of counting, and the procedure used for associating SWF values with a region of the material [2]. Figure 2 displays the results of examining a $[0,90_3]_S$ *E* glass laminated composite. The transducers were maintained at a separation of

3.125 cm and were translated in increments of 0.63 cm in order to examine the entire specimen. The value of SWF obtained has been plotted in two ways: the solid line displays the SWF value for the increment at the point directly between the two transducers, while the broken line records the data that is the average of all SWF values obtained while a .63 cm region remains between the two transducers. The latter procedure was developed to try to associate the SWF value with a small region (point) of the specimen. The schematic below the graph depicts the specimen: the grid indicates the various transducer positions and the shaded area the location of failure resulting from quasi-static tensile loading. The region of low SWF correlates quite well with the failure location.

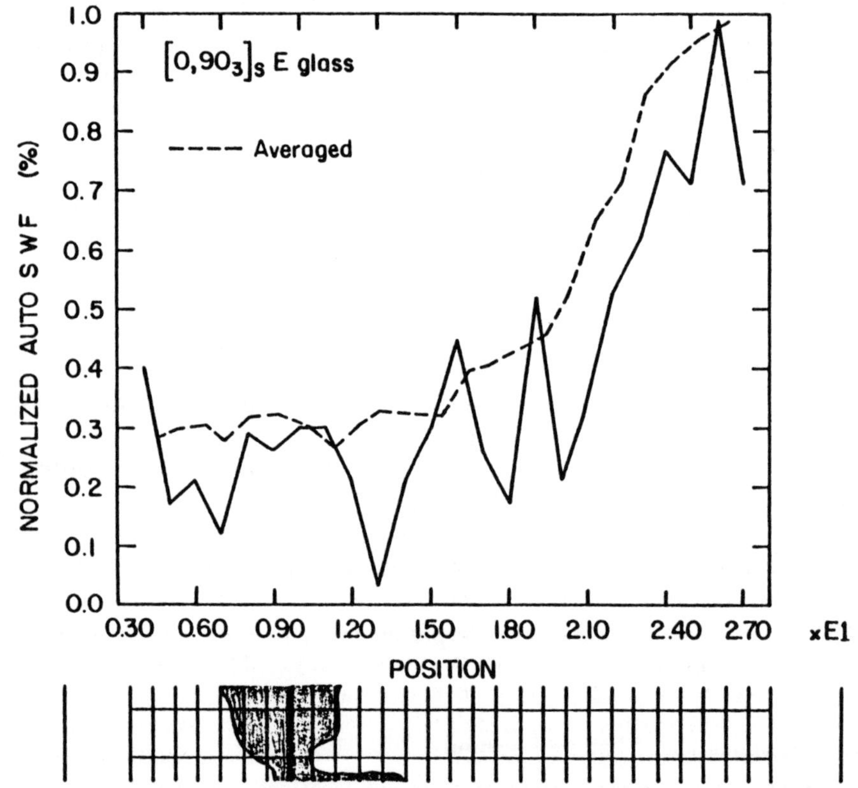

Figure 2: SWF values measured along the length of a E-glass epoxy laminate. The location of the failure site resulting from quasi-static tensile loading is also shown

4. Determination of Stiffness

Classically stiffness is considered to be a structure-insensitive property. However, because the very nature of fiber reinforced laminated composite material is that of a structure, the property of stiffness must of necessity be structure sensitive; this is not to say that the stiffnesses of the constituents of the composite material, i.e., the fiber and the matrix, are structure sensitive. Nevertheless it is necessary to recognize that, since laminated composite materials may in general continue to support load even after sustaining considerable damage and may exhibit property variations from point to point in even the undamaged condition, because of problems during manufacture, experimental determination of stiffness is inextricably linked with a gage length.

Several experiments were performed that were directed at comparing the stiffness of the material to the measured SWF. The technique of moiré interferometry was utilized to obtain the full field in-plane axial displacement during quasi-static tensile loading. Since the interference patterns obtained may be interpreted as displacement, it is possible to determine the stiffness at every point along the length of a straight sided coupon specimen, subjected to tensile load. Figure 3 displays a plot of measured SWF versus the axial displacement. For each region for which the SWF measurement was made, the axial displacement measured over the corresponding region has been plotted; each region measured 3.125 cm in length. The displacement values used were obtained at a load for which no detectable damage to the composite material resulted (76 MPa).

It is clear from the figure that these two measures are correlated. Since the stress wave used to make the measurement is not constrained to propagate solely in the axial direction, some variation due to nonuniform strains in the transverse directions might be expected. Nevertheless, within the experimental accuracy of the two measurement techniques, the correlation is quite good. Since the deformation at the loads employed is essentially linear and elastic, and the regions of material involved in the measurement are identical, the correlation of SWF with longitudinal stiffness is the same. However, the exact nature of the correlation is material and laminate dependent, Figure 4. As a result, in order to apply this technique to determining the stiffness from point to point in a structure, the correlation of the SWF with stiffness for the particular laminate and composite system would need to be known. In addition, the correlation of SWF with stiffness for damaged conditions is expected to be different.

To follow up on the experimental correlation we observed between the stiffness and stress wave factor parameters, a number of fatigue tests on graphite epoxy laminates of various stacking sequences were performed. A test set-up similar to that shown in Figure 1 was used. The transducers employed were matched broad band transducers, 1.27 cm in diameter and nominally 5 MHz. The transducers were held in a fixture so that the transducer centers were maintained a fixed distance apart, 3.81 cm. The received ultrasonic signals were amplified, digitized and stored in a computer for later spectral analysis. A Fast Fourier Transform algorithm was

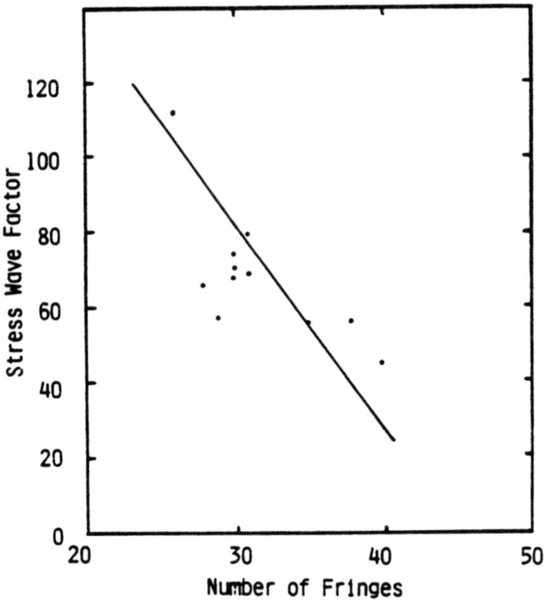

Figure 3: SWF versus the number of interference fringes measured by a moiré interferometric technique

used to perform the spectral analysis. For the fatigue tests, loading was halted at the mean load value after selected numbers of loading cycles and the acousto-ultrasonic tests were made.

The specimens tested were standard tensile coupons, nominally 25.4×20.32 mm of graphite epoxy. Several different laminate stacking sequences were investigated, $[0,90_2]_s$, $[0,\pm 45]_s$, and $[0,90,\pm 45,]_s$. All tests were run in a servo-hydraulic, closed-loop testing frame. Quasistatic tests were made during the first load cycle applied to each fatigue specimen. The first load cycle was applied very slowly to the maximum stress level to be used during fatigue, between 60-70% of the ultimate tensile strength of the laminate. The fatigue tests were tension-tension with a loading ratio of 0.1 and a rate of 10 Hz. An extensometer was used to measure the strain over a 25.4 mm gauge length.

To obtain a parameter from the acousto-ultrasonic test which is made more easily reproducible, the acousto-ultrasonic signals were analyzed by FFT to obtain the power spectral density. Talreja [3] has noted that three classes of parameters are needed to describe distribution functions, namely location, scale and shape parameters. He suggested that a convenient set of parameters to represent a power spectrum can be defined by considering the power spectrum to be a plane figure closed on the frequency axis. The location parameter for the power spectrum can then be taken to be the location of the centroid of this figure. The area of this figure forms a suitable scale parameter and the shape parameters can be described in terms of the various moments of the figure about a convenient axis. Talreja

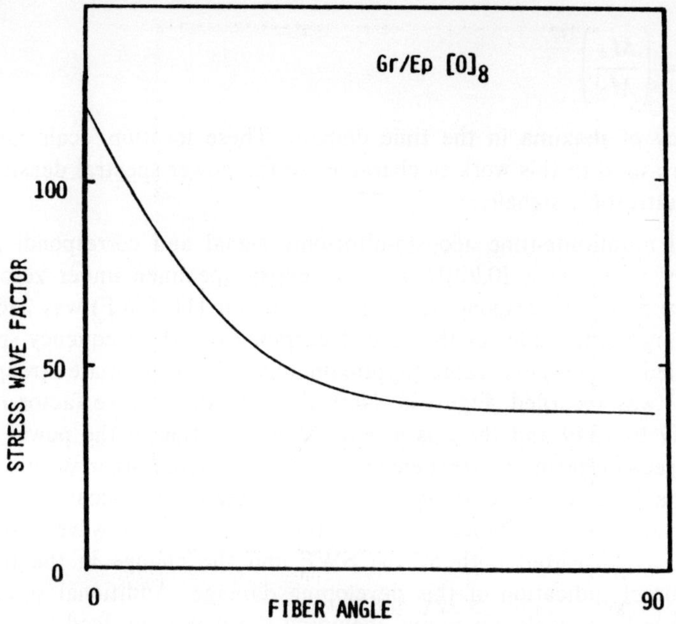

Figure 4: Variation of SWF with fiber angle in a undirectional graphite epoxy laminate

therefore proposed that a set of shape parameters be defined by

$$S_{r,k}=\frac{M_r}{M_{r-k}f_c^k} \quad \begin{array}{r} k=1,2,... \\ r=2,3,4... \\ r\geq k \end{array} \tag{1}$$

where

$$M_r=\int_{-\infty}^{\infty}S(f)f^r\,df \tag{2}$$

where $S(f)$ is the power spectral density, f is the frequency and f_c is the location of the centroid found by the following expression

$$f_c=\frac{M_1}{M_0} \tag{3}$$

Several of the moments and moment ratios can be given physical interpretations. For example, M_0 is the mean square value of the power spectral density. Also,

$$f_0=\left(\frac{M_2}{M_0}\right)^{1/2} \tag{4}$$

is the frequency of mean value crossings with positive slopes and

$$f_p = \left(\frac{M_4}{M_2} \right)^{1/2} \tag{5}$$

is the frequency of maxima in the time domain. These location, scale and shape parameters were used in this work to characterize the power spectral density curves of the acousto-ultrasonic signals.

The initial amplitude-time acousto-ultrasonic signal and corresponding power spectral density taken on a $[0,90_2]_s$ graphite epoxy specimen under zero load is shown in Figure 5. The corresponding stress wave factor [1] (SWF) was 2369. Note the relatively large amplitudes of the second harmonics in the frequency spectrum. Load was applied to the mean value (approximately 30% of ultimate strength) and a second signal was recorded, Figure 6. Note that the stress wave factor has been reduced slightly to 2349 and there is a noticeable reduction in the power spectral density at the second harmonic frequencies. It is known from other studies that for this laminate stacking sequence, transverse matrix cracking has occurred in the 90° plies at these load levels. Hence the reduction in the stress wave propagation characteristics, as measured both by the SWF and the change in the frequency spectra, is a direct indication of this developing damage. Additional signals were recorded at other load levels, up to and including the maximum load to be applied during fatigue.

After the specimen had been loaded to the maximum load level in the first cycle, the fatigue test was started. The fatigue test was halted at selected numbers of load cycles and acousto-ultrasonic signals recorded. Figure 7 is the signal and power spectral density for a signal recorded after 50,000 applied load cycles. Note the nearly complete disappearance of the higher frequency components. Also, the SWF has been reduced to 789.

In general, damage which occurs during fatigue causes a reduction in the laminate stiffness. Figure 8 presents the laminate stiffness, normalized with respect to the initial stiffness E_0, versus the number of applied load cycles. Quite typically, the stiffness shows a very large and rapid decrease during the first several hundred load cycles. During the large majority of the test, the stiffness changes only gradually. Sometimes, near the end of the test, and depending on the laminate stacking sequence, there is a second rapid decrease in stiffness. Also shown in Figure 8 is the root mean square of the power spectral density of the acousto-ultrasonic signal as defined by Equation (2) with $r=0$. While there is some scatter in this value during the early portion of the test, the general trend of this data is very close to that of the stiffness. Also note that while the stiffness degraded approximately 12% during the test, the root mean square decreased by 20%. Hence the root mean square of power spectral density is a more sensitive monitor of damage development than is stiffness. This conclusion has been borne out by a number of tested specimens of all three laminate types studied in our laboratory to date.

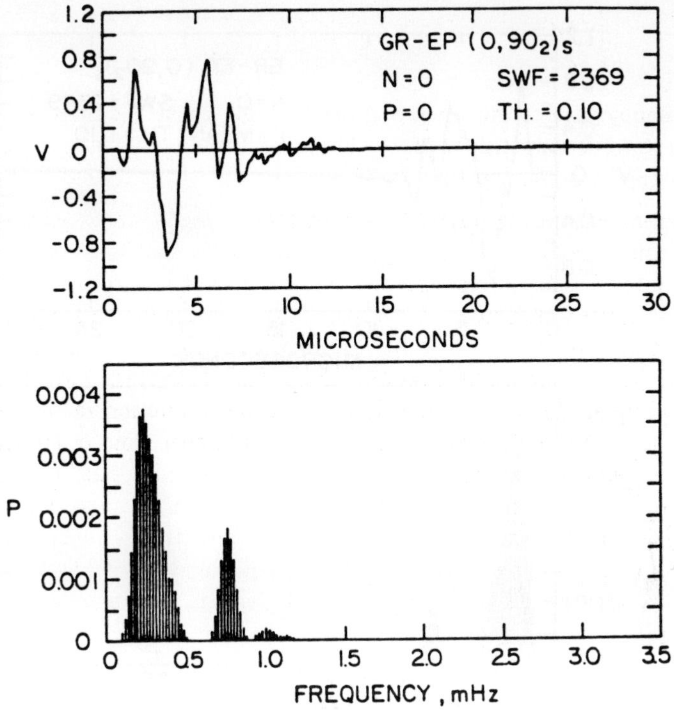

Figure 5: Amplitude-time and corresponding Fourier transform for graphite epoxy laminate in initial condition

5. Determination of Strength/Life

Although good agreement is seen between measured low values of SWF and failure location, some question exists regarding the exact relationship of the measured SWF and strength/life. In this regard, a fundamental difficulty exists because of the terminal nature of strength/life determining tests; the potential strength/life of regions outside of the failed region cannot be determined. It is important to recognize at this point in addition that the strength/life is dependent on the applied load history and future. That is to say, for a particular piece of composite material, the strength/life of any region is not unique. Consequently it is somewhat unrealistic to expect a relationship to exist in general between SWF and strength/life. Perhaps a particular relation may exist between the SWF, strength/life, and a specific load history. To fully investigate this aspect of SWF technique, additional clarification of the issue is required. First, as a composite material is subjected to mechanical loading the condition of the material becomes more and more damaged. Figure 8 demonstrates this for a specimen that has been cyclically loaded. Similarly such progressive damage development occurs for quasi-static, sustained or spectral loading. In general, a SWF measurement cannot be uniquely related to the material strength since as the material condition changes so does the SWF value; there

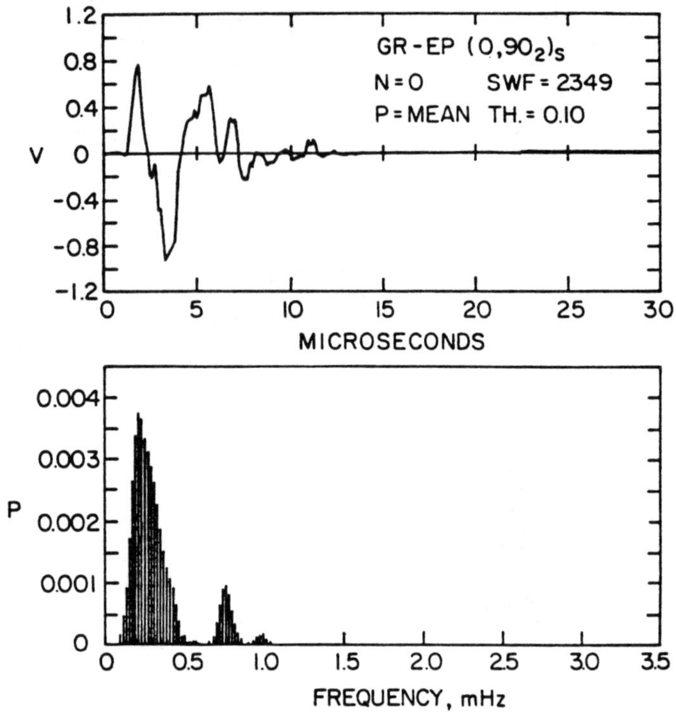

Figure 6: Amplitude-time and corresponding Fourier transform for graphite epoxy laminate after applying a tensile load equivalent to 30% of ultimate

would be only one value of strength.

Two areas of investigation need be considered: one directed at establishing the correlation between SWF and strength when the material condition being evaluated has resulted from cyclic loading; and the other is to determine the correlation between SWF and remaining life.

6. Nature of Acousto-Ultrasonic Waves

Experimental observations lead one to the hypothesis that the waves which are generated and received via the acousto-ultrasonic technique are plate (Lamb) waves. It is well known that an acoustic or electromagnetic waveguide will allow only propagation of a wave having a particular frequency or any of its harmonics for a given wave vector. For the case of an acoustic wave guide, the free surfaces cannot support a normal stress; thus acoustic waves propagating through a wave guide that have normal stress amplitudes other than zero at the free surface will be rapidly attenuated due to destructive interference. However, there will be one fundamental wave length or an integral multiple of this wave length that can propagate through

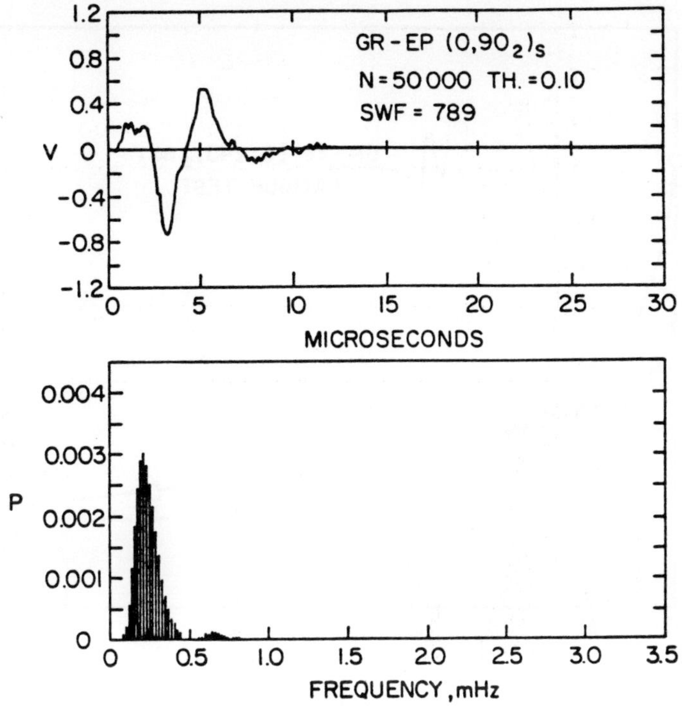

Figure 7: Amplitude-time and corresponding Fourier transform for graphite epoxy laminate after tension-tension cycling for 50 KHz at a maximum stress of 60% ultimate

the wave guide for a given wave vector and still maintain the stress free boundary conditions.

Wave propagation in plates can be separated into several regimes depending on the wavelength with respect to the thickness. If the wavelength is much larger than the thickness of the plate, then low frequency plate modes are set up. For the low frequency extensional modes, the wave vector is in the plane of the plate and there is no variation in displacement through the thickness. This mode can propagate for all frequencies, is nondispersive, and can be easily applied to angle-ply laminates. Because of its nature, it is perhaps easily established by the initial pulse applied to the generating transducer.

The results of a preliminary study described subsequently provide insight into the nature of the wave propagation that occurs in the stress wave factor technique. This work has included calculating the low frequency fundamental plate wave modes in composite materials. It should be pointed out that for the analysis described here, the composite plate has been treated as a macroscopically homogeneous, but anisotropic, material having the elastic constants of the composite material in question. This assumption is reasonable as long as the wave lengths are long in comparison with the fiber diameters, fiber spacing, and ply thicknesses,

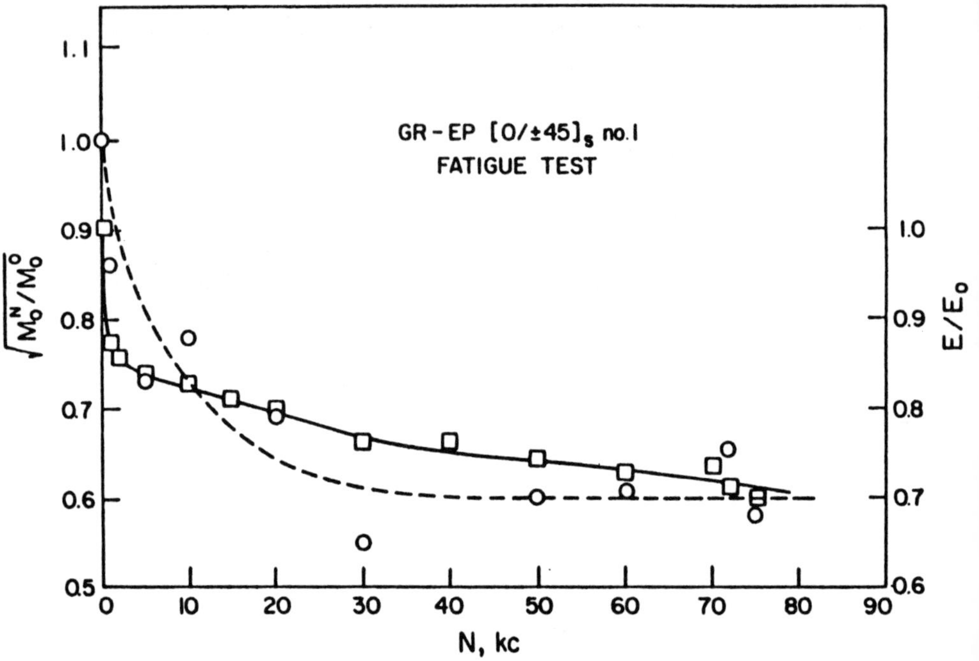

Figure 8: Laminate stiffness (square data points) and acousto-ultrasonic parameter (circular data points) versus fatigue cycles for graphite epoxy laminate

which is a reasonable assumption for frequencies less than those normally used for ultrasonic investigation (i.e., less than 10 MHz).

Preliminary experimental work has shown that the flexural phase velocity in the plate varies as a function of frequency due to the highly dispersive nature of the plate. The dispersion curves, i.e., curves for frequency versus velocity, are characteristic for a plate having a particular thickness and elastic properties. Once established, this analysis allowed good correlation of experimental dispersion relations with plate extensional stiffnesses, A_{11} and A_{22} and plate bending stiffnesses D_{11} and D_{22}. Analytical and experimental results, to date, show good correlation at the longer wave lengths. It appears from these results that fundamental plate wave modes are established by placing a transducer flat upon the surface of a plate. These plate wave modes propagate along the plane of the plate and can be detected by a similarly placed transducer "downstream" of the transmitter.

A condensed presentation of plate wave analysis is briefly reviewed before discussion of experimental results. The low frequency approximation described earlier gives several simplifying factors for formulating wave propagation in plates, namely:

1) For extensional and in-plane shear waves, the displacements are uniform across the thickness of the plate and out-of-plane inertial effects are negligible.

2) For flexural waves, rotary intertia and transverse shear are negligible.

With these approximations, classical laminate theory can be used as the basis for developing the low frequency wave equations.

Again, it should be noted that this two-dimensional development is valid only for wavelengths larger than the thickness of the plate. For short wavelengths, the thin plate assumption is no longer valid because the displacements will be a function of the thickness coordinate necessitating an analysis using three-dimensional elasticity.

For in-plane waves, the stress free surface condition is immediately satisfied by assuming a plane state of stress. The equations of motion for a plane state of stress are:

$$\frac{\partial \sigma_1}{\partial x_1}+\frac{\partial \sigma_6}{\partial x_2}=\rho \ddot{u}_1 \tag{6}$$

$$\frac{\partial \sigma_6}{\partial x_1}+\frac{\partial \sigma_2}{\partial x_2}=\rho \ddot{u}_2 \tag{7}$$

For a laminate that is symmetric with respect to its midplane, the stress-strain relationship is given by

$$\begin{bmatrix} \sigma_1 \\ \sigma_2 \\ \sigma_6 \end{bmatrix} = \frac{1}{d} \begin{bmatrix} A_{11} & A_{12} & A_{16} \\ A_{12} & A_{22} & A_{26} \\ A_{16} & A_{26} & A_{66} \end{bmatrix} \begin{bmatrix} \varepsilon_1 \\ \varepsilon_2 \\ \gamma_6 \end{bmatrix} \tag{8}$$

where A_{ij} are elements of the extensional stiffness matrix. The calculation of the extensional stiffness matrix is discussed in a number of books [4] on composite materials and is not discussed here.

If the displacements for a plane harmonic wave travelling in the plane of the plate are taken to be

$$\begin{aligned} u_1 &= A\alpha_1 \exp(i(k(x_1\ell_1+x_2\ell_2)-\omega t)) \\ u_2 &= A\alpha_2 \exp(i(k(x_1\ell_1+x_2\ell_2)\omega t)) \, , \end{aligned} \tag{9}$$

then we can obtain Christoffel's equation:

$$\begin{bmatrix} \ell_1^2 A_{11}+\ell_1\ell_2 A_{16}+\ell_2^2 A_{66}-\rho(\omega/k)^2 & \ell_1^2 A_{16}+\ell_1\ell_2(A_{12}+A_{66})+\ell_2^2 A_{12} \\ \ell_1^2 A_{16}+\ell_1\ell_2(A_{12}+A_{66})+\ell_2^2 A_{26} & \ell_1^2 A_{66}+2\ell_1\ell_2 A_{26}+\ell_2^2 A_{22}-\rho(\omega/k)^2 \end{bmatrix} \begin{bmatrix} \alpha_1 \\ \alpha_2 \end{bmatrix} = 0 \tag{10}$$

For a nontrivial solution, the determinant of the 2×2 matrix above must be equal to zero. The resulting characteristic equation yields the dispersion equations for the plate. Note that the phase speeds are dependent upon direction of propagation in the plate. Also, there are two solutions with appropriate directions of particle displacements.

For the experimental work, plate wave velocities were measured only for the x_1 and x_2 directions; thus only these directions will be considered in the subsequent analysis.

For propagation along the x_1 axis, the solution corresponding to an extensional wave has a velocity given by

$$V_1=\frac{\sqrt{2}}{2\rho}((A_{11}+A_{26})+\sqrt{(A_{11}+A_{66})^2+4(A_{11}A_{66}-A_{16}^2)})^{1/2} \tag{11}$$

For a unidirectional laminate or a laminate symmetric with respect to the x_1 axis, the extensional velocity reduces to

$$V_1=(\frac{A_{11}}{\rho})^{1/2} \tag{12}$$

for the extensional wave.

The derivation for flexural wave propagation is carried out in a fashion similar to that for extensional wave propagation. The governing equation for flexural wave propagation is

$$\frac{\partial^2 M_1}{\partial x_1^2}+\frac{2\partial^2 M_6}{\partial x_1 \partial x_2}+\frac{\partial^2 M_2}{\partial x_2}=-\frac{d}{\rho}\frac{\partial^2 u_3}{\partial t^2} \tag{13}$$

where

M_i bending moment about x_i coordinate axis

ρ mass density

d thickness of plate

The bending moments are related to the plate curvatures by

$$\begin{bmatrix} M_1 \\ M_2 \\ M_6 \end{bmatrix} = \begin{bmatrix} D_{11} & D_{12} & D_{16} \\ D_{12} & D_{22} & D_{26} \\ D_{16} & D_{26} & D_{66} \end{bmatrix} \begin{bmatrix} \kappa_1 \\ \kappa_2 \\ \kappa_6 \end{bmatrix} \tag{14}$$

where D_{ij} are the bending stiffnesses and κ_i are the curvatures of the plate.

If the displacement of a flexural wave, u_3, is taken to be

$$u_3=Ae^{i(k(\ell_1 x_1+\ell_2 x_2)-\omega t)} \tag{15}$$

the dispersion equation becomes

$$k^4(\ell_1^4 D_{11}+4\ell_1^3\ell_2 D_{16}+2\ell_1^2\ell_2^2(D_{12}+2D_{66})+4\ell_1\ell_2^3 D_{26}+\ell_2^4 D_{22})=\rho d\omega^2 \tag{16}$$

For flexural waves propagating along the x_1 axis, Equation (16) yields

$$k^4 D_{11} = \rho d \omega^2 , \tag{17}$$

and hence, the phase velocity is

$$V_p = \left(\frac{D_{11}\omega^2}{\rho d} \right)^{1/4} = \left(\frac{4\pi^2 D_{11} f^2}{\rho d} \right)^{1/4} \tag{18}$$

Table 1: Material properties

	E_{11} (MSI)	E_{22} (MSI)	G_{12} (MSI)	ν_{12}	ν_{21}	ρ (lbf-sec^2/in^4)
6061-T6 Aluminum	10.00	10.00	3.75	0.30	0.30	1.55×10^{-4}
AS-4(GR)/PR288 Graphite Epoxy	18.50	1.30	0.67	0.32	0.02	1.50×10^{-4}

In order to determine how well the analysis presented in the previous section represents the characteristics of the waves which are generated and detected during application of the acousto-ultrasonic technique, both analyses and experiments were performed on two different materials, aluminum and graphite epoxy. The aluminum plate was 6061-T6 aluminum, 0.065" thick × 12"×12". The graphite epoxy panels were positive pressure molded from AS-4(GR)/PR288 and consisted of $[0]_8$, $[0]_{22}$, $[0,90]_{2s}$, and $[0,+45,90\text{-}45]_s$ laminates. The mechanical properties of these materials can be found in Table 1. The laminate properties for the plates can be found in Table 2 and the low frequency extensional and flexural wave velocities can be found in Table 3.

A number of experiments utilizing low frequency acoustic waves were performed on several different laminates and aluminum to determine the applicability of this plate wave analysis. A piezoelectric transducer, which acted as the acoustic generator, was excited using a gated sine wave. The gated sine wave provided an excitation signal with a very narrow frequency band while still retaining a definite leading edge. A second piezoelectric transducer was used as the receiver. The position of the receiver was advanced a known distance and the change in time for a particular phase point noted from a digitizing oscilloscope. Dividing the displacement by the time yielded the phase velocity, from which the laminate stiffnesses in the direction of the wave are easily obtained.

To perform the experiment, a gated tone burst signal with the desired frequency, usually between 10 kHz and 1 MHz, was applied to the transmitting transducer. The receiver was positioned along the wave propagation direction relative to the transmitter, moved slowly so that proper phase point identification could be

Table 2: Laminate properties

Material	Lay-up	Thickness (inches)	Q_{11} (MSI)	Q_{22} (MSI)	A_{11} $\frac{lb}{in}x10^6$	A_{22} $\frac{lb}{in}x10^6$	D_{11} $\frac{in\text{-}lbs x10^3}{in}$	D_{22} $\frac{in\text{-}lbs x10^3}{in}$
6061-T6 Aluminum	--	0.0625	11.00	11.00	0.69	0.69	0.224	0.224
Graphite Epoxy	$[0]_8$	0.092	18.6	1.31	1.71	0.12	1.21	0.085
Graphite Epoxy	$[0]_{22}$	0.262	18.1	1.31	4.87	0.34	27.8	1.96
Graphite Epoxy	$[0,90]_{2s}$	0.095	--	--	0.940	0.940	.942	.480
Graphite Epoxy	$[0,45,90,-45]_s$	0.097	--	--	0.747	0.747	.891	.268

Table 3: Low frequency extensional and flexural wavelengths

		Extensional Waves		Flexural Waves	
Material	Stacking Sequence	V_1 (in/sec)	V_2 (in/sec)	V_1 (in/sec)	V_2 (in/sec)
6061-T6 Aluminum	--	207,695	207,695	$153\sqrt{f}$	$153\sqrt{f}$
Graphite Epoxy	$[0]_8$	352,136	93,452	$242\sqrt{f}$	$125\sqrt{f}$
Graphite Epoxy	$[0]_{22}$	352,136	93,452	$405\sqrt{f}$	$210\sqrt{f}$
Graphite Epoxy	$[0,90]_{2s}$	256,836	256,836	$226\sqrt{f}$	$191\sqrt{f}$
Graphite Epoxy	$[0,45,90,-45]_s$	226,583	226,583	$223\sqrt{f}$	$164\sqrt{f}$

maintained on the wave front, and the phase velocity measured. The wave number was calculated from the definition

$$k = V_p / 2\pi f \tag{19}$$

for each selected frequency. The dispersion curve was thus obtained for a range of frequencies between 10 kHz and 1 MHz.

Numerous tests were conducted on plates made from different materials and having a variety of thicknesses to determine the plausibility of applying the plate wave analysis to the wave modes transmitted in thin plates when one applies the stress wave factor technique. While obtaining data, it was noted that as the frequency of the excitation signal was increased, there existed a frequency at which there was an abrupt change in the phase velocity. Further investigation suggested that this was a change in mode from flexural wave propagation (the lower frequency signal) to extensional wave propagation (the higher frequency signal). Also, from observing the received signal, it was noted that the signals for the lower frequencies were dispersive while the signals at the higher frequencies were not.

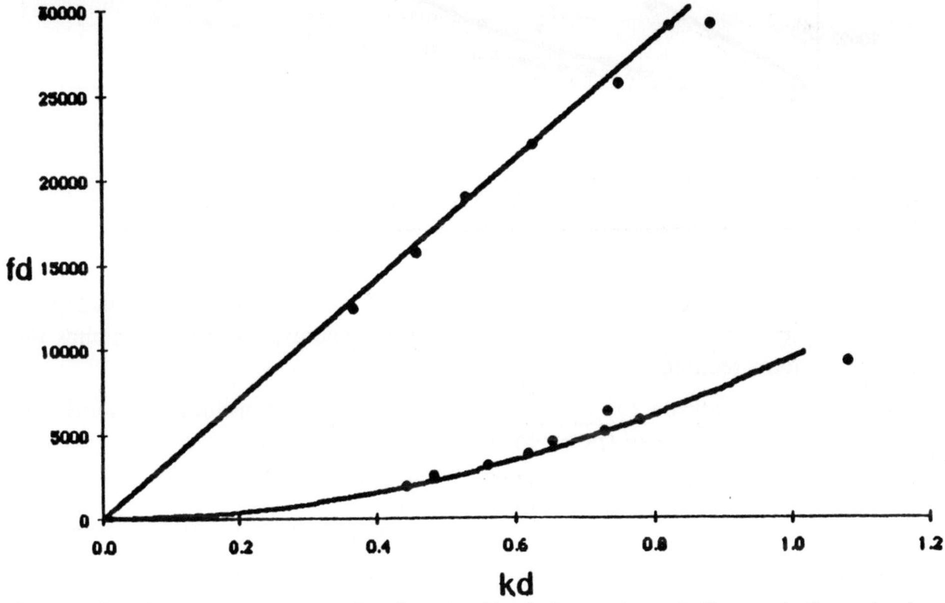

Figure 9: Dispersion curves for longitudinal (upper) and flexural (lower) plate modes in aluminum

Figures 9 and 10 are examples of the dispersion curves obtained. There is an excellent correlation between the experimental data and theoretical results for the higher (longitudinal) mode, for $kd > 0.5$, for all frequencies and materials tested. For the flexural mode, the correlation is good for small kd values, but as kd increases, the thin plate assumption is no longer valid for the composite laminates, and a better approximation is needed. Also, the flexural mode is noticeably dispersive and it is somewhat more difficult to measure the phase speed. From these plots, the extensional stiffnesses A_{11} and A_{22} can easily be found if the density of the material is known.

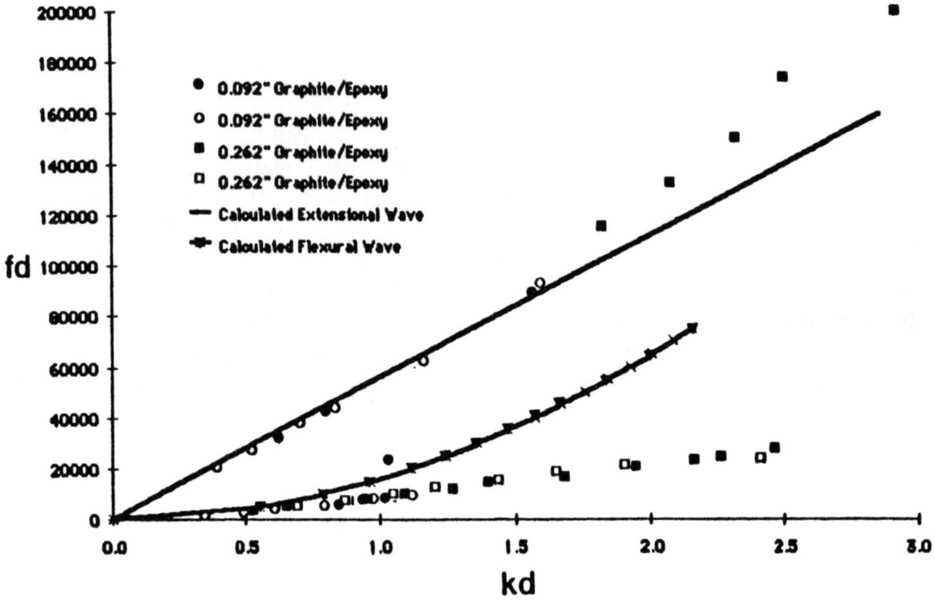

Figure 10: Dispersion curves for longitudinal (upper) and flexural (lower) plate modes in undirectional graphite epoxy laminate

The experimental data and analysis clearly indicates that at low frequencies acoustic energy is propagated by the fundamental extensional and flexural plate modes. This could explain the nature of the waves generated when one applies the stress wave factor technique.

For example, Govada [5] used a 5 MHz, wide-band transducer to launch an acoustic wave in a composite specimen. However, a spectral analysis of the received signal showed a frequency content much lower than 5 MHz. For an undamaged $[0,90_2]_s$ graphite epoxy specimen, Govada obtained a tri-modal frequency distribution with frequencies centered around 250 kHz, 750 kHz, and 1 MHz.

With the exception of the fundamental extensional and flexural mode, plate modes exhibit cut-off frequencies for which waves having a frequency content below the cut-off frequency will be attenuated. The lowest cut-off frequency for a plate, which occurs for an antisymmetric mode, is

$$f = \frac{V_s}{2d}$$

For the $[0,90_2]_s$ coupon, d = 0.030" and V_s = 60,000 in/sec gives the lowest cut-off frequency at 1 MHz. Plate waves having a frequency less than 1 MHz will propagate as one of the fundamental modes.

The reason higher modes do not propagate requires further investigation. It may simply be that higher modes are more rapidly attenuated and hence become lost in the noise level when one is making a measurement.

7. Conclusions

The acousto-ultrasonic technique is very sensitive to material and testing conditions. Because of this, a number of experimentalists have presupposed that the technique has limited usefulness for the interrogation of material condition. We have expended a considerable amount of time to develop the technique to the point at which we have been able to obtain reproducible data in our laboratory. Experimental correlations of data obtained from the technique have been found to exist for locating final failure sites and for determining laminate stiffness change with damage caused by mechanical loading. While such correlations are at the present time difficult to substantiate with physical models, they are nevertheless of extreme usefulness to engineers wishing to design and use composite materials in engineering structures.

Acknowledgement

The authors wish to thank Mr. Alex Vary and NASA-Lewis Research Center for their continuing support of the work reported herein.

References

[1] Vary, A. and Bowles, K.J., "An ultrasonic-acoustic technique for nondestructive evaluation of fiber composite quality", *Polymer Engng. Sci.*, 19, 1979, pp. 373-377.

[2] Henneke, E.G., II, Duke, J.C., Jr., Stinchcomb, W.W., Govada, A. and Lemascon, A., "A study of the stress wave factor technique for the characterization of composite materials", Interim Report of *NASA NAG 3-172*, June, 1982.

[3] Talreja, R., "On fatigue life under stationary Gaussian random loads", *Engineering Fracture Mechanics*, 5, 1973, pp. 993-1007.

[4] Jones, R.M., *Mechanics of Composite Materials*, McGraw Hill Book Co., New York, 1975.

[5] Govada, A.K., Duke, J.C., Henneke, E.G. and Stinchcomb, W.W., "A study of the stress wave factor technique of the characterization of composite materials", *NASA CR-174870*, February, 1985.

THERMAL WAVE IMAGING

THERMAL WAVE IMAGING FOR QUANTITATIVE NONDESTRUCTIVE EVALUATION

R.L. Thomas, L.D. Favro and P.K. Kuo
Department of Physics and Astronomy
Wayne State University
Detroit, Michigan

Abstract

Various techniques for making thermal wave images of defects and maps of material characteristics of opaque solids are reviewed and their potential applications are compared with those for other NDE techniques. Theory and experiment for thermal wave imaging are compared for the case of subsurface defects having simple geometries. The observed differences in sensitivity of various detection techniques to the presence of grain boundaries is discussed in the context of recent theoretical work. A variation of the technique which makes possible the contactless and local (~1mm) measurement of the thermal diffusivity of the solid is described and discussed. The application of this same technique for the study of coated materials is also described.

1. Introduction

Thermal wave imaging is an increasingly useful technique for the nondestructive evaluation (NDE) of subsurface features of opaque solids. In order to carry out this form of imaging, an intensity-modulated laser or particle beams is focused on the surface of the solid. The resulting spatial and temporal variations in the temperature of both the solid and the air above the heated surface region can be described in terms of thermal waves, whose propagation vectors, q, have equal real and imaginary components. In one dimension, the thermal wave can be written as

$$T(x,t)=T_o exp[i(qx-\omega t)] \ . \qquad (1)$$

Thus, the thermal waves have significant amplitudes only for a propagation distance of about one thermal wavelength, $\lambda_t=2\pi/Re(q)=2\pi(2k/\omega\rho c)^{1/2}$ from the periodic surface heat source. Here, k, ρ and c are the thermal conductivity, density and heat capacity, and $\omega/2\pi=f$, the modulation frequency of the heat source. For experimentally accessible modulation frequencies (about 1Hz to about 1 MHz) and thermal diffusivities ($\alpha=k/\rho c$) for typical solids and gases, the range of these waves can vary from a few microns to a few mm.

The objective of thermal wave imaging for NDE of solids is to deduce the position, shape, and other characteristics of a particular subsurface defect through its scattering of the surface-launched thermal waves. These scattered waves are then detected through their influence on the varying temperature distribution across the surface of the specimen, or by sound generated in the solid in the thermoelastic process. One obtains the imaging signal in some cases by phase-coherent observations of the surface temperature, in other cases by a suitable average of the near-surface temperature distribution in the air above the surface, and in still other cases through the resulting thermoelastic deformations of the solid. Typically, detectors such as microphones, ultrasonic transducers, infrared detectors, and laser probes are used to produce the imaging signal. An excellent review of the details of these techniques and their applications to NDE, together with an extensive set of references, may be found in the recent article by Birnbaum and White [1]. Common to these techniques is the fact that the source is localized. The techniques differ in that the detectors may be local or non-local to a greater or lesser degree. For example, the focused detector is a local point temperature detector; the mirage effect laser probe is a line detector; and the microphone is an area detector. In spite of the fact that the detector may be non-localized, the presence of the localized source gives all of these methods the potential for high spatial resolution. The symmetry of the non-locality of the detector, however, may limit detection of particular kinds of flaws.

For some of the detection schemes, comparisons between experiment and theory for imaging of flaws with simple geometry are straightforward, and good agreement has been achieved in most cases. By way of example, Figure 1 illustrates images obtained using the gas-cell detection experimental technique (to be described later) for a closed, planar, slanted crack at an angle of 45° with respect to the sample surface [2]. Theory [2] is in good agreement with the results shown in Figure 1. Similarly good agreement has been obtained [3] for gas-cell thermal wave imaging of spherical and cylindrical sub-surface defects. Other schemes, such as piezoelectric detection, are more complex in nature, and more difficult to treat theoretically. For example, the details of the conversion from thermal energy to acoustic energy may involve several different processes, and no quantitative three-dimensional theoretical model currently exists to assess the relative importance of these processes. The involvement of the elastic properties of the solid in such detection schemes may be responsible for the qualitatively different contrast seen by thermoacoustic microscopes as compared to strictly thermal wave microscopes.

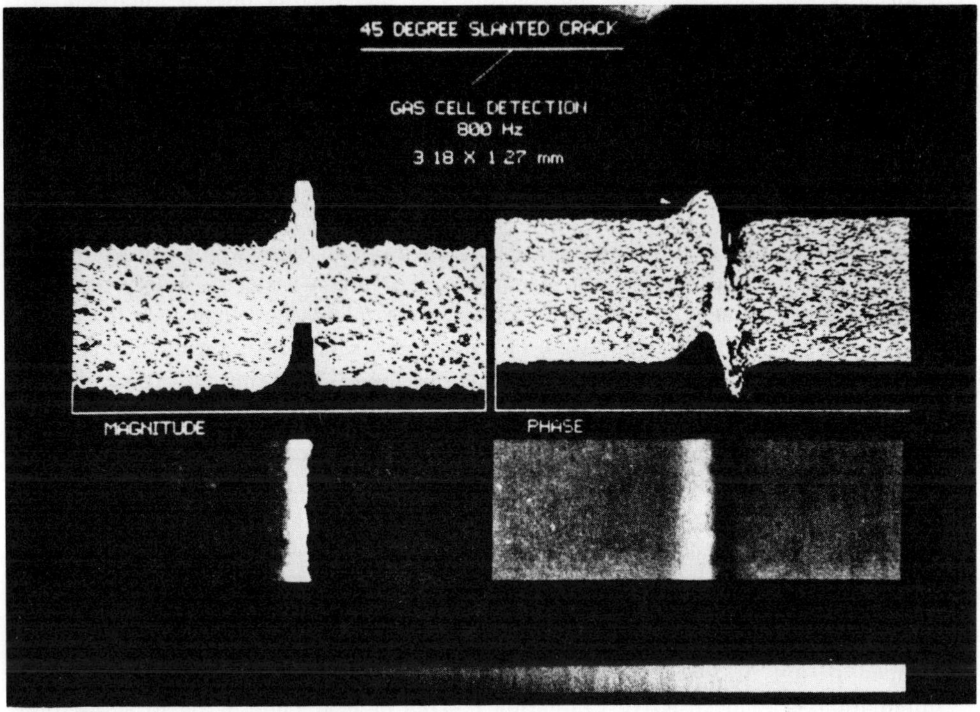

Figure 1: SPAM magnitude and phase perspective plots (top) and gray scale images (bottom) of a 45° slanted fabricated crack in an aluminum alloy. A detailed description is given in [2].

In this paper we review some general principles of imaging, as they pertain to the spatial resolution of thermal wave microscopes and discuss some of the advantages and disadvantages of various types of thermal wave imaging techniques for NDE.

2. Contrast and Resolution in Thermal Wave Imaging

The contrast of a microscope depends on the relative scattering efficiencies of neighboring parts of the scatterer and also on the microscope's ability to discriminate between the radiation scattered from adjacent parts of the object (spatial resolution). The resolution of a microscope depends on the ability either to localize the illumination at the scatterer (typical of scanning microscopes) or to localize the region of the scatterer to which the detector is sensitive (such as in conventional microscopes, which focus different points of the scatterer on different detectors). In microscopes which use lenses as focusing elements to achieve this localization, the localization is limited by the wavelength of the radiation being focussed (the Rayleigh criterion). For example, the resolution of high-quality optical microscopes is limited to about $\lambda/2$, where λ is the wavelength. On the other hand, if the source is localized to much better than a wavelength, *and* is brought into close (compared to a wavelength) proximity to the scatterer, resolution many times better than a wavelength is possible. This situation may be described as "the extreme-near-field limit" [4,5]. This limit was achieved with an optical microscope in 1984 by Pohl et al [6] and in the macroscopic domain using microwaves in 1972 by Ash and Nichols [7]. In the case of thermal wave microscopes, the sources are normally lasers or particle beams, which can be focused to dimensions which are small compared to the thermal wavelength. Therefore, it is much easier to satisfy the small source, close proximity criteria for thermal wave microscopy than it is for optical microscopy. The first demonstration of extreme-near-field imaging in thermal wave microscopy was made in 1983 by Inglehart et al [4,5], following an earlier observation by Kuo et al [8] in 1982.

Thermal wave imaging usually is carried out as scanned microscopy with a localized source. Since thermal waves decay in distances of the order of a wavelength, thermal wave lenses do not appear to be practical. However, this same exponential damping also means that the contrast of thermal wave images is dominated by scatterers located within a fraction of a thermal wavelength from the source. Thus, by varying the thermal wavelength one can vary the region of the specimen which contributes to the image. Another consequence of the heavily damped nature of these waves is that thermal wave imaging is especially well suited for the nondestructive evaluation (NDE) of near (submicron to a few mm) subsurface defects in opaque solids. This region is extremely important because very small flaws near the surface often lead to catastrophic failure. Furthermore, alternative NDE techniques for the most part either have poorer contrast, poorer resolution, or are adversely effected by the shallow depth of the defects.

Thermal wave microscopes take on a number of different forms with corresponding advantages and disadvantages, depending on the nature of the source and the detector. Gas-cell detection [2,9-13] uses a microphone to detect the pressure variations in an enclosed volume of gas as a focused, intensity-modulated laser beam is scanned over the surface of a sample. Conceptually, this can be thought of as imaging with a point source and an *area* detector (the acoustic wavelength is typically larger than the dimensions of the cell, so that the microphone responds equally to all parts of the cell). For this type of thermal wave microscope, because the source is a laser and the solid is normally opaque, the source is localized on the surface to a region whose lateral dimensions are limited by the *optical* wavelength. The proximity of the source to the scatterers is determined by the depth of the scatterers, and thus the resolution is limited by that depth. An advantage of this imaging technique is that it lends itself readily to theoretical analysis. This ease of theoretical analysis originates in the planar symmetry of the detection scheme, which allows one to utilize plane-wave scattering theory [13]. This same symmetry can be a disadvantage, however, since it precludes the detection of an important class of defects, closed vertical cracks [2]. This result, illustrated for a fabricated vertical crack in Figure 2, is easily understood if one considers the mathematically identical problem for describing the gas-cell/vertical crack geometry: an equivalent geometry in which source and detector are imagined to be interchanged, so that the scattering problem becomes one with a planar, temporally periodic temperature source and with a point detector of the resulting temperature. Since the closed vertical crack cannot scatter plane waves which strike its edge, the crack cannot be seen (see Figure 2) with gas-cell detection (point source, area detector). Nevertheless, fatigue cracks in metal alloys can be observed using this technique, even though they are nearly vertical, as illustrated in Figure 3. Another disadvantage of the gas-cell technique is its limited range (audio) of modulation frequencies and hence the limited range of thermal wavelengths. Perhaps a more serious disadvantage is that it usually involves some contact with the sample surface.

A thermal wave imaging technique which avoids the disadvantage of surface contact, and which reduces the symmetry of the detection scheme, utilizes the mirage effect or optical beam deflection (OBD) detection [14-16]. Again, the localization is achieved by focusing the source, but the detection is along a *line* rather than over an area. This method, in addition to being contactless, can operate over a much wider frequency range (1 Hz to a few hundred kHz). In contrast to the gas cell technique, which cannot detect vertical cracks, the mirage effect provides an excellent method for imaging such cracks [2,17]. To illustrate, Figure 4 shows a mirage effect thermal wave image for the same vertical crack which produced the negative gas-cell detection result shown previously in Figure 2. The improvement in detection of fatigue cracks is demonstrated by comparison of the two methods for imaging a fatigue crack on the inside surface of a fabricated bolt hole in and aluminum alloy, shown in Figure 5 [19]. An obvious disadvantage to this technique is the requirement of flat or cylindrical sample surfaces. Furthermore, there are the practical difficulties of maintaining *two* laser positions, as well as the height of the probe beam, during the scan. Also, the *line* symmetry of the detector means that a theoretical analysis requires, at the very least, considerations of scattering with cylindrical waves, and hence is considerably more difficult than that for gas-cell imaging.

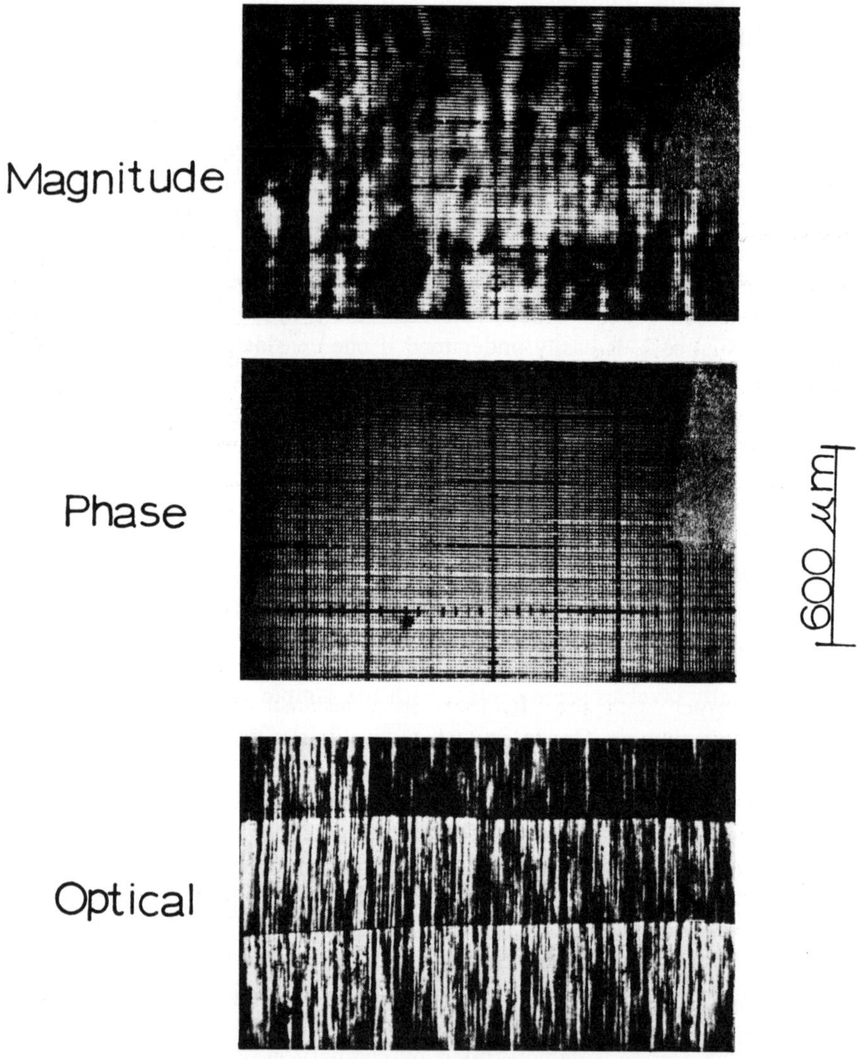

Figure 2: SPAM gas-cell images [2]: [magnitude of the signal (top); phase of the signal (middle); composite optical image (bottom)]. The crack (not observable in these images) runs from the top center to the bottom center of each image. The horizontal lines in the optical image are an artifact of the photographic composition.

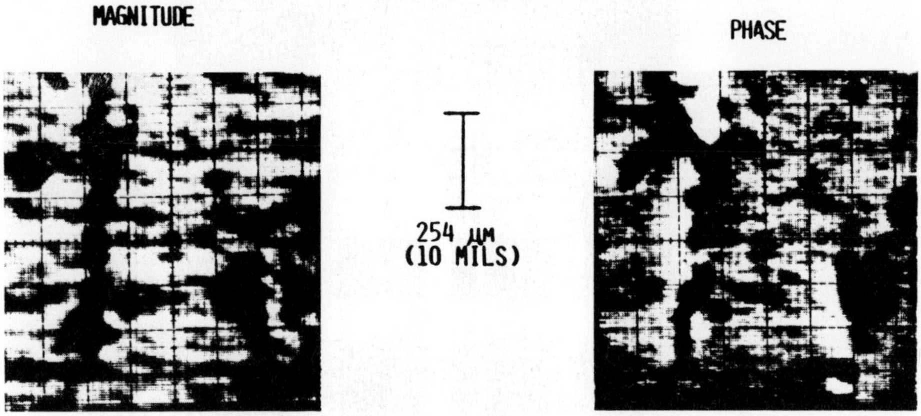

Figure 3: SPAM magnitude and phase thermal wave images at 7.5 kHz of fatigue cracks in an aluminum alloy test specimen [18].

Another contactless thermal wave imaging technique is photothermal radiometry [20-22], in which variations in the local surface temperature are measured with a focused infrared (IR) detector. The source is again a modulated laser beam, focused on the surface. In this case, one has a point source and a *point* detector, with the ultimate resolution in the extreme-near-field limit being provided by the optical, rather than the IR, focal spot, because of the shorter optical wavelength. An advantage of this technique is that, in contrast to both the gas-cell and mirage effect techniques, it does not depend upon heat flow from the solid to the air, and therefore does not have the complications associated with the additional phase delays and magnitude changes of the signal due to the presence of the air. This also permits the application of the technique in vacuum. A disadvantage of the technique is that variations in the emissivity of the surface can obscure the thermal wave image. However, the theoretical analysis is facilitated by the absence of complications associated with the presence of the air.

There are several other thermal wave techniques which also utilize a point source on the surface of the sample (a chopped laser beam) but where the symmetries of the detection schemes vary. One is the photothermal deflection scheme [21-23] which, like the mirage effect, also uses a probe laser beam. The probe beam is reflected from the surface in the immediate vicinity of the heating beam to detect the slope of the bump produced by the thermal expansion of the surface. The periodic variations in the slope cause periodic deflection of the reflected beam. Unless the experiment is performed in vacuum, there will also be mirage-type deflections included in the signal. Advantages of this scheme are that it can be used on a concave surface, that the technique is applicable to experiments in ultra-high vacuum (UHV), and that it is non-contacting. A disadvantage is that the signal-to-noise ratio is strongly dependent on the surface roughness and coefficient of reflectivity. Another disadvantage is the large number of experimental

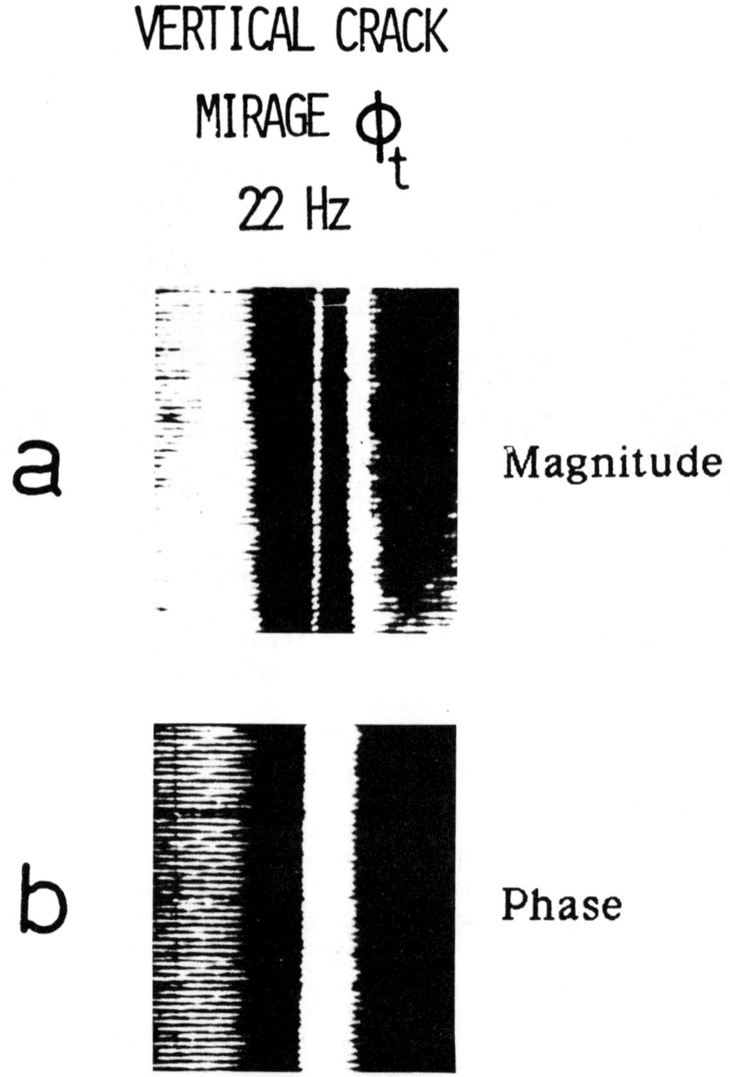

Figure 4: Mirage effect thermal wave images of the same fabricated vertical crack as in Figure 2, using the variations in magnitude (top) and phase (bottom) of the transverse probe beam deflection [2].

parameters, e.g., the orientations and positions of the two beams with respect to each other as well as to the surface. Also, the theoretical analysis is quite complicated, because one has to solve the thermo-elastic equations to calculate the shape of the thermal bump and to calculate the mirage-effect contribution due to the heating of the air. A recent variation of this scheme [27] takes advantage of the very small but detectable change in reflectivity induced by the heating beam. In another detection scheme, the height of the thermal bump is measured

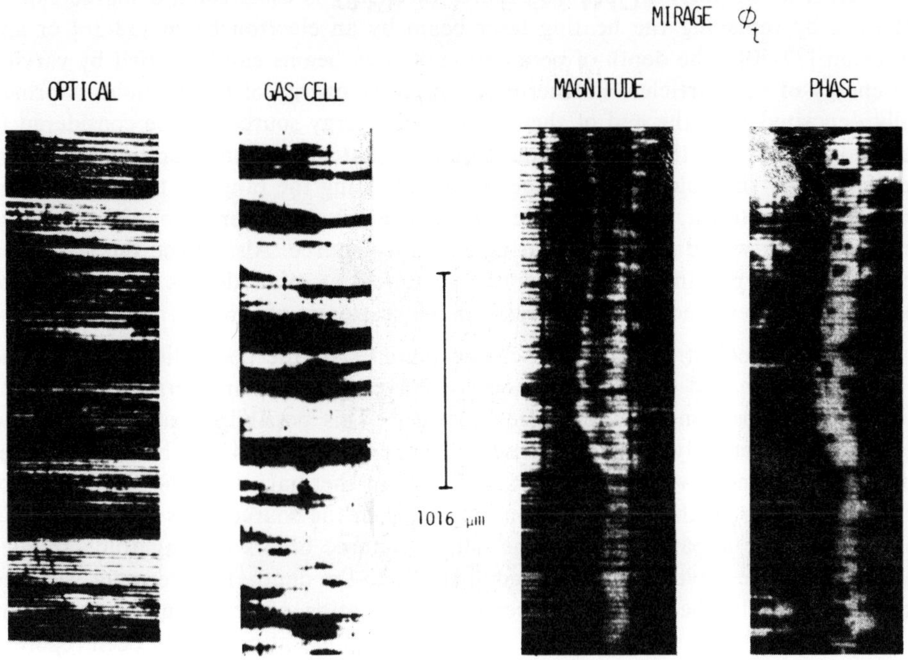

Figure 5: Comparison of optical, gas-cell thermal wave, and mirage effect (magnitude and phase of the transverse probe beam deflection) for a fatigue crack on the inner surface of a cylindrical bolt hole [19].

interferometrically [28]. Yet another scheme uses an ultrasonic beam to measure the change in the ultrasonic phase velocity due to the localized change in temperature, either in the air [29], or on the surface of the sample [30]. The ultrasound generated by the heating can also be directly detected in the air with a focused ultrasonic probe [31]. This ultrasonic detection technique is essentially a contactless version of an earlier technique, in which the ultrasonic vibrations in the solid are detected by means of a peizoelectric transducer bonded elsewhere on the sample [32]. In addition to having the disadvantage of requiring bonding between the sample and transducer, the analysis of the latter scheme is greatly complicated by the presence of many possible modes of elastic vibration of the solid.

In all of the techniques described above, the source of the waves, whether they are thermal or elastic, is the focal spot of the laser on the surface of the sample. The resolution, therefore, is limited by the size of that spot and the closest approach of that spot to the subsurface scatterer. Thus, even though the intrinsic resolution can be much better than either the thermal or acoustic wavelength, it must deteriorate with increasing depth of the scatterers. Consequently, if submicron resolution is required, only scatterers which are much less than a micron in depth can be investigated. This restriction can be circumvented to some extent, however, if a *sub-surface*, localized source of energy is used. Such a source can be achieved by replacing the heating laser beam by an electron beam [33-36] or any ion beam [37-40]. The depth of penetration of such beams can be varied by varying the energy of the particles. Furthermore, since the energy of the particles is principally deposited near the end of their range, the energy source can be considered to be localized at a depth approximately equal to that range and laterally to a region determined by the "bloom" of the beam. By adjusting the range to the depth of the scatterers of interest, it is possible to achieve much better resolution of those scatterers than would be possible using a surface source. Also, since the resolution depends on the proximity of the source, scatterers at other depths will be "out of focus", thereby enhancing the contrast for the scatterers of interest.

In the piezoelectric detection scheme there exists the possibility that it is the direct generation and scattering of acoustic waves, rather than thermal waves, that is responsible for some of the images obtained. This possibility has recently been described theoretically, [41] and observed experimentally [42]. The experimental observation is unequivocally acoustic, rather than thermal, since the observed subsurface defect is 26 thermal diffusion lengths from the source. It is also interesting to note that the apparent size of the defect is three orders of magnitude smaller than the acoustic wavelength and is comparable to the depth, in agreement with the extreme-near-field theory. An independent observation of scattering from subsurface defects up to 20 thermal diffusion lengths deep has also recently been reported [43]. The vivid grain boundary contrast observed in some electron acoustic images [44,45], is conceivably due to the same acoustic scattering mechanism, or perhaps due to the mode conversion of thermal waves to acoustic waves at the boundary [41].

3. Thermal Wave Imaging of Coatings

Thermal wave measurements of the transverse deflection of an optical probe beam (Mirage effect) as a function of the transverse offset between the probe beam and the heating source have been shown experimentally [46,47] and theoretically [48,49] to provide a direct, local and contactless method for determining the thermal diffusivities of pure materials. Furthermore, if such materials are coated with a layer of another material of different thermal diffusivity, the characteristics of the coating and its adhesion to the substrate can significantly change the transverse deflection signal as a function of transverse offset. This effect is illustrated in Figure 6, which shows the frequency dependence of the transverse offset separation, x, between the 90 degree phase points of the transverse mirage deflection for a coated nickel-based (713C) alloy turbine blade, a portion of which had a 1-2 mil coating of AEP 32 [46].

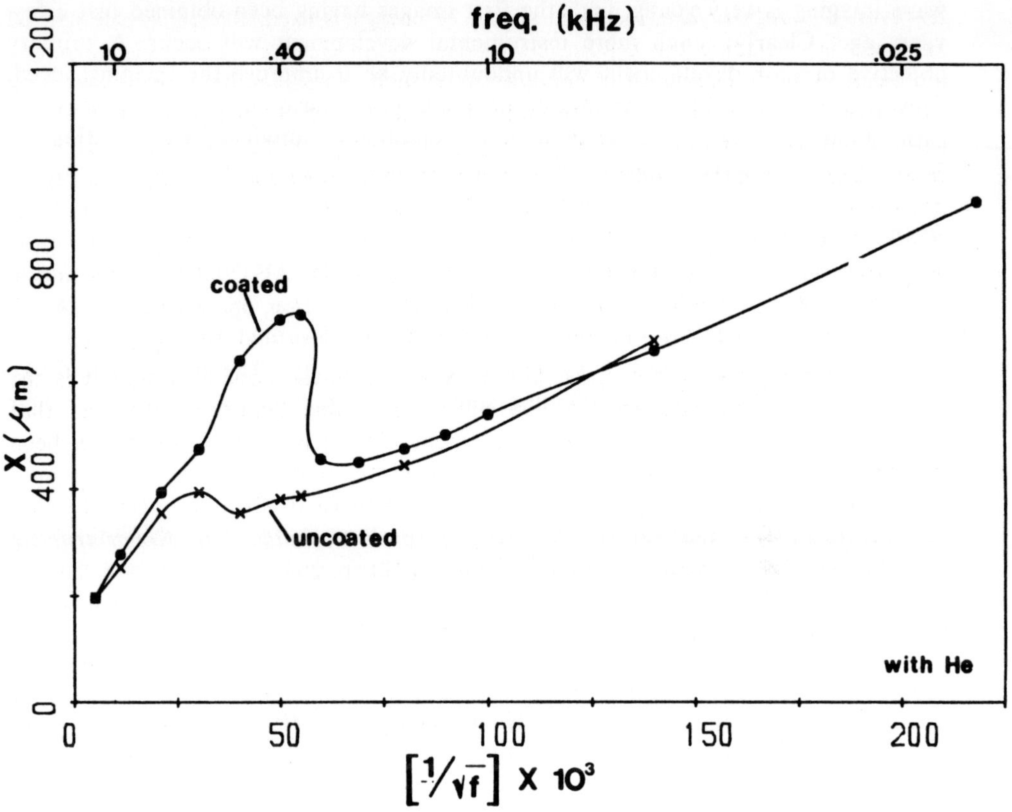

Figure 6: A graph of the measured zero crossings (x) for a coated nickel-based super-alloy (IN 738) plotted as a function of inverse-square-root frequency, for both the coated and uncoated portions of the sample [46].

4. Discussion and Conclusions

The general principles of thermal wave imaging have now been established experimentally and theoretically. At least for the simpler detection schemes it is possible to carry out detailed calculations of images for subsurface defects having simple geometries, and quantitative agreement with experiments has been obtained. All of these thermal wave techniques are capable of producing high resolution images under the right experimental conditions. Different detection techniques have different degrees of symmetry, and this symmetry, or lack of symmetry, can be exploited for the detection of flaws with particular geometries. A number of the imaging schemes involve the generation and detection of acoustic waves in addition to the thermal waves. Interpretation of such images is necessarily more complicated, but such microscopes present the possibility of making high resolution images with both thermal waves and acoustic waves using the same instrument. It should be noted that as compared to optical, and even acoustic imaging, thermal wave imaging is very young, with the first images having been obtained just a few years ago. Clearly, much more instrumental development will occur. A primary objective of such development will undoubtedly be to improve the imaging speed. This probably will be accomplished through some form of parallel processing, rather than point by point scanning. The potential for production applications of NDE of thermal wave imaging would then be established.

Acknowledgements

This work was supported by ARO under contract No. DAAG 29-84-K-0173, and in part by the Center for Advanced Nondestructive Evaluation, operated by the Ames Laboratory, USDOE, for the Air force Wright Aeronautical Laboratories under Contract No. W-7405-ENG-82 with Iowa State University. The authors would like to thank Dr. D.N. Rose for providing some of the thermal wave images of their data.

References

[1] Birnbaum, G. and White, G.S., in *Research Techniques in Nondestructive Testing*, Vol. 7, Academic Press, London, 1984, p. 259.

[2] Grice, K.R., Inglehart, L.J., Favro, L.D., Kuo, P.K. and Thomas, R.L., *J. Appl. Phys. 54*, 1983, p. 6245.

[3] Inglehart, L.J., Thomas, D.J., Lin, M.J., Favro, L.D., Kuo, P.K. and Thomas, R.L., in *Review of Progress in Quantitative Nondestructive Evaluation*, Vol. 4B, D.O. Thompson and D.E. Chimenti (Eds.), Plenum, New York, 1985, p. 753.

[4] Inglehart, L.J., Grice, K.R., Favro, L.D., Kuo, P.K. and Thomas, R.L., *Appl. Phys. Lett. 43*, 1983, p. 446.

[5] Inglehart, L.J., Lin, M.J., Favro, L.D. and Thomas, R.L., *Proceedings of the 1983 IEEE Ultrasonics Symposium*, B.R. McAvoy (Ed.), IEEE, New York, 1983, p. 668.

[6] Pohl, D.W., Denk, W. and Lanz, M., *Appl. Phys. Lett. 44*, 1984, p. 651.

[7] Ash, E.H. and Nichols, G., *Nature, 237*, 1972, p. 510.

[8] Kuo, P.K., Favro, L.D., Inglehart, L.J., Thomas, R.L. and Srinivasan, M., *J. Appl. Phys. 53*, 1982, p. 1258.

[9] Wong, Y.H., Thomas, R.L. and Hawkins, G.F., *Appl. Phys. Lett. 32*, 1978, p. 538.

[10] Busse, G., *Appl. Phys. Lett. 35*, 1979, p. 759.

[11] Luukkala, M. and Penttinen, A., *Electron. Lett. 15*, 1979, p. 326.

[12] Thomas, R.L. Pouch, J.J., Wong, Y.H., Favro, L.D., Kuo, P.K. and Rosencwaig, A., *J. Appl. Phys. 51*, 1980, p. 1152.

[13] Kuo, P.K. and Favro, L.D., *Appl. Phys. Lett. 40*, 1982, p. 1012.

[14] Boccara, A.C., Fournier, D. and Badoz, J., *Appl. Phys. Lett. 36*, 1980, p. 130.

[15] Murphy, J.C. and Aamodt, L.C., *J. Appl. Phys. 51*, 1980, p. 4580.

[16] Aamodt, L.C. and Murphy, J.C., *J. Appl. Phys. 54*, 1983, p. 581.

[17] Lin, M.J., Inglehart, L.J., Favro, L.D., Kuo, P.K. and Thomas, R.L., in *Review of Progress in Quantitative Nondestructive Evaluation*, Vol. 4B, D.O. Thompson and D.E. Chimenti (Eds.), Plenum, New York, 1985, p. 739.

[18] Grice, K.R., Favro, L.D., Kuo, P.K. and Thomas, R.L., in *Review of Progress in Quantitative Nondestructive Evaluation*, Vol. 2B, D.O. Thompson and D.E. Chimenti, (Eds.), Plenum, New York, 1983, p. 1019.

[19] Grice, K.R., Inglehart, L.J., Favro, L.D., Kuo, P.K. and Thomas, R.L., in *Review of Progress in Quantitative Nondestructive Evaluation*, Vol. 3B, D.O. Thompson and D.E. Chimenti (Eds.), Plenum, New York, 1984, p. 763.

[20] Nordal, P.E. and Kanstad, S.O., *Phys. Scr. 20*, 1979, p. 659.

[21] Busse, G., *Appl. Opt. 21*, 1982, p. 107.

[22] Luukkala, M., Lehto, A., Jaarinen, J. and Jokinnen, M., *Proceedings of the 1982 IEEE Ultrasonics Symposium*, B.R. McAvoy (Ed.), IEEE, New York, 1982, p. 591.

[23] Almond, D.P., Patel, P.M. and Reiter, H., *J. de Physique Colloque C6*, 1983, p. 491.

[24] Olmstead, M.A. and Amer, N.M., *J. Vac. Sci. Technol. B1*, 1983, p. 751.

[25] Olmstead, M.A., Amer, N.M., Kohn, S., Fournier, D. and Boccara, A.C., *Appl. Phys. A32*, 1983, p. 141.

[26] Rosencwaig, A., Opsal, J. and Willenborg, D.L., *Appl. Phys. Lett. 43*, 1983, p. 166.

[27] Rosencwaig, A., Opsal, J., Smith, W.L. and Willenborg, D.L., *Appl. Phys. Lett. 46*, 1985, p. 1013.

[28] Wickramasinghe, H.K., Martin, Y., Spear, D.A.H. and Ash, E.A., *J. de Physique Colloque C6*, 1983, p. 191.

[29] Stearns, R.G., Khuri-Yakub, B.T., Kino, G.S., in *Review of Progress in Quantative Nondestructive Evaluation*, Vol. 4B, D.O. Thompson and D.E. Chimenti (Eds.), Plenum, New York, 1985, p. 761.

[30] Stearns, R.G., Khuri-Yakub, B.T. and Kino, G.S., *Proceedings of the 1983 IEEE Ultrasonics Symposium*, B.R. McAvoy (Ed.), IEEE, New York, 1983, p. 649.

[31] Stearns, R.G. and Kino, G.S., *Proc. 4th Int. Topical Meeting on Photoacoustic, Thermal, and Related Sciences*, Laurentides, August 4-8, 1985 (to be published).

[32] Busse, G. and Rosencwaig, A., *Appl. Phys. Lett. 36*, 1980, p. 815.

[33] Rosencwaig, A., *J. Appl. Phys. 51*, 1980, p. 2210.

[34] Rosencwaig, A., *Proceedings of the 1980 IEEE Ultrasonics Symposium*, B.R. McAvoy (Ed.), IEEE, New York, 1980, p. 600.

[35] Cargill, G.S. II, *Nature, 286*, London, 1980, p. 691.

[36] Ringermacher, H.I., *Proceedings of the 1982 IEEE Ultrasonics Symposium*, B.R. McAvoy (Ed.), IEEE, New York, 1982, p. 576.

[37] Hawkins, G.R. (private communication).

[38] Murphy, J.C. Satkiewcz, F.G. and Aamodt, L.C., *Bull. Am. Sphys. Soc. 30*, 1985, p. 474.

[39] Satkiewcz, F.G., Murphy, J.C. and Aamodt, L.C., *Proc. 4th Int. Topical Meeting on Photoacoustic, Thermal and Related Sciences*, Laurentides, August 4-8, 1985 (to be published).

[40] Rose, D.N., Turner, H.R. and Legg, K.O., *Proc. 4th Int. Topical Meeting on Photoacoustic, Thermal and Related Sciences*, Laurentides, August 4-8, 1985 (to be published).

[41] Favro, L.D., Kuo, P.K., Lin, M.J., Inglehart, L.J., and Thomas, R.L., *Proceedings of the 1984 IEEE Ultrasonics Symposium*, B.R. McAvoy (Ed.), IEEE, New York, 1984, p. 629.

[42] Ringermacher, H.I. and Jackman, L., in *Review of Progress in Quantitative Nondestructive Evaluation*, Vol. 5, D.O. Thompson and D.E. Chimenti (Eds.), Plenum, New York (to be published).

[43] Murphy, J.C., Satkiewicz, F.G. and Aamodt, L.C., in *Review of Progress in Quantitative Nondestructive Evaluation*, Vol. 5, D.O. Thompson and D.E. Chimenti (Eds.), Plenum, New York (to be published).

[44] Rosencwaig, A., *Science, 218*, 1982, p. 223.

[45] Wetsel, G.C. Jr., Maclachlan, J.W. and Murphy, J.C., in *Review of Progress in Quantitative Nondestructive Evaluation*, Vol. 5, D.O. Thompson and D.E. Chimenti (Eds.), Plenum, New York (to be published).

[46] Thomas, R.L., Inglehart, L.J., Lin, M.J., Favro, L.D. and Kuo, P.K., in *Review of Progress in Quantitative Nondestructive Evaluation*, Vol. 4B, D.O. Thompson and D.E. Chimenti (Eds.), Plenum, New York, 1985, p. 859.

[47] Thomas, R.L., Kuo, P.K., Reyes, C.B., Favro, L.D. and Zhang, Shu-yi, in *Review of Progress in Quantitative Nondestructive Evaluation*, Vol. 5B, D.O. Thompson and Chimenti, D.E., Plenum, New York (to be published).

[48] Kuo, P.K., Inglehart, K.J., Sendler, E.D., Lin, M.J., Favro, L.D. and Thomas, R.L., in *Review of Progress in Quantitative Nondestructive Evaluation*, Vol. 4B, D.O. Thompson and D.E. Chimenti (Eds.), Plenum, New York, 1985, p. 745.

[49] Kuo, P.K., Reyes, C.B., Favro, L.D., Thomas, R.L. and Zhang, Shu-yi, in *Review of Progress in Quantitative Nondestructive Evaluation*, Vol. 5B, D.O. Thompson and D.E. Chimenti (Eds.), Plenum, New York (to be published).

QNDE OF MATERIAL PROPERTIES AND RESIDUAL STRESS STATES I

THEORY OF ACOUSTOELASTICITY AND ACOUSTOPLASTICITY

Yih-Hsing Pao
Department of Theoretical and Applied Mechanics
Cornell University
Ithaca, New York, U.S.A.
and
Institute of Applied Mechanics
National Taiwan University
Taipei, Taiwan, Republic of China

Abstract

The theory of acoustoelasticity and several recently developed theories of acoustoplasticity are reviewed in this article. For a body with plastic deformation, the acoustoelastic birefringence is affected by the initial stress and the texture, and the latter is related to plastic strains and inherent anisotropy of the body. The birefringence formula may be applied to measure residual stresses by ultrasonic techniques.

1. Introduction

Acoustoelasticity is the interaction of acoustic waves with elastic deformation in a solid. An interesting phenomenon of such an interaction is the change of wave speeds due to initial stresses or initial strains in a predeformed body, generally known as the acoustoelastic effect.

The relative change of wave speeds, however, is very small, about 10^{-5} per MPa for steel or aluminum. Only in recent years, accurate measurements of such a small change were possible in ultrasonic techniques. Acoustoelasticity has thus been a basis for a new method of experimental stress analysis.

The theory of acoustoelasticity was well developed in the 1960's (Toupin and Bernstein, [18]; Thurston and Brugger [15]). The theory predicts, for an originally isotropic solid, that the difference of two transverse wave speeds v_{T1} and V_{T2}, polarized in two perpendicular directions, is proportional to the difference of two principal stresses, σ_1 and σ_2,

$$\Delta v/v_0=(v_{T1}-v_{T2})/v_{T0}=C_A(\sigma_1-\sigma_2) \tag{A}$$

The birefringent coefficient C_A, is related to the second and third order elastic constants of the undeformed body in which the transverse wave speed is v_{T0}. This theory substantiated earlier discovery of acoustoelastic birefringence made by Benson and Raelson [1].

For a body that is slightly orthotropic, the birefringence is given by (Tokuoka and Saito, [17]),

$$\frac{\Delta v}{v_0}=\alpha_0+C_A(\sigma_1-\sigma_2) \tag{B}$$

where α_0 depends on original anistropy. As reviewed in the article by Pao, Sachse, and Fukuoka [13], these formulae are valid for measuring stresses in an elastically deformed body, but they might be in large error for measuring residual stresses.

Residual stresses are developed in a body as a result of inhomogeneous plastic deformation. The permanent (plastic) strains of unequal magnitude at various points prevent the total relaxation of stresses in the body even when all external loads are removed. If ultrasonic waves are propagated through the body, the wave speeds are affected by the stress as well as plastic strains. We call the interaction of acoustic waves with plastic deformation in a solid the acoustoplasticity.

Although the acoustoplastic effect was mentioned in earlier work, detailed studies were not made until recent years (see the review article by Pao et al. [13]). The study is yet incomplete as currently none of the theories is generally accepted, nor are there enough experimental data to substantiate one of them. In this article, we summarize the theories and discuss the salient features that must be considered in the study of acoustoplasticity.

2. Applied Stresses and Residual Stresses

To determine residual stresses in a pre-deformed body by ultrasonic techniques, we must distinguish carefully four states (configurations) of the body: the natural (undeformed) state, the initial (pre-deformed) state, the final (current) state, and the relaxed (sectioned) state.

At the *natural state*, a body is free from any stress and strain. Under the action of external forces, thermal loading, or chemical reaction, the body is deformed elastically or plastically to the *initial state*, at which an ultrasonic experiment is performed. The superimposed ultrasonic wave, although very weak, deforms the body further to the *final state*. These are the three states usually considered in the theory of acoustoelasticity.

If the process of predeformation is elastic, the body recovers its original configuration upon the total removal of external forces, and return to its natural state. The initial stress in the body is called the *applied* (elastic) *stress*.

On the other hand, if the process of predeformation is inelastic, plastic (permanent) strains are developed in the body. After the total removal of external loading, a part of the body is prevented from returning to its natural configuration due to a distribution of unequal plastic strains. The initial stress that is locked inside the region with the unrecoverable deformation is called the *residual stress*.

After loading and unloading, the residual stresses are usually measured by destructive methods. The body with inhomogeneous plastic strains is sectioned into small pieces so as to remove all constraints on deformation. The plastic part of the strain within each piece remains after sectioning, but the elastic part is relaxed to zero magnitude. The stress corresponding to the relaxed elastic strain is the residual stress in the piece before sectioning. We call the state of the body after sectioning the *relaxed state*.

The choice of the natural state for a body must be clearly designated. For example, consider that a steel block (an ingot) is hot rolled into a straight beam, and the beam is then bent by cold working into a curved bar. Residual stresses are developed in the straight beam due to rolling, and are amplified in the curved bar due to plastic bending. There are three choices of the natural and final states, depending on which one of the residual stresses is to be measured. In each case, the final state is always the state after the superposing of ultrasonic waves.

(i) Residual stress in the beam relative to the steel block – The beam is the initial state, and the block the natural state (isotropic).

(ii) Residual stress in the curved bar relative to the steel block – The bar is the initial state, and the block natural state (isotropic).

(iii) Residual stress in the curved bar relative to the straight beam – The bar is the initial state, and the beam the natural state (slightly anisotropic due to rolling).

3. Deformation, Strain and Stress

The kinematics and kinetics of the deformation in three states are described in the theory of small motion superposed on a finitely deformed body (Eringen and Suhubi [3]). Three vectors $\boldsymbol{\xi}$, $\mathbf{X}$, $\mathbf{x}$ are used to denote the positions of a material point (a particle) in a body at three different states, all being referred to a common Cartesian frame of reference. At the natural state, the particle is identified by the coordinates $\boldsymbol{\xi}(\xi_1,\xi_2,\xi_3)$. At the initial and final state, the positions of the same particle are described by coordinates $\mathbf{X}(X_1,X_2,X_3)$ and $\mathbf{x}(x_1,x_2,x_3)$ respectively (Figure 1).

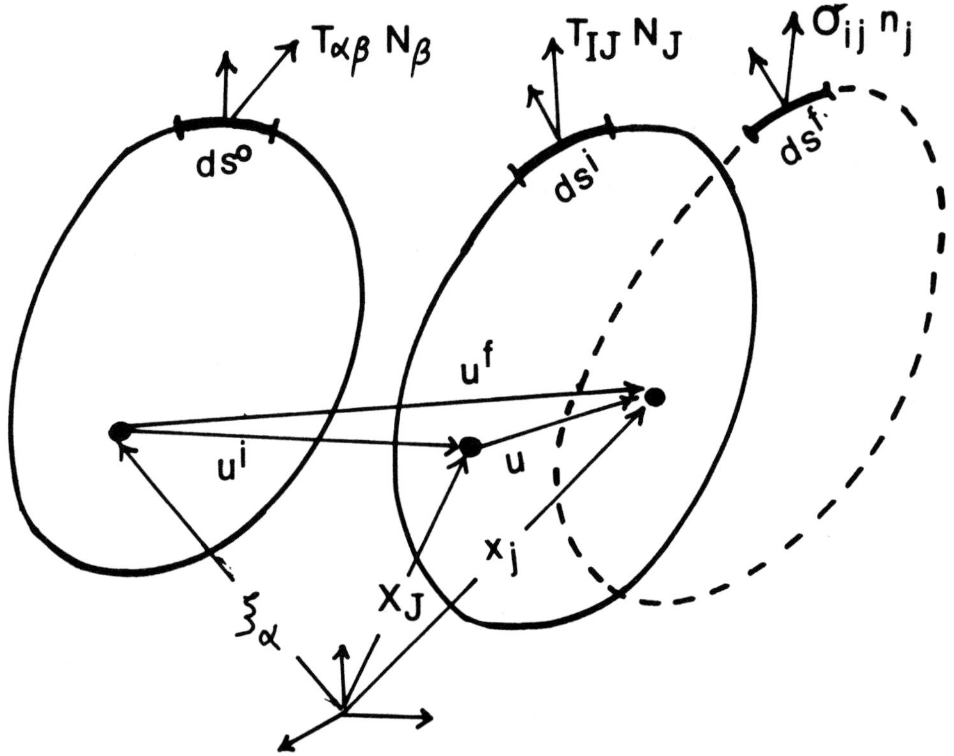

Figure 1: Coordinates, displacements and stress vectors at natural, initial and final state

The deformations are characterized by transformations of coordinates. From the natural state to the initial state, the transformation for static deformation is defined by

$$\mathbf{X}=\hat{\mathbf{X}}(\xi_1,\xi_2,\xi_3) \tag{1}$$

From the initial state to the final state, the transformation is

$$\mathbf{x}=\hat{\mathbf{x}}(x_1,x_2,x_3,t) \tag{2}$$

where $\hat{\mathbf{X}}$ and $\hat{\mathbf{x}}$ are two continuous vector functions. The inverse transformations are assumed to exist.

From the natural state, the displacements to the initial and final states are defined respectively as

$$\mathbf{u}^i(\xi)=\mathbf{X}-\xi$$

$$\mathbf{u}^f(\boldsymbol{\xi},t)=\mathbf{x}-\boldsymbol{\xi} \tag{3}$$

From the initial to the final state, the increment of displacement is given by

$$\mathbf{u}(\boldsymbol{\xi},t)=\mathbf{x}-\mathbf{X}=\mathbf{u}^f-\mathbf{u}^i \tag{4}$$

In these equations and the sequel, we indicate all physical variables in the final state by the superscript f; those in the initial state by i; and those in the natural state by o. All incremental variables are written without superscripts. The components of a vector or tensor in the natural coordinates, $\boldsymbol{\xi}$, are indicated by the Greek subscripts, α, β, γ,...; those in the initial coordinates $\mathbf{X}$, by the upper case Roman subscripts I, J, K,...; and those in the final coordinates, $\mathbf{x}$, by lower case subscripts i, j, k,....

The Lagrangian strain tensors of the body in the initial and final state are given by

$$E^i_{\alpha\beta}=\frac{1}{2}\left\{\frac{\partial X_J}{\partial\xi_\alpha}\frac{\partial X_J}{\partial\xi_\beta}-\delta_{\alpha\beta}\right\}$$

$$E^f_{\alpha\beta}=\frac{1}{2}\left\{\frac{\partial x_j}{\partial\xi_\alpha}\frac{\partial x_j}{\partial\xi_\beta}-\delta_{\alpha\beta}\right\} \tag{5}$$

where $\delta_{\alpha\beta}$ is the Kronecker symbol. In terms of the displacement gradients, they are

$$E^{i,f}_{\alpha\beta}=\frac{1}{2}\left\{\frac{\partial u^{i,f}_\alpha}{\partial\xi_\beta}+\frac{\partial u^{i,f}_\beta}{\partial\xi_\alpha}+\frac{\partial u^{i,f}_\gamma}{\partial\xi_\alpha}\frac{\partial u^{i,f}_\gamma}{\partial\xi_\beta}\right\} \tag{6}$$

The difference of these two strains is defined as the incremental strain $\mathbf{E}$,

$$\mathbf{E}=\mathbf{E}^f-\mathbf{E}^i \tag{7}$$

Substituting Equation (5) into the above and replacing $\mathbf{u}^f$ by $\mathbf{u}^i+\mathbf{u}$, we obtain

$$E_{\alpha\beta}=\frac{1}{2}\left(\frac{\partial u_\alpha}{\partial\xi_\beta}+\frac{\partial u_\beta}{\partial\xi_\alpha}+\frac{\partial u^i_\lambda}{\partial\xi_\alpha}\frac{\partial u_\lambda}{\partial\xi_\beta}+\frac{\partial u_\lambda}{\partial\xi_\alpha}\frac{\partial u^i_\lambda}{\partial\xi_\beta}\right) \tag{8}$$

In the preceding equation, the product $(\partial u_\lambda/\partial\xi_\alpha)$ $(\partial u_\lambda/\partial\xi_\beta)$ has been omitted, based on the assumption that the superimposed disturbance is small.

When expressed in the initial coordinates, the incremental Lagrangian strain appears extremely simple,

$$E_{IJ}=\frac{1}{2}\left(\frac{\partial u_I}{\partial X_J}+\frac{\partial u_J}{\partial X_I}+\frac{\partial u_K}{\partial X_I}\frac{\partial u_K}{\partial X_J}\right)\simeq e_{IJ} \tag{9}$$

where

$$e_{IJ}=\frac{1}{2}\left(\frac{\partial u_I}{\partial X_J}+\frac{\partial u_J}{\partial X_I}\right) \tag{10}$$

is the infinitesimal strain.

The Eulerian form of the incremental strain is

$$\varepsilon_{ij}=\frac{1}{2}\left(\frac{\partial u_i}{\partial x_j}+\frac{\partial u_j}{\partial x_i}-\frac{\partial u_k}{\partial x_i}\frac{\partial u_k}{\partial x_j}\right) \tag{11}$$

It equals E_{IJ} when the superimposed motion is small. The tensorial components of $\boldsymbol{\varepsilon}$, however, cannot be easily calculated from the difference of Eulerian strains, $\boldsymbol{\varepsilon}=\boldsymbol{\varepsilon}^f-\boldsymbol{\varepsilon}^i$. Therefore, we shall not use Eulerian strain tensors here.

At the initial state, the Cauchy stress tensor is denoted by $\boldsymbol{\sigma}^i$. It is the very stress to be measured by ultrasonic methods. In the theory, however, it is more convenient to deal with the second Piola-Kirchhoff stress tensor, $\mathbf{T}^i$, which is related to the former by

$$\sigma^i_{JK}=\left|\frac{\partial \mathbf{X}}{\partial \boldsymbol{\xi}}\right|^{-1}\frac{\partial X_J}{\partial \xi_\alpha}\frac{\partial X_K}{\partial \xi_\beta}T_{\alpha\beta}{}^i \tag{12}$$

After the superposition of ultrasonic waves, the stress in the body is changed to $\boldsymbol{\sigma}^f$ at the final state, which can also be converted to $\mathbf{T}^f$ expressed in either initial or natural coordinates,

$$\sigma^f_{jk}=\left|\frac{\partial \mathbf{x}}{\partial \boldsymbol{\xi}}\right|^{-1}\frac{\partial x_j}{\partial \xi_\alpha}\frac{\partial x_k}{\partial \xi_\beta}T^f_{\alpha\beta}=\left|\frac{\partial \mathbf{x}}{\partial \mathbf{X}}\right|^{-1}\frac{\partial x_j}{\partial X_M}\frac{\partial x_k}{\partial X_N}T^f_{MN} \tag{13}$$

In previous equations, $|\partial \mathbf{X}/\partial \boldsymbol{\xi}|$ is the determinant of the matrix $\partial X_J/\partial \xi_\alpha$ $(J,\alpha=1,2,3)$ etc.

In analogous to Equation (7), the incremental stress tensor is defined by $\mathbf{T}=\mathbf{T}^f-\mathbf{T}^i$, or

$$T_{JK}=T^f_{JK}-\sigma_{JK}{}^i \quad , \quad T_{\alpha\beta}=T^f_{\alpha\beta}-T^i_{\alpha\beta} \tag{14}$$

The calculation of the incremental Cauchy stress components from the definition $\boldsymbol{\sigma}=\boldsymbol{\sigma}^f-\boldsymbol{\sigma}^i$, however, is not convenient.

4. Equations of Motion

In the previous sections, we have defined three states of a deformed body, and five stress tensors σ^i_{JK}, $T^i_{\alpha\beta}$, σ^f_{jk}, T^f_{JK} and $T^f_{\alpha\beta}$. Corresponding to each type of stress, there is an equation of equilibrium for the initial state, or an equation of motion for the final state.

The equation for σ^f_{jk} is the simplest,

$$\frac{\partial \sigma^f_{jk}}{\partial x_k} \simeq \rho^f \ddot{x}_j = \rho^f \frac{\partial^2 u_j}{\partial t^2} \tag{15}$$

where ρ is the mass density. The approximation for $\ddot{x}_j$ is based on the assumption of a small disturbance being superposed on a statically predeformed body,

$$\ddot{\mathbf{x}} = \dot{\mathbf{v}}^f = d\mathbf{v}^f/dt = \partial \mathbf{v}^f/\partial t + \mathbf{v}^f \cdot \nabla \mathbf{v}^f/\partial t \simeq \partial \mathbf{u}^2/\partial t^2 \tag{16}$$

where $\mathbf{v}^f = \dot{\mathbf{x}} = \partial \mathbf{x}/\partial t = \partial\ (\mathbf{X}+\mathbf{u})/\partial t = \partial \mathbf{u}/\partial t$. We shall not start from this equation because the final deformation is composed of two parts, and each part may require a different constitutive equation.

The remaining two equations of motion for T^f_{JK} and $T^i_{\alpha\beta}$ are respectively given by

$$\frac{\partial}{\partial X_K}\left[T^f_{JK} + T^f_{KL}\frac{\partial u_J}{\partial X_L}\right] = \rho^i \frac{\partial^2 u^f_J}{\partial t^2} = \rho^i \frac{\partial^2 u_J}{\partial t^2} \tag{17}$$

$$\frac{\partial}{\partial \xi_\beta}\left[T^f_{\alpha\beta} + T^f_{\beta\gamma}\frac{\partial u^f_\alpha}{\partial \xi_\gamma}\right] = \rho^o \frac{\partial^2 u^f_\alpha}{\partial t^2} = \rho^o \frac{\partial^2 u_\alpha}{\partial t^2} \tag{18}$$

The equations of equilibrium for σ^i_{JK} and $T_{\alpha\beta}i$ are respectively

$$\frac{\partial \sigma^i_{JK}}{\partial X_K} = 0 \tag{19}$$

$$\frac{\partial}{\partial \xi_\beta}\left[T_{\alpha\beta}i + T_{\beta\gamma}i\frac{\partial u^i_\alpha}{\partial \xi_\gamma}\right] = 0 \tag{20}$$

All four equations are exact.

Subtracting Equation (19) from (17), and Equation (20) from (18), we obtain the equation of motion for the incremental stress in two coordinates systems:

$$\frac{\partial}{\partial X_K}\left[T_{JK} + \sigma^i_{KL}\frac{\partial u_J}{\partial X_L}\right] = \rho^i \frac{\partial^2 u_J}{\partial t^2} \tag{21}$$

$$\frac{\partial}{\partial \xi_\beta}\left[T_{\alpha\beta} + T_{\beta\gamma}i\frac{\partial u_\alpha}{\partial \xi_\gamma} + T_{\beta\gamma}\frac{\partial u^i_\alpha}{\partial \xi_\gamma}\right] = \rho^o \frac{\partial^2 u_\alpha}{\partial t^2} \tag{22}$$

In the derivation, the products $T_{KL}(\partial u_J/\partial X_L)$ and $T_{\beta\gamma}(\partial u_\alpha/\partial \xi_\gamma)$ have been neglected for small incremental motion.

Equation (21) and (22) are derived without invoking any constitutive relation for the initial, final, or incremental stresses. The only assumptions made so far are that the predeformed body at the initial state is in equilibrium, and the superimposed wave motion is small. The initial deformation can be infinitesimal or finite, elastic or inelastic.

5. Theory of Acoustoelasticity

The equation of motion for the incremental stress **T** and dynamic disturbance **u**, either in the initial coordinates of Equation (21) or in the natural coordinates of Equation (22), is one of two key elements for a theory of acoustoelasticity or acoustoplasticity. The second element is the constitutive equation for the incremental stress **T**.

5.1. *Equations in Natural Coordinates*

The theory of acoustoelasticity assumes that both initial deformation and final deformation are elastic or, more restrictively, hyperelastic. For a hyperelastic body, there exists an internal energy function, $\psi(\mathbf{E})$, such that in an isentropic or adiabatic process,

$$T^{i,f}_{\alpha\beta}=\rho^o\frac{\partial\psi(\mathbf{E}^{i,f})}{\partial E^{i,f}_{\alpha\beta}} \tag{23}$$

The energy function can be expanded into a Taylor series about the natural state with zero strain,

$$\rho^o\psi(\mathbf{E}^{i,f})=\frac{1}{2}c^o_{\alpha\beta\gamma\delta}E^{i,f}_{\alpha\beta}E^{i,f}_{\gamma\delta}+\frac{1}{6}c^o_{\alpha\beta\gamma\delta\epsilon\eta}E^{i,f}_{\alpha\beta}E^{i,f}_{\gamma\delta}E^{i,f}_{\epsilon\eta}+\cdots \tag{24}$$

where second and third order moduli are given by

$$c^o_{\alpha\beta\gamma\delta}=\rho^o\left[\frac{\partial^2\psi(\mathbf{E}^{i,f})}{\partial E^{i,f}_{\alpha\beta}\,\partial E^{i,f}_{\gamma\delta}}\right]^o \quad \ldots\ldots \tag{25}$$

They remain constant throughout the deformation.

Substituting the series, truncated after two terms, into Equation (23), we obtain the constitutive equation for $T^i_{\alpha\beta}$, or $T^f_{\alpha\beta}$ as follows:

$$T^i_{\alpha\beta}=c^o_{\alpha\beta\gamma\delta}E^i_{\gamma\delta}+c^o_{\alpha\beta\gamma\delta\epsilon\eta}E^i_{\gamma\delta}E^i_{\epsilon\eta} \tag{26}$$

$$T^f_{\alpha\beta}=c^o_{\alpha\beta\gamma\delta}E^f_{\gamma\delta}+c^o_{\alpha\beta\gamma\delta\epsilon\eta}E^f_{\gamma\delta}E^f_{\epsilon\eta} \tag{27}$$

A constitutive equation for the incremental stress, **T**, is thus derived by subtracting $T^i_{\alpha\beta}$ from $T^f_{\alpha\beta}$;

$$T_{\alpha\beta}=c^o_{\alpha\beta\gamma\delta}E_{\gamma\delta}+c^o_{\alpha\beta\gamma\delta\epsilon\eta}E^i_{\gamma\delta}e_{\epsilon\eta} \tag{28}$$

$$\simeq c^o_{\alpha\beta\gamma\delta}\left(\delta_{\rho\gamma}+\frac{\partial u^i_\rho}{\partial\xi_\gamma}\right)\frac{\partial u_\rho}{\partial\xi_\delta}+c^o_{\alpha\beta\gamma\delta\epsilon\eta}\frac{\partial u^i_\gamma}{\partial\xi_\delta}\frac{\partial u_\epsilon}{\partial\xi_\eta} \tag{29}$$

where $e_{\epsilon\eta}$ is the linear part of $E_{\epsilon\eta}$.

The constitutive equation (29) can be rewritten as

$$T_{\alpha\beta}=C^o_{\alpha\beta\gamma\delta}\frac{\partial u_\gamma}{\partial\xi_\delta}$$

where $\mathbf{C}^o$ is the "effective" elastic moduli of the predeformed body,

$$C^o_{\alpha\beta\gamma\delta}=c^o_{\alpha\beta\gamma\delta}+c^o_{\alpha\beta\rho\delta}\frac{\partial u^i_\gamma}{\partial\xi_\rho}+c^o_{\alpha\beta\gamma\delta\varepsilon\eta}\frac{\partial u^i_\varepsilon}{\partial\xi_\eta} \tag{30}$$

Substituting the above into Equation (22), we obtain

$$\frac{\partial}{\partial\xi_\beta}\left[\Gamma^o_{\alpha\beta\gamma\delta}\frac{\partial u_\gamma}{\partial\xi_\delta}+T^i_{\gamma\beta}\frac{\partial u_\alpha}{\partial\xi_\gamma}\right]=\rho^o\frac{\partial^2 u_\alpha}{\partial t^2} \tag{31}$$

where

$$\Gamma^o_{\alpha\beta\gamma\delta}=C^o_{\alpha\beta\gamma\delta}+c^o_{\rho\beta\gamma\delta}\frac{\partial u^i_\alpha}{\partial\xi_\rho} \tag{32}$$

Equation (31) is the governing equation for acoustic waves propagating in an elastically deformed medium in natural coordinates. The initial stress $\mathbf{T}^i$ and initial displacement gradient $\partial\mathbf{u}^i/\partial\boldsymbol{\xi}$ are related by Equation (26), and one of them can be eliminated from the equation. The T^i_α should be converted to σ^i_{JK} according to Equation (12) as the latter is the initial stress to be measured by ultrasonic method. The difference between them is neglected for most of the applications when the predeformation is also assumed small.

The equations and results discussed in this section are well documented (Thurston and Brugger, [15]; Pao and Gamer, [11]). We repeat them here for the purpose of emphasizing the assumptions and approximations made in the derivations. Some of the assumptions must be changed when the predeformation is inelastic.

5.2. *Equations in Initial Coordinates*

In terms of the initial coordinates X_J, the constitutive equation for T_{JK} may be derived for small $\mathbf{u}$, from the transformation,

$$T_{JK}=\left|\frac{\partial\mathbf{X}}{\partial\boldsymbol{\xi}}\right|^{-1}\frac{\partial X_J}{\partial\xi_\alpha}\frac{\partial X_K}{\partial\xi_\beta}T_{\alpha\beta} \tag{33}$$

The result is

$$T_{IJ}=C_{IJKL}e_{KL}=C_{IJKL}\frac{\partial u_L}{\partial X_K} \tag{34}$$

where

$$C_{IJKL}=c^o_{IJKL}(1-e^i_{NN})+c^o_{IJKLMN}e^i_{MN}$$

$$+c^o_{MJKL}\frac{\partial u^i_I}{\partial X_M}+c^o_{IMKL}\frac{\partial u^i_J}{\partial X_M}+c^o_{IJML}\frac{\partial u^i_K}{\partial X_M}+c^o_{IJKM}\frac{\partial u^i_L}{\partial X_M} \tag{35}$$

and

$$c^o_{IJKLMN}=\delta_{I\alpha}\delta_{J\beta}\delta_{K\gamma}\delta_{L\delta}c^o_{\alpha\beta\gamma\delta} \tag{36}$$

Substituting Equation (35) into Equation (21), we obtain the complete equation for acoustoelastic waves in initial coordinates,

$$\frac{\partial}{\partial X_J}\left[(C_{IJKL}+\sigma^i_{JL}\delta_{IK})\frac{\partial u_K}{\partial X_L}\right]=\rho^i\frac{\partial^2 u_I}{\partial t^2} \tag{37}$$

It is equivalent to Equation (31) in natural coordinates.

For an isotropic and homogeneous body at the natural state, we have

$$c^o_{IJKL}=\lambda\delta_{IJ}\delta_{KL}+\mu(\delta_{IK}\delta_{JL}+\delta_{IL}\delta_{JK}) \tag{38}$$

where λ and μ are Lamé constants. The third order moduli, c^o_{IJKLMN}, are reduced to three constants ν_1, ν_2, ν_3. The natural isotropy is, however, destroyed by the initial stress, and the body at the initial state is anisotropic, and inhomogeneous.

We note that the constitutive Equation (34) can also be dreived from the assumed energy function $\psi(\mathbf{E})$,

$$T^f_{IJ}=\left|\frac{\partial \mathbf{X}}{\partial \xi}\right|^{-1}\frac{\partial X_I}{\partial \xi_\alpha}\frac{\partial X_J}{\partial \xi_\beta}\frac{\partial \psi(\mathbf{E})}{\partial E^f_{\alpha\beta}} \tag{39}$$

and an expansion of the previous result about the initial state (Eq. (2.20) of Toupin and Bernstein, [18]). This approach is not easy to modify when the predeformation is inelastic, unless another internal energy function is assumed to exist also for elasto-plastic deformation (Johnson, [7]).

5.3. *Acoustoelastic Birefringence*

Equation (31) or (37) may be applied to analyze the propagation of plane waves. The inhomogeneity of the predeformed body, however, causes difficulty in the calculation of wave speeds (Tokuoka and Iwashimizu, [16]). This difficulty is removed by the assumption that the initial strains are homogeneous in the predeformation. This assumption is valid in applications when a small ultrasonic transducer is used to measure the average value of initial stresses across the thickness of a specimen. It implies that the initial strain is homogeneous within a small area of the transducer, but it does not require homogeneous strain over the entire specimen.

Based on the assumptions for small and homogeneous initial strains, the birefringent formula is found as

$$\Delta v/v_0=(1+\nu_3/\mu)(e^i_1-e^i_2)$$

$$=(2\mu)^{-1}(1+\nu_3/\mu)(\sigma_1^i-\sigma_2^i) \tag{40}$$

where $\Delta v = v_{T1} - v_{T2}$ and $v_0^2 = \mu/\rho^o$. This is the formula (A) mention in the Introduction. When the body is slightly orthotropic at the natural state, the acoustoelastic birefringence is given by (Tokuoka and Saito, [17]),

$$\frac{\Delta v}{v_0} = \frac{v_{T1}^0 - v_{T2}^0}{v_0} + \frac{1}{2\mu}(1+\frac{\nu^3}{\mu})(\sigma_1^i - \sigma_2^i) \tag{41}$$

where v_{T1}^0 and v_{T2}^0 are the speeds of waves at the natural state, polarized in the directions X_1 and X_2 respectively of the principal stresses σ_1^i and σ_2^i. This is the formula (B) mentioned in the Introduction.

6. Theory of Acoustoplasticity

The birefringent formula, Equation (40), was applied to measure residual stresses and was found to be inadequate in early investigations (Crecraft, [2]). It was suggested that the acoustoelastic birefingence should be decomposed into two parts:

$$\frac{\Delta v}{v_0} = \left(\frac{\Delta v}{v_0}\right)_{\text{texture}} + \left(\frac{\Delta v}{v_0}\right)_{\text{stress}} \tag{42}$$

Whereas the second term on the right hand side is given by Equation (40), the first term, which is due to microstructural change of materials accompanying residual stresses, has not been explicitly given.

Before discussing the effect of texture on the wave speeds, we recall the basic assumptions in the theory of acoustoelasticity:

(i) The predeformed body is in equilibrium (Equation (19) or (20)).

(ii) The superposed motion is small (Equation (8), (21) or (22)).

(iii) The predeformed body is hyperelastic with the constitutive Equation (26).

(iv) The superposed dynamic deformation is also hyperelastic with a linear constitutive Equation (30) or (34).

It is obvious that assumptions (iii) and (iv) must be modified for a body with plastic deformation.

Residual stresses in polycrystalline metals are induced mainly by inhomogeneous plastic deformation. A metal in uniaxial loading tests often exhibits an yield point, beyond which the strain increases rapidly. The incremental strain may be decomposed into two parts, the elastic strain and plastic strain,

$$d\mathbf{E}^i = d\mathbf{E}^e + d\mathbf{E}^p \tag{43}$$

The total initial strain is given by an integral along a certain path of loading and unloading,

$$\mathbf{E}^i = \int_{\text{path}} d\mathbf{E}^i$$
$$= \mathbf{E}^e + \mathbf{E}^p \tag{44}$$

Upon the removal of all loading, the elastic strain vanishes but the plastic strain remains, resulting in a permanent set.

In a state of triaxial loading, the plasticity is defined by a yield condition

$$f(\boldsymbol{\sigma}', \mathbf{E}^p) = \kappa \tag{45}$$

where κ is the work hardening parameter, and $\boldsymbol{\sigma}'$ the deviatoric part of the stress tensor. Metal deforms plastically when the yield condition is reached.

Based on two different yield conditions for the predeformed body, Johnson [7], [8] and Kobayashi [10] proposed two theories of acoustoplasticity, resulting in two different birefringent formulas. A third theory, independent of any yield condition, was proposed by Pao and Gamer [11], [12]. These theories are discussed in the following.

6.1. *Constitutive Relations for the Initial Deformation*

It was noted by Pao and Gamer [11] that Equation (21) or (22) was still valid for inelastic predeformation. It contains both initial stress, $\boldsymbol{\sigma}^i$ or $\mathbf{T}^i$, and initial displacement gradient, $\partial \mathbf{u}^i / \partial \boldsymbol{\xi}$ or $\partial \mathbf{u}^i / \partial \mathbf{X}$. Their determination depends not only on the yield condition and constitutive equation but also on the loading and unloading history, geometry, and boundary conditions of the specimen. Therefore, instead of replacing the assumption (iii) with a new constitutive equation to relate initial stresses and initial strain, both $\boldsymbol{\sigma}^i$, and $\mathbf{E}^i$ are retained as two independent quantities in the theory.

As given by Equation (44), the plastic part of the total strain is unknown because the path of integration is unspecified. But the elastic part, $\mathbf{E}^e$, may be assumed to follow a hyperelastic constitutive equation. The equation of motion, (21) or (22), thus contains $\boldsymbol{\sigma}^i$ or $\mathbf{T}^i$, and $\mathbf{E}^p$ as two independent quantities for homogeneous deformations.

6.2. *Constitutive Relation for the Incremental Deformation*

The modification of the assumption (iv) is a difficult but crucial step. There have been several approaches reported in literature. Johnson [7] assumed an elastic-plastic function $\psi(\mathbf{E}, \mathbf{E}^p, \kappa)$ to replace $\psi(\mathbf{E})$ in Equation (39). Kobayashi [10] adopted the incremental theory of plastic deformation, $d\mathbf{E}^p = \lambda d\mathbf{T}'$, where λ is the plastic potential, and $\mathbf{T}'$ the stress deviator, and introduced a new constitutive relation for the incremental stress.

In place of Equation (34), Pao and Gamer [12] postulated the following constitutive relation for infinitesimal strains,

$$T_{IJ}=\overline{C}_{IJKL}e_{KL} \tag{46}$$

where the effective moduli $\overline{C}$ are

$$\overline{C}_{IJKL}=c^r_{IJKL}(1-e^e_{NN})+c^r_{IJKLMN}e^e_{MN}+c^r_{MJKL}\frac{\partial u^e_I}{\partial X_M}$$

$$+c^r_{IMKL}\frac{\partial u^e_J}{\partial X_M}+c^r_{IJML}\frac{\partial u^e_K}{\partial X_M}+c^r_{IJKM}\frac{\partial u^e_L}{\partial X_M} \tag{47}$$

The c^r_{IJKL} and c^r_{IJKLMN} are the elastic moduli at the relaxed state. For a body that is isotropic at the natural state, they assumed

$$c^r_{IJKL}=\lambda\delta_{IJ}\delta_{KL}+\mu(\delta_{IK}\delta_{JL}+\delta_{IL}\delta_{JK})+\lambda'(e^p_{IJ}\delta_{KL}+\delta_{IJ}e^p_{KL}) \tag{48}$$

$$+\mu'(e^p_{IK}\delta_{JL}+\delta_{IK}e^p_{JL}+e^p_{IL}\delta_{JK}+\delta_{IL}e^p_{JK}) \tag{48}$$

where λ' and μ' are two new material constants. The third order constants c^r_{IJKLMN} can be approximated by those at the natural state, c^0_{IJKLMN}. In essence, they assumed that the superimposed dynamic deformation is *linearly elastic*, but the elastic moduli at the initial state are affected by the plastic strains.

Substitution of Equations (46), (47), (48) into Equation (21) completes the equation of motion for acousto-plasticity. Based on this equation, the birefringence formula in a naturally isotropic body with plastic deformation is given by (Pao and Gamer, [12]),

$$\frac{\Delta v}{v_0}=\frac{\mu'}{2\mu}(e^p_1-e^p_2)+(1+\frac{v_3}{\mu})(e^e_1-e^e_2) \tag{49}$$

$$=\frac{\mu'}{2\mu}(e^p_1-e^p_2)+\frac{1}{2\mu}(1+\frac{v_3}{\mu})(\sigma^i_1-\sigma^i_2) \tag{50}$$

On the right hand side, the second term, which is the same as that in Equation (41), represents the stress effect; and the first term represents the texture effect in Equation (42).

We note that the theory by Kobayashi [10] leads to a birefringent equation similar to Equation (50), but with a different coefficient for the first term. The result obtained by Johnson [8], however, is different. It can be expressed as

$$\Delta v/v_0=(a_2+b_3)(e^e_1-e^e_2) \tag{51}$$

where a_2 and b_3 are two elasto-plastic moduli depending on e^p and κ.

6.3. *Effects of Texture and Anisotropy*

The inelastic behavior of polycrystalline metals is very complex, and is by no measns well understood. Within a grain in the matrix, the deformation is composed of a recoverable part, represented by the elastic strain $\mathbf{E}^e$ and an irrecoverable part, represented by the plastic strain $\mathbf{E}^p$. Even at room temperature, the plastic strain rate is coupled to a nonlinear viscous flow within a grain (Hart, [5]). In between grains, there is the anelastic deformation due to grain-boundary sliding. In addition, there are the inherent anisotropy for a givn crystalline structure, and the induced anisotropy along a preferred orientation of plastic flow. The latter is called the texture effect.

Excluding the time varying behavior of anelasticity and nonlinear viscosity, Pao and Gamer considered the effect of plastic deformation by keeping both initial stress and initial strain in the equation, and decomposing the latter into an elastic strain and a plastic strain (Equation (44)). The effect of induced anisotropy is incorporated in the effective moduli of Equation (47), and that of inherent anisotropy can be treated by replacing λ and μ with c^0_{IJKL} in Equation (48), and λ' and μ' with similar moduli.

To illustrate these key features in their theory, we consider two material points, A and B, inside a curved bar mentioned in Section 2. The bar is bent from a straight beam by cold working, the point A being in the elastic core region of the bar, and B in the outer region with plastic deformation. The stress-strain histories for both points are shown in Figure 2.

In the figure, the initial stresses at A and B are denoted by σ^i_A and σ^i_B respectively, the initial strains by E^i_A and E^i_B, where $E^i_B = E^e_B + E^p_B$. The two initial stresses, which are assumed equal, are locked inside the bar by non-uniform plastic strains generated by cold working. The straight beam is in the natural state which is indicated by the point O in Figure 2.

Around the material point A, the deformation is elastic and the initial stress σ^i_A is determinable from the acoustoelastic formula, Equation (40). The orthotropy of the straight beam with texture is treated as an inherent anisotropy at the natural state in this case (Hirao and Pao, [6]).

The region around the material point B has undergone plastic deformation. The path of loading and unloading is marked by OYNB on the stress-strain diagram, where O is the nature state, Y the yield point, N the unloading point, and B the initial state. If a small block around the material point B is cut off from the curved bar, the stress at B drops to zero and the path follows the unloading curve of uniaxial test to point O′, which is in the relaxed state.

The unloading (from N to O′) and reloading (from O′ to N) processes are known to be elastic. If the path OYNB is replaced by OO′B, the residual stress, σ^i_B, relative to the natural state O may be regarded as an applied stress relative to the relaxed state O′. On this basis, Pao and Gamer [12] postulated the elastic constitutive equation (46) for the incremental stress at the initial state.

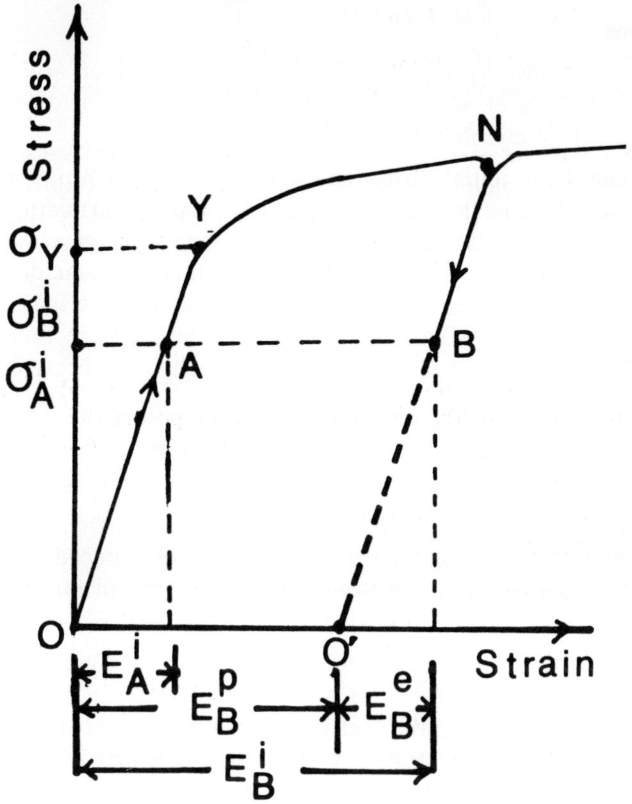

Figure 2: Initial stresses and initial strains of two material points A and B

The moduli of the elastic (or hyperelastic) deformation along the path O′BN, however, are differnt from those along the path OAY. This is evidenced by ultrasonic wave speed data (Fukuoka [4]). The change of c^r_{IJKL} is attributed to preferred orientations of crystalline grains during plastic flow, the texture effect.

One method of evaluating the change of moduli is to assume a probabilistic distribution function for the orientations of all grains, and then to calculate the averaged moduli for the aggregate by applying a statistical theory (Sayers, [14]; Johnson, [9]). Although the distribution function is strongly affected by the plastic deformation, the dependence of the function on plastic strains has not been specified in this method.

In the theory by Pao and Gamer discussed in the preceding section, they simply assume that the moduli at the natural state c^0_{IJKL} are changed to those at the relaxed state c^r_{IJKL}, and the change is a function of strains as given by Equation (48). The linear form is guided by that an isotropic material is changed to an orthotropic material when stretched in three principal directions of strains.

7. Conclusions

In this paper, we have reviewed the well known theory of acoustoelasticity and several recent theories of acoustoplasticity for the purpose of determining residual stresses by ultrasonic method.

By keeping both initial stress and the plastic part of initial strain as two independent quantities in the equation of motion, and by postulating a linear elastic constitutive equation for the incremental stress, the theory of acoustoelasticity was extended by Pao and Gamer [12] to include the effect of plastic deformation on the ultrasonic wave speeds. Their theory gives rise to an acoustoelastic birefringent formula (Equation (50)) for a naturally isotropic material. With proper modification for anisotropic materials, the birefringent formula may be expressed as

$$\Delta v/v_0=\alpha_0+C_B(e_1^p-e_2^p)+C_A(\sigma_1^i-\sigma_2^i) \qquad \text{(C)}$$

where the constant α_0 equals $(v_{T1}^0-v_{T2}^0)/v_0$ as given by Equation (41); C_B and C_A are given by the corresponding coefficients in Equation (50). The first term, α_0, on the right hand side of formula (C) is the effect of inherent anisotropy (at the natural state), and the second term is the effect of induced anisotropy (at the relaxed state). Together, they represent the texture induced birefringence in Equation (42).

The formula (C) which reduces to formula (B) in the absence of plastic strains contains three coefficients, α_0, C_A, C_B and two unknown variables $(e_1^p-e_2^p)$, and $(\sigma_1^i-\sigma_2^i)$. Assuming that all three coefficients can be calibrated accurately, one still cannot determine both unknowns by a single measurement of Δv. Therefore, successful applications of the formula (C), or the equivalent, depend on supplemental information or additional measurement on the initial stress or plastic strains.

References

[1] Benson, R.W. and Raelson, V.J.. "Acoustoelasticity", *Product Engineering*, 30, 1959, pp. 56-59.

[2] Crecraft, D.I., "The Measurement of Applied and Residual Stresses in Metals Using Ultrasonic Waves", *J. Sound Vib.*, 5, 1967, pp. 173-192.

[3] Eringen, A.C. and Suhubi, E.S., *Elastodynamics, Vol. 1*, (Chapter 4), Academic Press, New York, 1974.

[4] Fukuoka, H., An article in this volume, 1986.

[5] Hart, E.W., "Constitutive Relations for the Nonelastic Deformation of Metals", *J. Eng. Materials and Technology*, (Trans. ASME), 98, 1976, pp. 193-202.

[6] Hirao, M. and Pao, Y.H., "Dependence of Acoustoelastic Birefringence on Plastic Strains in a Beam," *J. Acoust. Soc. Am.*, 77, 1985, pp. 1659-1664.

[7] Johnson, G.C., "Acoustoelastic Theory for Elastic-Plastic Materials", *J. Acoust. Soc. Am.*, 70, 1981, pp. 591-595.

[8] Johnson, G.C., "The Effect of Plastic Deformation on the Acoustoelastic Response of Metals", *J. Appl. Mech.*, 50, 1983, pp. 689-691.

[9] Johnson, G.C., "Acoustoelastic Response of a Polycrystalline Aggregate with Orthotropic Texture," in *Wave Propagation in Inhomogeneous Media and Ultrasonic NDE*, G.C. Johnson (Ed.), *AMD, Vol. 62*, ASME, New York, 1984, pp. 175-184.

[10] Kobayashi, M., "Analysis of Acoustic Phenomena in Elasto-Plastically Deformed Condition", (in Japanese) *Transactions of Japan Soc. Mech. Engr.*, A48, No. 432, 1982, pp. 1072-1981.

[11] Pao, Y.H. and Gamer, U., "Acoustoelastic Waves in Orthotropic Media", *J. Acoust. Soc. Am., 77*, 1985, pp. 806-812.

[12] Pao, Y.H. and Gamer, U., "Theory of Acoustoelasticity for Elastic-Plastic Solids", Materials Science Center Report No. 5121, 1983, Cornell University, Ithaca, New York (Revised in 1985).

[13] Pao, Y.H., Sachse, W. and Fukuoka, H., "Acoustoelasticity and Ultrasonic Measurements of Residual Stresses", in *Physical Acoustics*, Vol. 17, W.P. Mason and R.N. Thurston (Eds.), Academic Press, New York, 1984, pp. 62-143.

[14] Sayers, C.M., An article in this volume. 1986.

[15] Thurston, R.N. and Brugger, K., "Third-Order Elastic Constants and the Velocity of Small Amplitude Elastic Waves in Homogeneously Stressed Media", *Phys. Rev.*, 133, 1964, pp. A1604-1610.

[16] Tokuoka, T. and Iwashimizu, Y., "Acoustical Birefringence of Ultrasonic Waves in Deformed Isotropic Elastic Materials," *Int. J. Solids and Structures*, 4, 1968, pp. 383-389.

[17] Tokuoka, T. and Saito, M., "Elastic Wave Propagations and Acoustical Birefringence in Stressed Crystals", *J. Acoust. Soc. Am.*, 45, 1969, pp. 1241-1246.

[18] Toupin, R.A. and Bernstein, B., "Sound Waves in Deformed Perfectly Elastic Materials. Acoustoelastic Effect", *J. Acoust. Soc. Am.*, 33, 1961, pp. 216-225.

[19] Hogar, A., "On the Determination of Residual Stress in an Elastic Body", to appear in *J. Elasticity*.

[20] Man, C.S. and Lu, W.Y., "Towards an Acoustoelastic Theory for Measurement of Residual Stress", to appear *J. Elasticity*.

ULTRASONIC MEASUREMENT OF RESIDUAL STRESS

H. Fukuoka
Osaka University
Toyonaka, Japan

Abstract

Our main interest in this paper is a practical application of acoustoelastic measurement of residual stress in industry.

1. Review of Birefringent Law

Although several modes of wave can be used in acoustoelasticity, a birefringent method is most convenient for practical use because it requires the measurement of transit time difference of two shear waves and it is free from errors in measurement of length.

In engineering metallic materials, a texture-induced acoustic birefringence is the same order as a stress-induced one. Therefore, acoustoelasticity theory should be based on intrinsically anisotropic elasticity. Anisotropy of rolled metal is typically modeled by an orthorhombic symmetry in which seven second order and 20 third order elastic constants are independent.

We summarize the results of birefringent law of orthotropically anisotropic material. Let the coordinate axes x_1, x_2, x_3 be oriented along the axes of orthotropic symmetry. Shear waves propagate along x_3-axis which is also the direction of principal stress σ_3 as well as one of the axes of orthotropic symmetry.

First, we consider a simple case in which directions of principal stresses σ_1 and σ_2 coincide with x_1 and x_2 axes, respectively. Then, to first order in displacement gradients, we have [1]

$$\rho V_{31}^2 = C_{55} + \sigma_3 + C_{55}(\varepsilon_1 - \varepsilon_2 + \varepsilon_3) + C_{551}\varepsilon_1 + C_{552}\varepsilon_2 + C_{553}\varepsilon_3 \quad (1)$$

$$\rho V_{32}^2 = C_{44} + \sigma_3 + C_{44}(\varepsilon_2 - \varepsilon_1 + \varepsilon_3) + C_{441}\varepsilon_1 + C_{442}\varepsilon_2 + C_{443}\varepsilon_3 \quad (2)$$

for velocities V_{31} and V_{32} of shear waves polarized in x_1 and x_2 directions, respectively. ε_1, ε_2 and ε_3 are infinitesimal principal strains and ρ is the material density in initial stressed state. C's are second and third order elastic constants in the contracted Voigt's notation. Second and third terms of the right hand sides of these equations are originated in geometrical non-linearity, while fourth to sixth terms are in material non-linearity.

From Equations (1) and (2), we have

$$\rho(V_{31}^2 - V_{32}^2) = C_{55} - C_{44} + (C_{55} - C_{44})\varepsilon_3 + (C_{55} + C_{44})(\varepsilon_1 - \varepsilon_2) + (C_{55} - C_{441})\varepsilon_1 + (C_{552} - C_{441})\varepsilon_2 + (C_{553} - C_{443})\varepsilon_3 \quad (3)$$

For weak anisotropy, we can neglect the product of anisotropy terms in elastic stiffness and initial strain.

Then, we obtain

$$\frac{V_{31} - V_{32}}{V_{30}} = \frac{C_{55} - C_{44}}{\mu} + \frac{\mu + \nu_3}{\mu}(\varepsilon_1 - \varepsilon_2) \quad (4)$$

where $V_{30} = (\mu/\rho)^{1/2}$ is the shear wave velocity of isotropic material which may be replaced by $(V_{31} + V_{32})/2$.

As we assume weak anisotropy, we can neglect the difference between principal directions of stress and those of strain and use the relation

$$\varepsilon_1 - \varepsilon_2 = \frac{1}{2\mu}(\sigma_1 - \sigma_2) \quad (5)$$

into Equation (4).

Then, the birefringent acoustoelasticity law for this simple case is

$$\frac{V_{31} - V_{32}}{V_{30}} = \alpha + C_A(\sigma_1 - \sigma_2) \quad (6)$$

where

$$\alpha = \frac{C_{55} - C_{44}}{2\mu} \quad (7)$$

is the texture-induced acoustic birefringence and

$$C_A=\frac{1}{2\mu}(1+\frac{\nu_3}{\mu}) \tag{8}$$

is the acoustoelasticity constant.

For the measurement of applied stress, α can be obtained by acoustoelastic measurement at stress free state. In case of nondestructive residual stress measurement, however, the measurement at stress free state is unavailable in general and α should be separated from total acoustic birefringence in order to obtain the stress-induced acoustic birefringence. The separation problem is the point at issue in nondestructive residual stress measurement.

Next, we consider the case where the directions of principal stresses σ_1 and σ_2 do not coincide with axes of orthotropic symmetry. In this case, Equation (3) is replaced by

$$\begin{aligned}\rho(V_{3I}^2-V_{3II}^2)=[\{&C_{55}-C_{44}+(C_{55}+C_{44})(\varepsilon_{11}-\varepsilon_{22})\\&+(C_{55}-C_{44})\varepsilon_3+(C_{155}-C_{144})\varepsilon_{11}+(C_{255}-C_{244})\varepsilon_{22}\\&+(C_{355}-C_{344})\varepsilon_3\}^2+4\{(C_{55}+C_{44}+2C_{456})\varepsilon_{12}\\&+(C_{44}-C_{55})\omega_{12}\}^2]^{1/2}\end{aligned} \tag{9}$$

where V_{3I} and V_{3II} are the velocities of shear waves polarized perpendicularly, and ωij is an infinitesimal rigid body rotation.

For weak anisotropy, the birefringent law is obtained as [2]

$$\frac{V_{3I}-V_{3II}}{V_{30}}=\{\alpha^2+2\alpha C_A(\sigma_1-\sigma_2)cos\,2\phi+C_A^2(\sigma_1-\sigma_2)^2\}^{1/2} \tag{10}$$

where ϕ is the angle between the direction of principal stress σ_1 and the x_1-axis.

The angle θ between one of the polarization directions of shear waves and the x_1-axis is given by

$$\tan 2\theta=\frac{C_A(\sigma_1-\sigma_2)\sin 2\phi}{\alpha+C_A(\sigma_1-\sigma_2)\cos 2\phi} \tag{11}$$

2. Texture Effect Versus Stress Effect

Several methods have been proposed for the separation problem.

In a method of oblique incidence of the SH wave presented by King and Fortunko [1], relative difference of the velocities of two SH waves being polarized perpendicularly to each other and obliquely incident at the same angle relative to the normal to the surface was measured. Measurement at two different values of incident angle were required to evaluate the principal stress difference separated from the texture effect.

Thompson et al proposed a method [3] of measuring the difference in the velocities of two shear waves whose directions of propagation and polarization have been interchanged for the material of orthotropic anisotropy. The velocity difference gives principal stress difference independent of texture anisotropy.

One approach to the separation problem is to combine several wave velocities in such a way as to be independent of texture. Allen and Sayers [4] used a relationship between longitudinal wave velocity and birefringence both normalized by the sum of squares of three velocities which is a texture independent quantity. For a plate with homogeneous texture, Sayers et al [5] gave a texture independent determination of the principal stress difference in the plate by combining the Rayleigh wave measurements with the normal shear wave birefringence method. Also, they suggested the importance of the fundamental SH waves for the texture independent measurement of residual stresses in the rolled plate. Toda et al [6] used a ratio of a longitudinal wave velocity to a mean value of two shear wave velocities to obtain the principal stress sum.

Salama et al [7] made use of the temperature dependence of longitudinal wave velocity. They found that the velocity varies linearly with temperature and the slope of the linear relationship depends on the stress.

Mahadevan [8] proposed to make use of the frequency dependent effect for the separation problem and Hirsekorn [9] gave a theoretical explanation to this effect.

3. Measurement of Birefringences of Plastically Deformed Materials

There are two effects of plastic deformation on birefringent acoustoelasticity; one is the effect on the acoustoelasticity constant C_A, and the other is on the texture acoustic birefringence α.

One approach to this problem is a series of experiment in which an acoustic birefringence is measured during a cycle of uniaxial test on tension and compression specimen of the material under investigation.

A typical experiment was done by Hirao and Pao [10] on an aluminum 6061-T6 rolled plate by making use of the echo-overlap method for relative velocity measurement. In Figure 1, the results of tension and compression tests are shown in the same diagram. The acoustic birefringence is plotted against the total strain during loading (denoted by solid circle) and unloading (denoted by hollow circle). The slope of the curve divided by twice the rigidity modulus is the acoustoelasticity constant C_A and the acoustic birefringence at the stress-free state (denoted by triangle) is the texture acoustic birefringence. The acoustoelasticity constants C_A are almost constant in the tested range of plastic strain and the calibrated value for C_A of this material is $-4.0\times10^{-5} MPa^{-1}$. On the other hand, the texture acoustic birefringence α varies with plastic strain and Figure 2 shows the variations of α against the plastic strain for two tensile and two compression specimens.

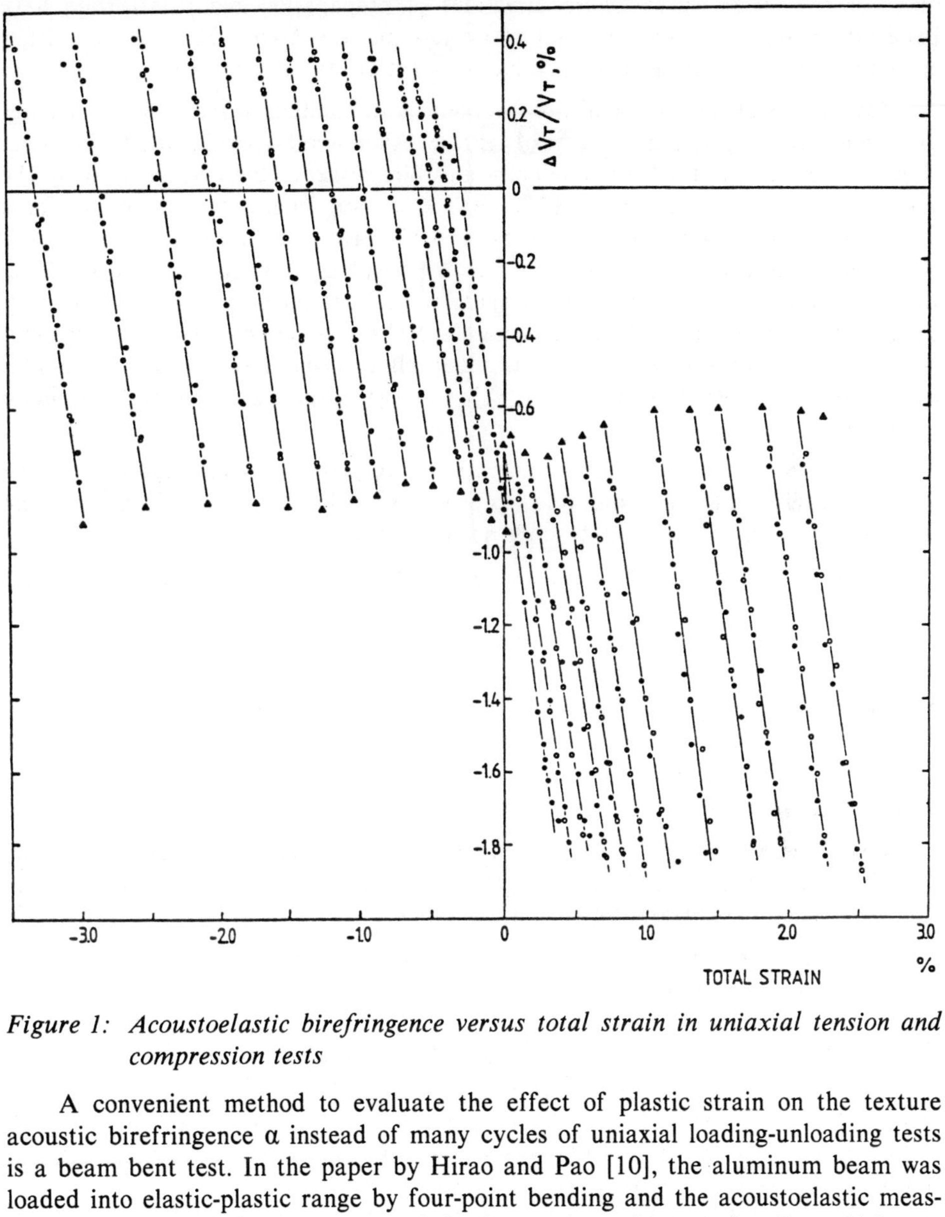

Figure 1: Acoustoelastic birefringence versus total strain in uniaxial tension and compression tests

A convenient method to evaluate the effect of plastic strain on the texture acoustic birefringence α instead of many cycles of uniaxial loading-unloading tests is a beam bent test. In the paper by Hirao and Pao [10], the aluminum beam was loaded into elastic-plastic range by four-point bending and the acoustoelastic measurements were done at the center of the beam along the scanning line which was perpendicular to the beam axis. The difference of acoustic birefringence before bending and after sectioning the beam to release the residual stress represents the change of the texture acoustic birefringence α due to plastic strain. Plastic strain at the measuring point along the scanning line was measured by the strain gauge.

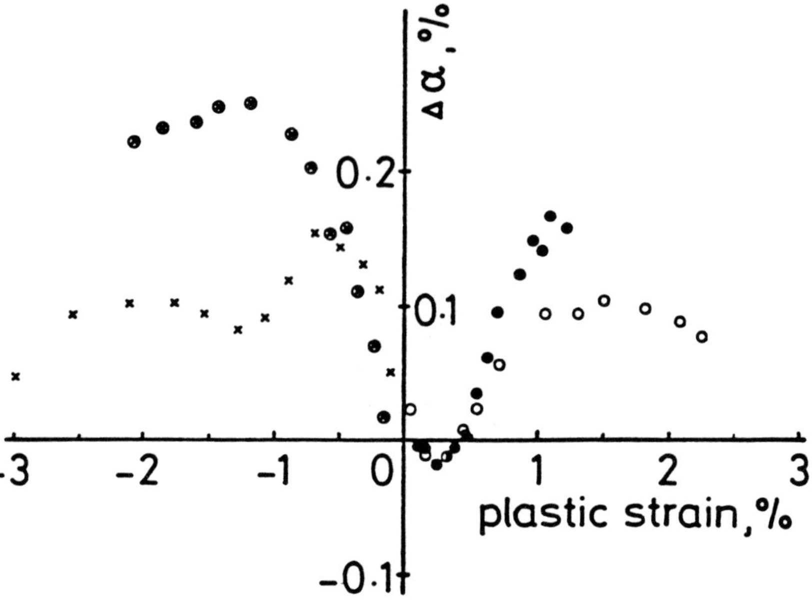

Figure 2: Change of texture birefringence as a function of plastic strain obtained from uniaxial tension and compression tests

The results are shown in Figure 3.

4. Measurement of Residual Stresses

4.1. *Wide-Flanged Beam*

Residual stresses in as-rolled wide-flanged beam of 1018 steel were measured [11] by making use of the sing-around method for relative transit time measurement.

The separation of the stress and texture-induced acoustic birefringence was done by the following consideration. Residual stress in the wide-flanged beam was known to be predominantly longitudinal from experience and the both ends of the beam were considered to be free from residual stress. We assumed that the texture anisotropy was constant along a rolling line. Then, the mesurement of acoustic birefringence at the both ends along transversal directions should yield the distribution of the texture acoustic birefringence along the transversal direction for the entire beam. Subtracting the texture acoustic birefringence thus obtained, the distribution of residual stress at the middle region of the beam was obtained from the acoustic birefringence measured there.

Solid circles in Figure 4(a) and (b) show the residual stress in the flange and web, respectively, obtained by this procedure. The abscissas of these figures are transversal distances from the center of the flange or web. For comparison, residual stresses were measured by a conventional destructive method. Measurements of the gauge length of a contact strain gauge were done before and after sectioning

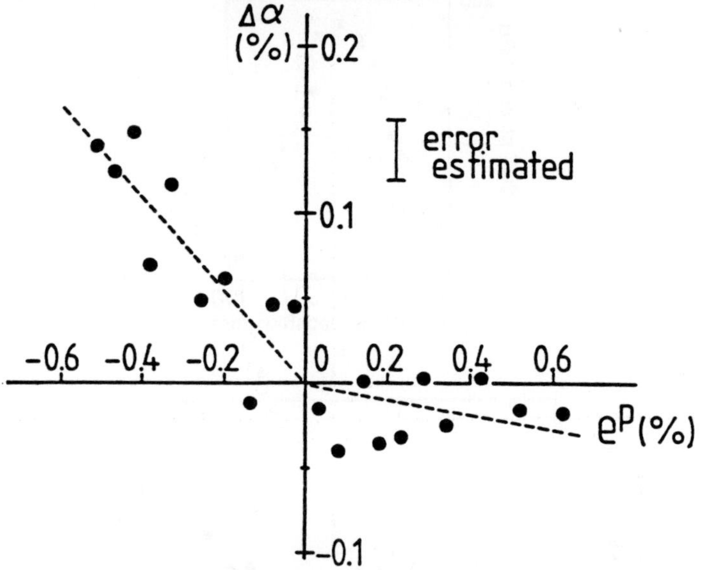

Figure 3: Change of texture birefringence as a function of plastic strain obtained from beam bending tests

the beam. Residual stresses obtained by this destructive method are shown by cross marks in Figure 4(a) and (b). Although the nondestructive acoustoelasticity evaluation gives good agreement for the flange, this is not the case for the web.

To investigate the origin of this discrepancy for the web, the assumption used for the separation of two effects was checked by measuring acoustic birefringence of the sectioned pieces of the beam. By this measurement, distributions of texture acoustic birefringence were obtained and are shown in Figure 5(a) and (b). The acoustic birefringences at the both ends were essentially the same for both before and after sectioning the beam as was assumed. The distribution of the texture acoustic birefringence in the flange was nearly constant along a rolling line, and the magnitudes were the same order as the stress acoustic birefringence. On the other hand, the distribution of the texture acoustic birefringence in the web fluctuated considerably along a rolling line, and the magnitudes were several times larger than those of stress acoustic birefringence, both of which caused the large discrepancies in Figure 4(b).

By subtracting the texture acoustic birefringence of each measuring point shown in Figure 5(a) and (b) from the acoustic birefringence measured at the middle regions before the sectioning, the calculated residual stresses so obtained are shown in Figure 6(a) and (b) by solid circles together with those obtained by the conventional destructive method. This example illustrates the importance of the separation problem for nondestructive residual stress measurement.

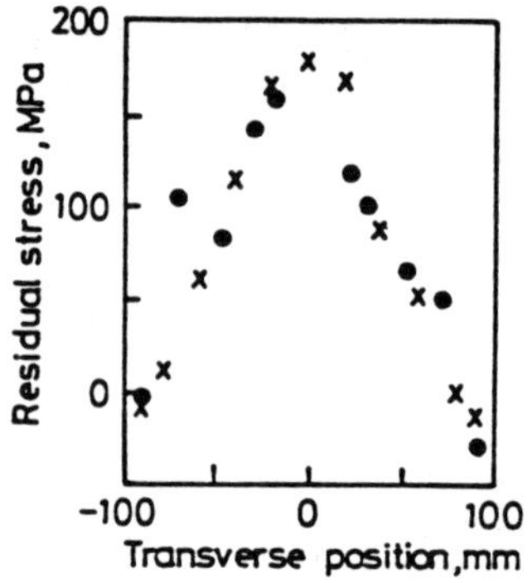

(a)

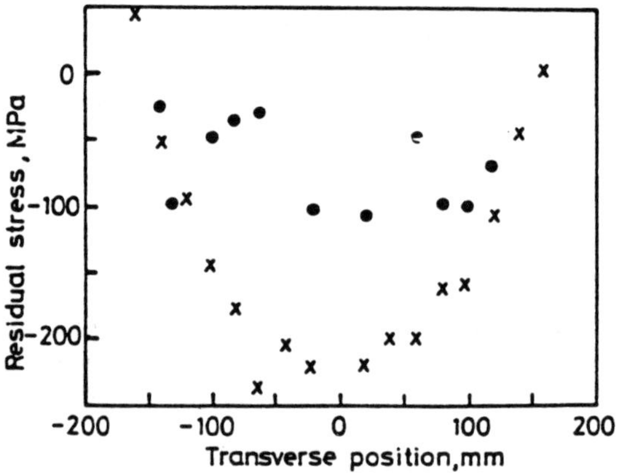

(b)

Figure 4: *Residual stresses in (a) the flange and (b) the web of the wide-flange beam. Solid circles and cross marks are those obtained by acoustoelasticity (assuming uniform texture birefringence along a rolling line) and conventional destructive method, respectively.*

4.2. *Railroad Wheel Rim*

Under abnormal drag braking condition the circumferential tensile residual stress develops in the rim of railroad wheel and may cause a fracture if the residual stress becomes large enough such that the stress intensity factor at the crack tip exceeds the fracture toughness. In this respect the evaluation of the residual stress in the rim of railroad wheel is of importance to avoid such catastrophic failure. The residual stress in the rim of railroad solid wheel was measured by birefringent acoustoelasticity method [12]. Seven wheels were tested. The excess drag braking were applied by the brake testing machine to four wheels, Nos. 1, 2, 3 and 7 with the level of thermal input of 65.3KW, 32.7KW, 54.4KW and 43.6KW, respectively.

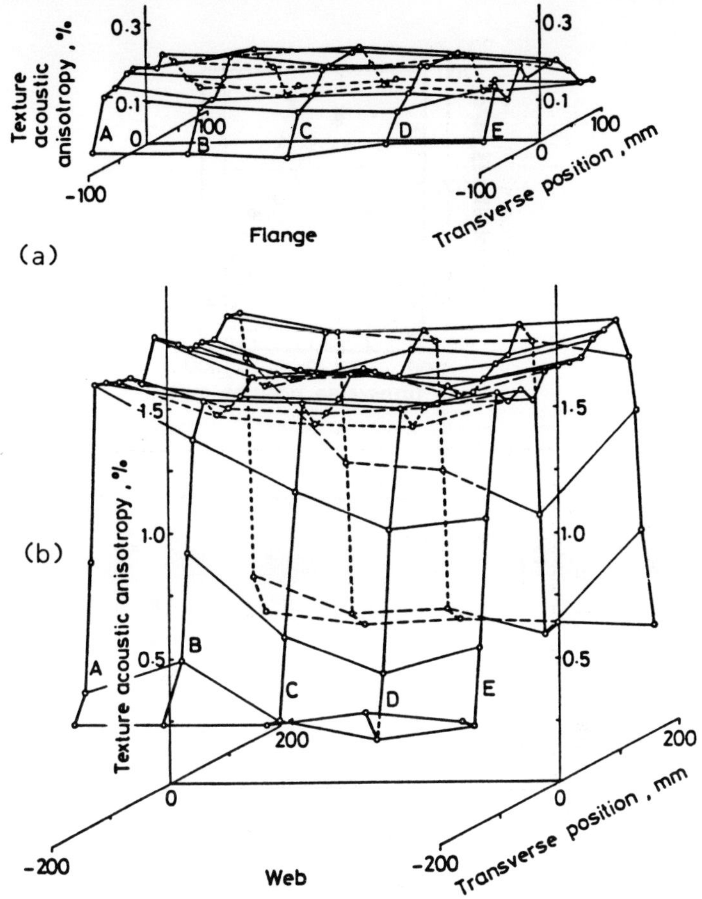

Figure 5: Texture birefringence in (a) the flange and (b) the web of the wide-flange beam.

Acoustic birefringence was given as $(T_r - T_\theta)/T_0$, where T_r and T_θ were the transit time of the shear waves polarized in the radial and circumferential directions of the wheel, respectively, and T_0 was the average transit time. The transit time was measured by a sing-around method with a shear type 5MHz transducer of 7mm square size.

Acoustoelastic measurements of a wheel rim were made at the four regions I, II, III and IV as shown in Figure 7. In each region, radial positions A, B, C and D which were 15mm, 40mm, 50mm and 60mm from the inner edge of the rim were selected as measuring positions. For each position, measurements were done at three points a, b and c, and the maximum, minimum and average values were recorded as the values of the relevant radial position. The acoustoelastic measurements were done at the time as manufactured, after braking test, and as cut into rim blocks to release the residual stress.

(a)

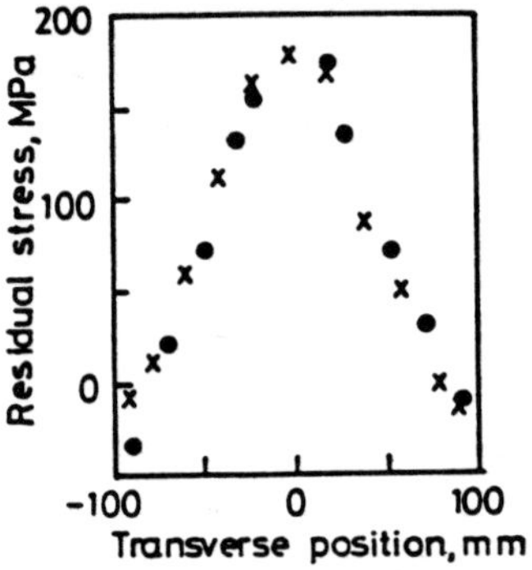

(b)

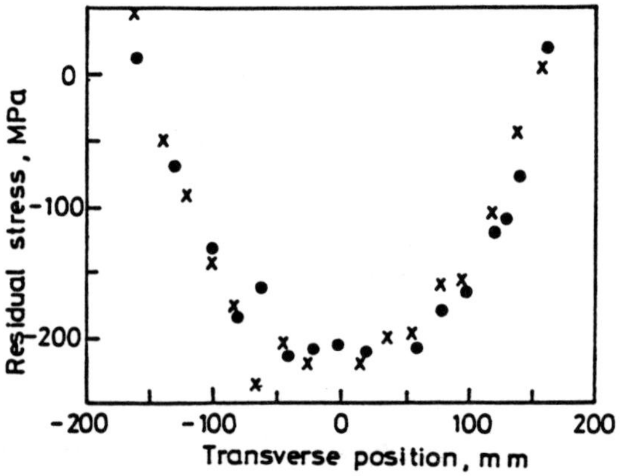

Figure 6: Residual stresses in (a) the flange and (b) the web of the wide-flange beam. Solid circles and cross marks are those obtained by acoustoelasticity (based on actual texture birefringence) and conventional destructive method, respectively.

The acoustoelastic constant C_A was calibrated directly from uniaxial tests. Three tension and three compression test pieces cut from the wheel were used for the calibration test. The acoustoelastic constant was obtained by a slope of the relation between the acoustic birefringence and the applied stress. Figure 8 shows the examples of the uniaxial tests. The results are summarized in Table 1. As the acoustoelastic constant of the wheel material, the mean value of $-7.6\times10^{-6} 1/MPa$ was used for conversion of the acoustic birefringence to stress in the wheel rim.

The acoustic birefringences of the wheel rims measured at three stages are shown in Figures 9 to 11. The conversion from acoustic birefringence to stress in Figures 9 and 10 is made by neglecting the texture birefringence which is shown in

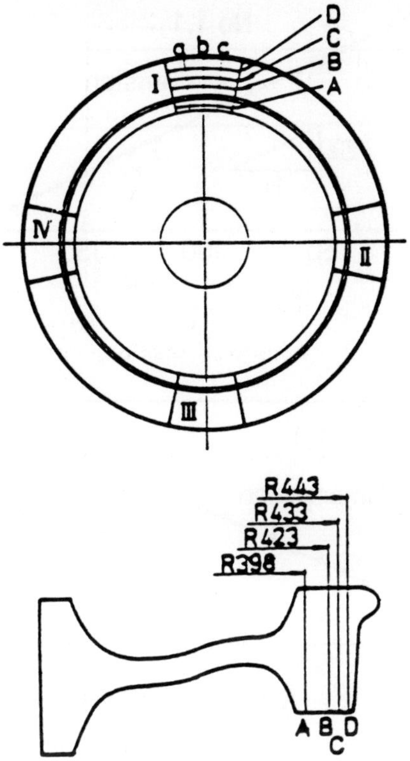

Figure 7: Locations of acoustoelastic measurement of the wheel rim

Figure 11. The stress state in the wheel rim is assumed to be uniaxial circumferential stress.

The texture birefringence of the wheels as shown in Figure 11 are small compared with those of the wide-flanged beam. Such small texture birefringence of the wheel material enables a rough estimation of the residual stress in the wheel rim by a single mesurement of acoustoelasticity as shown in Figures 9 and 10. However, more precise estimation of residual stress can be done by knowing the texture acoustic birefringence.

As the axes of texture anisotropy and principal stress direction coincide in this case, we can make use of Equation (6). The acoustic birefringence for each stage of the measurement can be written as

(as manufactured, Figure 9)

$$\frac{T_r - T_\theta}{T_0} = \alpha^0 + C_A^0 \sigma^0 \tag{12}$$

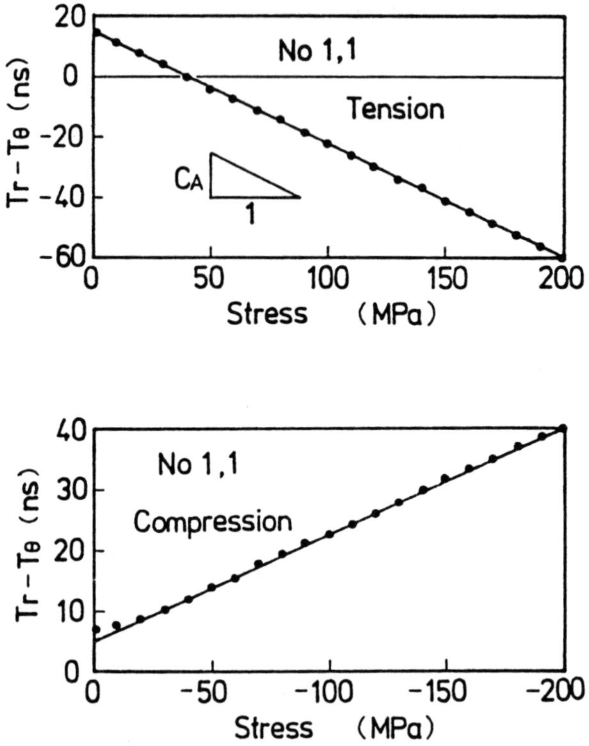

Figure 8: Transit time difference versus stress in calibration tests.

(after breaking test, Figure 10)

$$\frac{T_r - T_\theta}{T_0} = \alpha + C_A \sigma \tag{13}$$

(as rim blocks, Figure 11)

$$\frac{T_r - T_\theta}{T_0} = \alpha \tag{14}$$

where σ^0 and σ are the residual stresses at the time of manufacture and after braking test and α^0 is the initial, as manufactured texture acoustic birefringence. C_A^0 is the acoustoelastic constant at the time of manufacture. The change of C_A due to braking test is found to be negligible as shown in Table 1. The same is true for the change of α as is seen from Figure 11.

The method obtaining the residual stress from Figures 10 and 11 can be said as "destructive" acoustoelasticity. For "nondestructive" acoustoelasticity, we have to assume a priori the likelihood values of texture acoustic birefringence or as manufactured residual stress of the wheel rim. One possibility is to measure either of them for a sample wheel chosen from one lot of wheels and use them as the likelihood values. This method cannot be successful unless the scatter of the texture

Table 1: Acoustoelastic constants obtained from uniaxial tension and compression tests.

($\times 10^{-6}$/MPa)

	Tension			Compression		
	1	2	3	1	2	3
NO.1	-7.66	-7.49	-7.66	-7.43	-7.56	-7.59
NO.3	-7.85	-7.79	-7.39	-7.43	-7.50	-7.56
NO.3	-8.12	-7.75	-7.66	-7.99	-7.59	-7.28
NO.4	-8.21	-8.05	-7.69	-7.27	-7.36	-7.99
NO.5	-7.59	-7.66	-7.49	-7.07	-7.26	-7.27
NO.6	-7.69	-7.62	-7.56	-7.66	-7.13	-7.73
NO.7	-7.82	-	-	-7.46	-7.59	-7.00
Average	-7.72			-7.46		
	-7.59					

acoustic birefringence or as manufactured residual stress of individual wheel of the lot from the assumed value is small. Now, if the average value of the texture acoustic birefringence in Figure 11 or as manufactured residual stress in Figure 12 is used as the assumed value, the residual stresses can be estimated from the data in Figure 10 and are shown in Figure 14.

The residual stress in the No. 1 wheel rim estimated by acoustoelasticity is compared with those obtained by plate-cut-out method and elastic-plastic finite element computation which are shown in Figure 15. Distributions of the circumferential residual stress along axial direction at the radial positions A, B, C and D are taken from Figure 15 and are shown in Figure 16. As the stress obtained by acoustoelasticity is an averaged value through the axial thickness, the results in Figure 16 are averaged at the points A, B, C and D. In Figure 17, they are compared with acoustoelasticity values. Acoustoelastic analysis shows a good agreement within 40MPa with the result of the plate-cut-out method, which is at present considered to be the most reliable destructive method.

As for the wheels of No. 2, 3 and 7, the acoustoelastic analysis which used the assumed texture acoustic birefringence is also compared with the results obtained by the sectioning method and are shown in Figure 18. The results obtained by the nondestructive acoustoelasticity agree well within 40MPa with those obtained by the conventional destructive method.

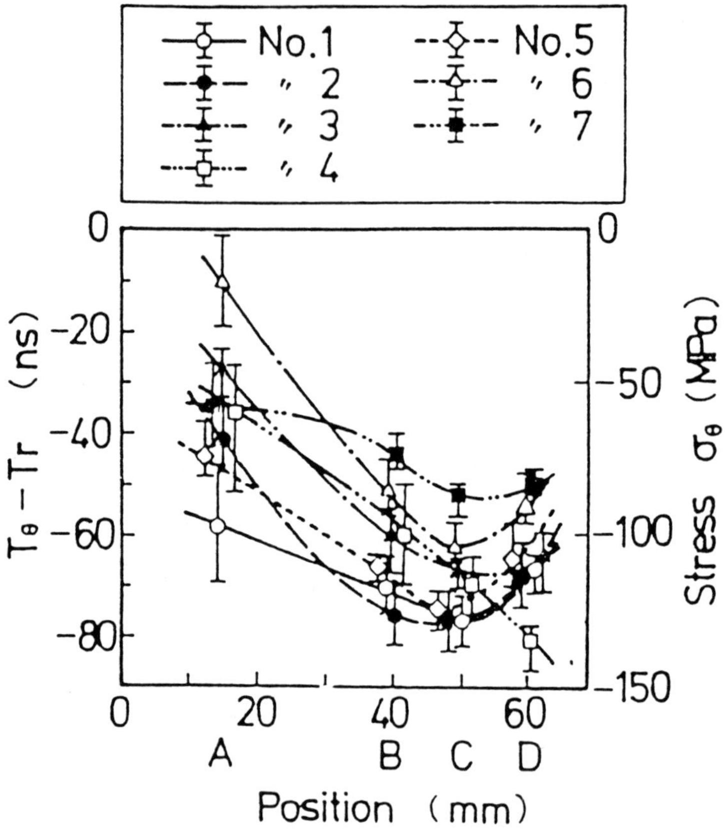

Figure 9: Acoustic birefringence of the wheel rim as manufactured

4.3. *Water Pipe*

The stress in the water pipe for cooling the condenser in a nuclear power plant caused by uneven settlement of the ground was measured by the birefringent acoustoelasticity. The pipe was made of SS41 steel (equivalent to 1018) and the main dimension and the measuring positions A, B, C and D on the three circumferential lines 1, 2 and 3 are shown in Figure 19. The pipe was laid about 7m under the ground.

Stresses acting on the buried water pipe are the following. The axial stress σ_z consists of the stress σ_b by uneven settlement and the thermal stress σ_t caused by the temperature difference at the time of installation and measurement. The circumferential stress σ_θ consists of the bending stress σ_{tb} by earth pressure and the residual stress σ_R caused by welding. As the stress measured by acoustoelasticity is an average value through the thickness of the pipe, the bending stress σ_{tb} by earth pressure has no effect to the acoustoelastic stress measurement.

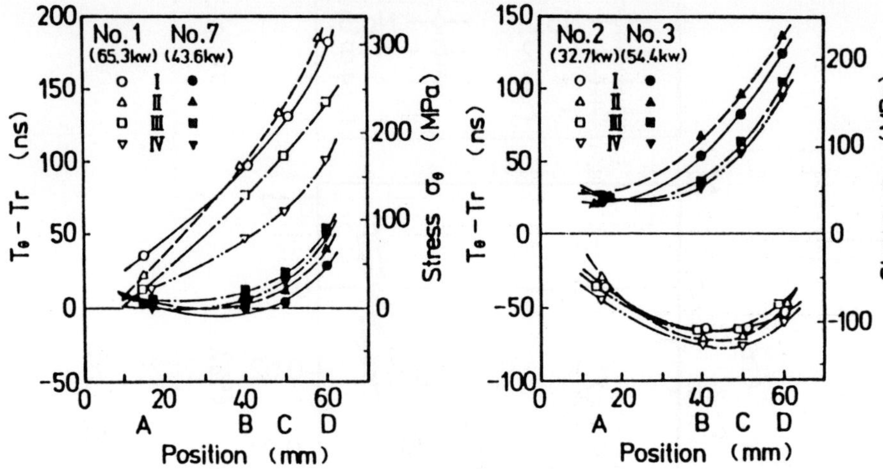

Figure 10: Acoustic birefringence of the wheel rim after braking tests

On the same circumferential line, the value of the residual stress σ_R caused by welding is the same for all measuring points and so is the case for the thermal stress σ_t caused by the temperature difference. For the stress σ_b by uneven settlement, the values at the top point A and the bottom point B are of the opposite sign on the same circumferential line. The stresses σ_b for the side points C_1 and C_2 are zero. Therefore, Equation (6) for the buried water pipe can be written as,

at the top point A:

$$\frac{T_z^A - T_\theta^A}{1/2(T_Z^A + T_\theta^A)} = \alpha + C_A(\sigma_\theta^A - \sigma_z^A)$$

$$= \alpha + C_A\{\sigma_R - (-\sigma_b + \sigma_t)\} \tag{15}$$

at the bottom point B:

$$\frac{T_z^B - T_\theta^B}{1/2(T_z spC - T_\theta^C)} = \alpha + C_A(\sigma_\theta^B - \sigma_z^A)$$

$$= \alpha + C_A\{\sigma_R - (\sigma_b + \sigma_t)\} \tag{16}$$

at the side point C_1, C_2:

$$\frac{T_z^C - T_\theta^C}{1/2(T_z^C + T_\theta^C)} = \alpha + C_A(\sigma_\theta^C - \sigma_z^C)$$

$$= \alpha + C_A(\sigma_R - \sigma_t) \tag{17}$$

The stress caused by uneven settlement can be obtained from Equations (15) and (16) as

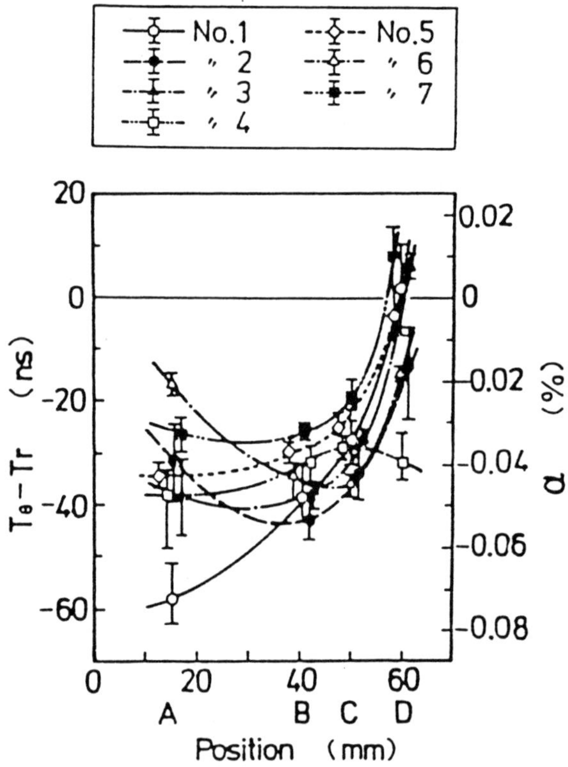

Figure 11: Acoustic birefringence of the wheel rim as rim blocks

$$\sigma_b = \frac{1}{2C_A}\left\{\frac{T_z^B - T_\theta^B}{1/2(T_z^B + T_\theta^B)} - \frac{T_z^A - T_\theta^A}{1/2(T_z^A + T_\theta^A)}\right\} \tag{18}$$

The acoustic birefringence measured at the side point C_1 or C_2 includes the effects of texture acoustic birefringence α, the residual stress σ_R by welding and the thermal stress σ_t by temperature difference.

The acoustoelastic measurement was done by a sing-around method with a shear type 5MHz PZT transducer of 4mm square size. The acoustoelastic constant of the pipe material was calibrated as

$$-7.1 \times 10^{-6} 1/MPa \ .$$

The results of the measurement are summarized in Table 2.

The stress obtained by the birefringent acoustoelasticity was compared with calculated value derived from the measured settlement and is shown in Figure 20. The acoustoelastic measurement gives a good agreement with the calculated value derived from the settlement.

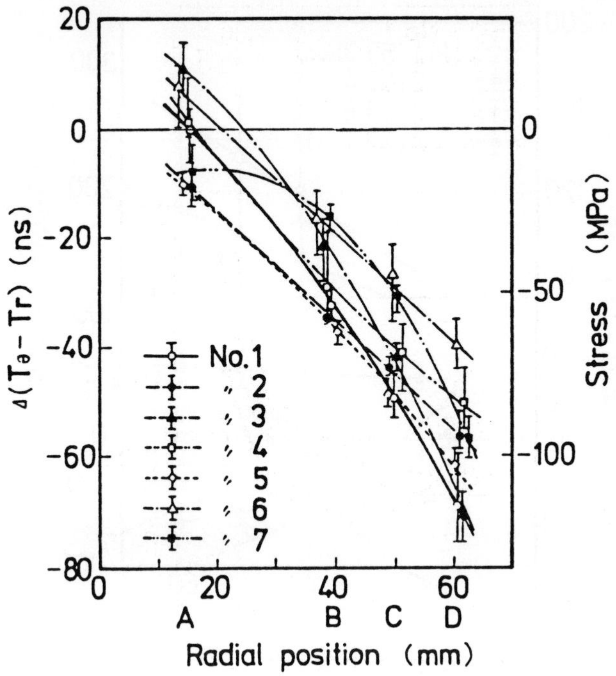

Figure 12: As manufactured residual stresses using the data of rim blocks

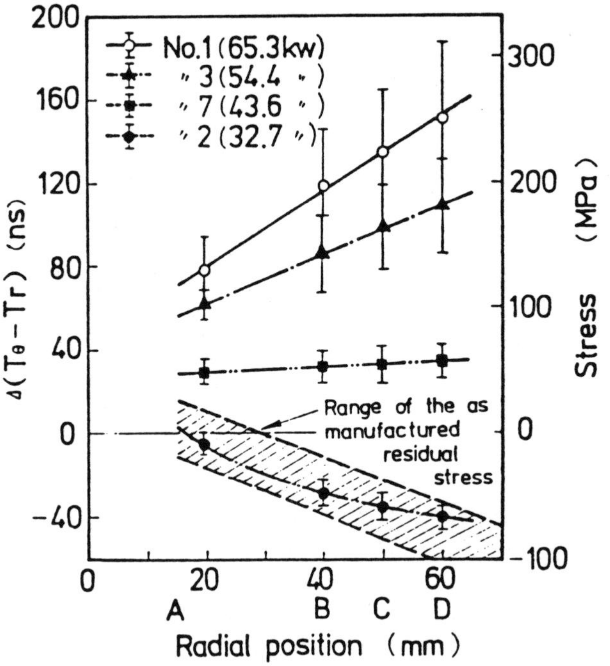

Figure 13: After braking residual stresses using the data of rim blocks

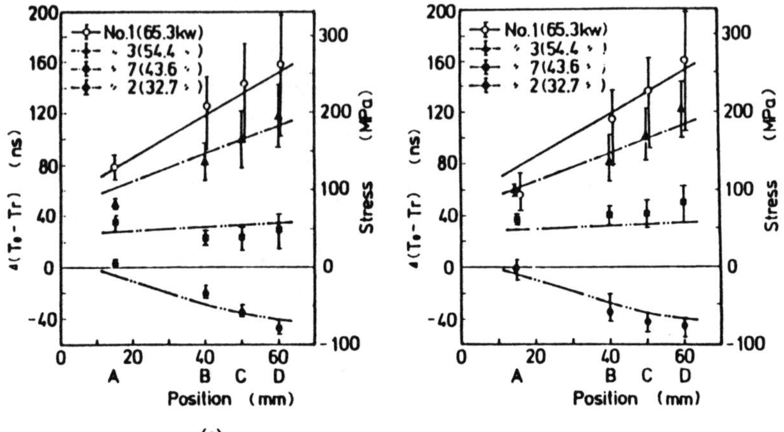

Figure 14: Residual stress estimation by nondestructive acoustoelasticity by (a) assuming the average as manufactured residual stress and (b) assuming the texture birefringence

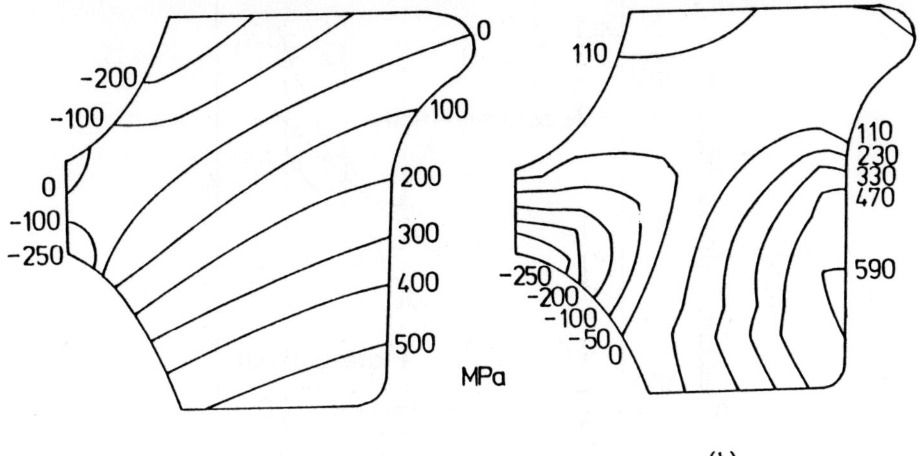

Figure 15: Residual stresses obtained by (a) plate-cut-out method and (b) elastic-plastic finite element method

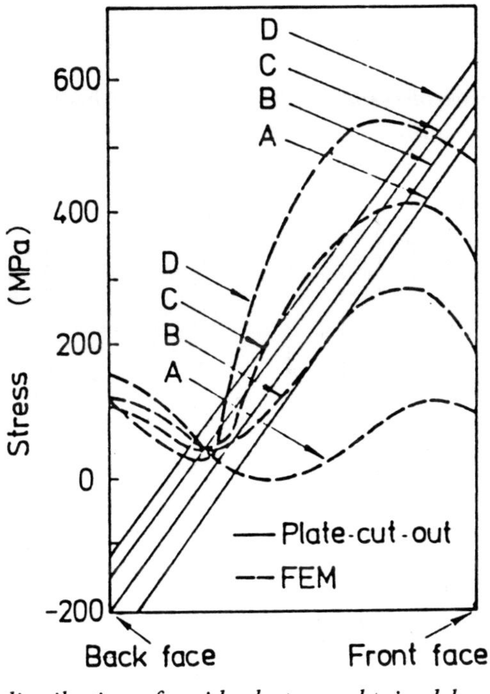

Figure 16: Axial distribution of residual stress obtained by plate-cut-out method and finite element method

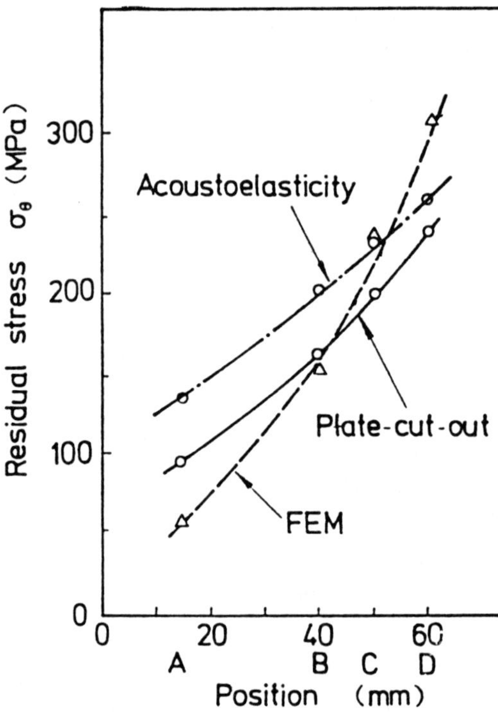

Figure 17: Comparison of residual stresses obtained by acoustoelasticity analysis with plate-cut-out and FEM for No. 1 wheel

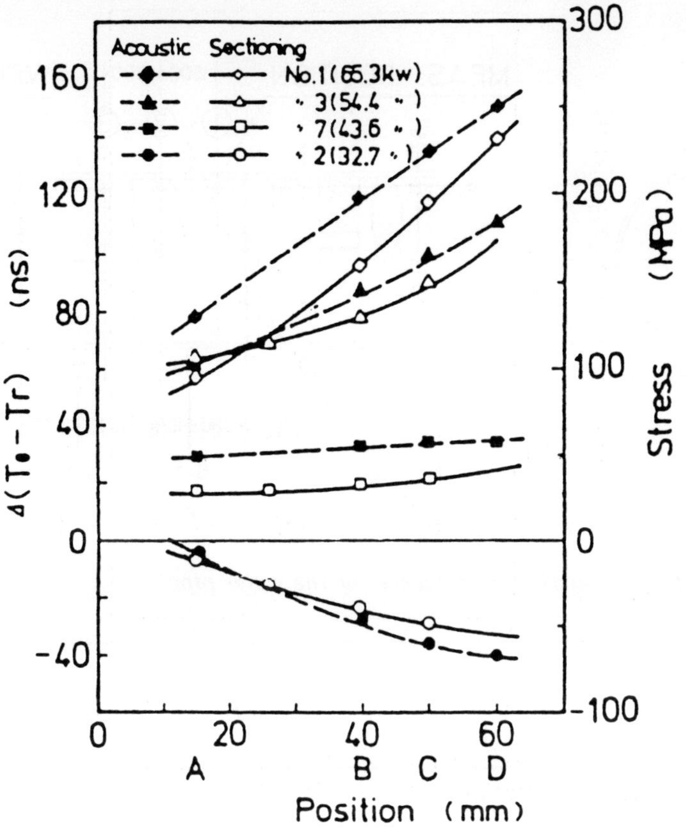

Figure 18: Comparison of residual stresses obtained by acoustoelasticity with sectioning method

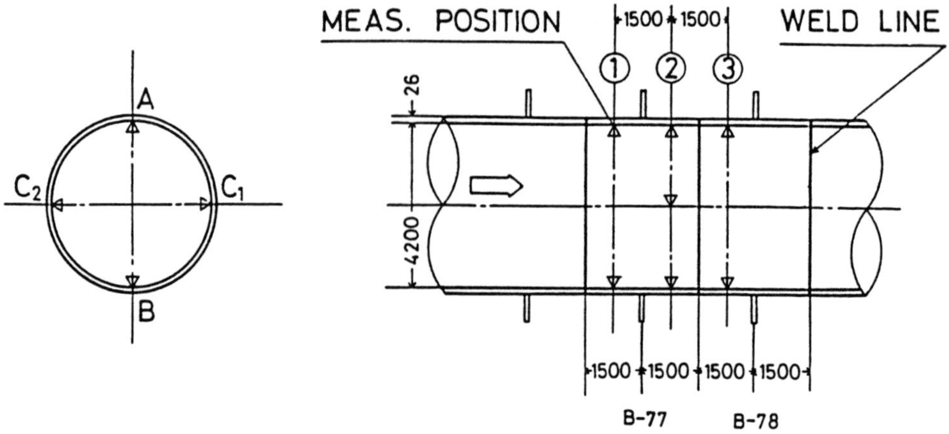

Figure 19: Locations of acoustoelasticity measurement for the water pipe.

Table 2: Data of acoustic birefringence for the water pipe

	$\frac{T_Z - T_\theta}{\frac{1}{2}(T_Z + T_\theta)}$ (%)				σ_b (MPa)
	TOP A	BOTTOM B	SIDE C_1	SIDE C_2	
①	0.025	0.184	–	–	112
②	0.037	0.174	0.099	0.081	96
③	0.029	0.181	–	–	107

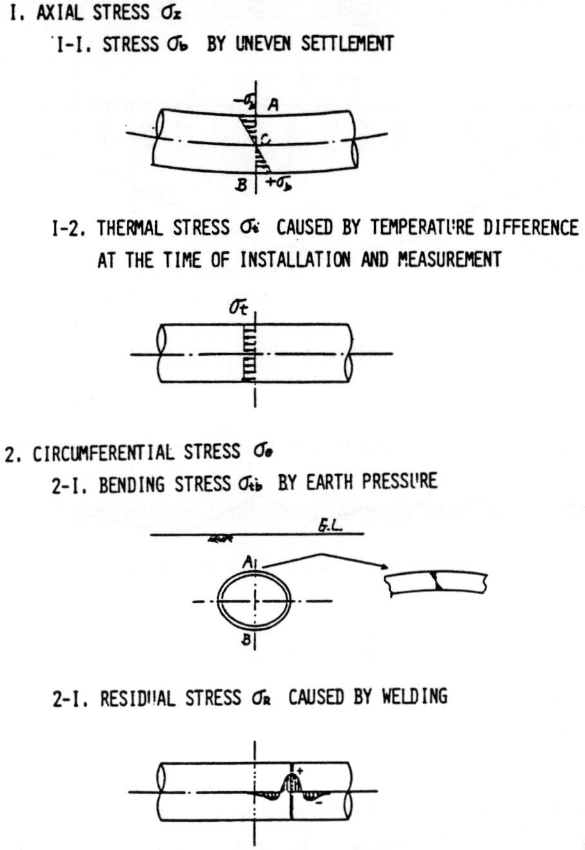

Figure 20 Stresses acting on the buried water pipe

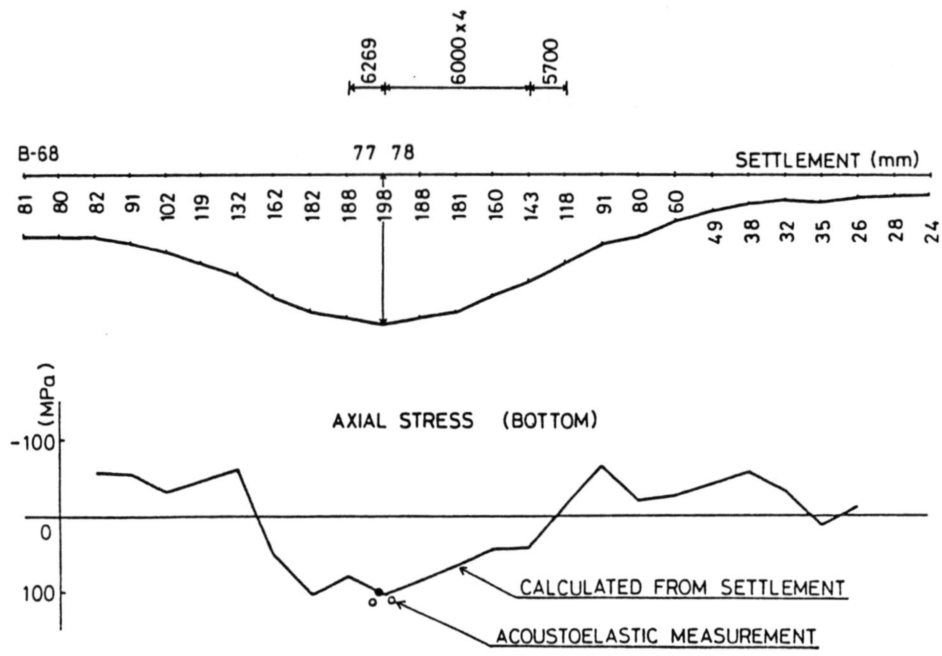

Figure 21 *Stresses due to uneven settlement obtained by acoustoelasticity and calculated from the data of settlement*

References

[1] King, R.B. and Fortunko, C.M., *J. Appl. Phys., 54*, 1983, p. 3027.

[2] Iwashimizu, Y. and Kubomura, K., *Int. J. Solids Structures*, 9, 1973, p. 99.

[3] Thompson, R.B., Smith, J.F. and Lee, S.S., in *Review of Progress in Quantitative NDE, 2B*, 1983, p. 1339.

[4] Allen, D.R. and Sayers, C.M., *Ultrasonics, 22*, 1984, p. 179.

[5] Sayers, C.M., Allen, D.R., Haines, G.E. and Proudfoot, G.G., *AERE-R11860*, July, 1985.

[6] Toda, H., Fukuoka, H. and Aoki, Y., *Jap. J. Appl. Phys., 23*, Suppl. 23-1, 1984, p. 86.

[7] Salama, K., Wang, J.J. and Barber, G.C., *Review of Progress in Quantitative NDE, 2B*, 1983, p. 1355.

[8] Mahadevan, P., *Nature, 211*, 1966, p. 621.

[9] Hirsekorn, S., *J. Acoust. Soc. Am., 77*, 1985, p. 832.

[10] Hirao, M. and Pao, Y.H., *J. Acoust. Soc. Am., 77*, 1985, p. 1659.

[11] Fukuoka, H., Toda, H. and Naka, H., *Exp. Mech., 23*, 1983, p. 120.

[12] Fukuoka, H., Toda, H., Hirakawa, K., Sakamoto, H. and Toya, Y., in *Wave Propagation in Inhomogeneous Media and Ultrasonic Nondestructive Evaluation*, G.C. Johnson (Ed.), ASME, AMD-Vol. 62, 1984, p. 185.

LOW-FIELD MAGNETOACOUSTIC RESIDUAL STRESS MEASUREMENT IN STEEL

M. Namkung
Department of Physics
College of William and Mary
Williamsburg, VA 23185

D. Utrata
Association of American Railroads
3140 S. Federal Street
Chicago, IL 60616

J.S. Heyman
S.G. Allison
NASA-Langley Research Center
Hampton, VA 23665

Abstract

Much of the recent development of the magnetoacoustic technique has been devoted to refine this technique as a reliable and practical tool for measuring bulk residual stress in steel components. For this, the effects of structural and metallurgical properties on the magnetoacoustic interaction have been studied for various types of steel samples. Also, progress is being made to obtain quantitative residual stress measurements in railroad wheels. This paper reviews the physical basis and the experimental results of the magnetoacoustic stress measurements in steels.

1. Introduction

One of the most persisting difficulties in nondestructive bulk stress measurements is to obtain a calibration standard. This is because material inhomogeneity and structural anisotropy, both significantly affecting ultrasonic wave velocities, can easily be introduced to individual metallic components during the fabrication process. As a result, various ultrasonic techniques cannot determine reliably even the sign of bulk residual stress.

Recent developments of the low-field magnetoacoustic technique [1-3] have shown that it is possible to differentiate the sign of residual stress in steel alloys without reference calibration data. This is because the initial slope of acoustic natural velocity variation, with respect to net magnetic induction in the stress axis, is negative in steel samples under static uniaxial compression and is positive otherwise. The method is based mainly on two ferromagnetic properties: the stress-dependent magnetic domain structure and the domain structure-dependent elastic modulus.

The ferromagnetic and stress states in a material are closely related to each other through the magnetoelastic interaction [4,5]. The interaction energy density of a particular domain depends on the orientations of domain magnetization and uniaxial stress axes with respect to the lattice unit cell axes. For the two types of magnetic domain walls, 90- and 180-degree walls, in iron base alloys having the body-centered cubic (BCC) crystal structure, only the former respond to uniaxial stress. This is because two adjacent domains separated by a 90-degree wall, in general, have different magnetoelastic energy densities under uniaxial stress. This difference in energy density is the net pressure acting on the domain wall [6]. Hence, domain walls move to expand the volume of domains with lower magnetoelastic energy densities at the expense of others. The anisotropy in the domain structure thus produced is called the stress-induced domain alignment.

The elastic modulus of a ferromagnetic solid changes during magnetization. More generally, the elastic modulus depends on the domain structure in the material [5,7]. The initial domain structure in the demagnetized state is determined according to the stress state, and the subsequent change in domains due to the application of magnetic field proceeds distinctively at each stress level. Consequently, the change in elastic modulus accompanying the magnetization process should depend on the stress state in the material.

A typical magnetization curve is divided into three regions according to different contributing mechanisms; reversible domain wall motion, irreversible domain wall motion and domain rotation [5]. Upon the completion of domain rotation, the material reaches the state of technical saturation. The change in elastic modulus, hence the acoustic wave velocity, up to saturation has been known as the ΔE-effect, which is most pronounced during the domain wall motion. Further change in the physical properties beyond saturation is called the pure para-process [8] which includes the magnetic field dependence of acoustic wave velocity.

The uniaxial stress-induced domain structure change in pure iron is possible by domain wall motion only. This is because the crystalline anisotropy energy in iron is unusually high and the rotation of magnetization vectors away from the easy axes requires a uniaxial stress far greater than the ultimate strength of iron [4]. This is assumed to be true in many iron base alloys, including steels, unless their magnetic properties are drastically altered by impurities. Hence, stress information can be obtained only by the changes of acoustic wave velocity during the domain wall motion process when magnetic field is applied. This is why the present method is described as low-field magnetoacoustics, as opposed to the magnetoacoustic interaction phenomena observed in the pure para-process [9].

2. Uniaxial Stress Effects on Domain Structure

The introduction of the ferromagnetic state into a crystal causes a reduction of its lattice symmetry. For example, the BCC unit cells of ferromagnetic iron do not have the exact cubic symmetry which is recovered by a magnetic phase transition into the paramagnetic state at the temperature called the Curie point (770 °C). The lowered symmetry in the ferromagnetic state reduces the crystalline anisotropy energy by deforming the crystal with respect to the cubic lattice [10]. This lattice strain is called spontaneous magnetostriction. The distortion in ferromagnetic unit cells of iron is such that they are tetragonal with the longer edges aligned parallel to one of the six equivalent directions of <100>, i.e., $\pm x, \pm y$ and $\pm z$ - directions, called the easy axes of magnetization.

The effect of uniaxial stress on the ferromagnetic state can be described by the following expression of the magnetoacoustic interaction [5,10]:

$$f_{k,me} = K_1(\alpha_1^2\alpha_2^2 + \alpha_2^2\alpha_3^2 + \alpha_3^2\alpha_1^2) - 3/2\lambda_{100}\sigma(\alpha_1^2\gamma_1^2 + \alpha_2^2\gamma_2^2 + \alpha_3^2\gamma_3^2)$$
$$- 3\lambda_{111}\sigma(\alpha_1\alpha_2\gamma_1\gamma_2 + \alpha_2\alpha_3\gamma_2\gamma_3 + \alpha_3\alpha_1\gamma_3\gamma_1)$$

where K_1 is the first anisotropy constant, λ_{100} is the saturation magnetostriction along the <100>-axis, λ_{111} is that along the <111>-axis and α's and γ's are the direction cosines of the domain magnetization vector and uniaxial stress axis, respectively, with respect to the cube axes. Since uniaxial stress cannot rotate the magnetization away from the easy axes, the first and third terms in the above equation are zero before the application of magnetic field. In iron where λ_{100} is positive, the energy density is negative under tension for the domain aligned in the stress axis and it is zero in domains aligned perpendicular to the stress axis. A difference in energy density exists only at a domain wall separating two domains perpendicular to each other. Hence, 90-degree walls move to reduce the total interaction energy in the crystal, while the positions of 180-degree walls are unaffected by uniaxial stress. Figure 1 illustrates the domain alignment of iron-like materials, i.e., having positive λ_{100}, by uniaxial tension and compression.

During their motion in a crystal, domain walls interact with various lattice defects. This is because they "see" each other through local lattice strains they create [11]. It is also obvious that lattice strain is severe at 90-degree walls and is

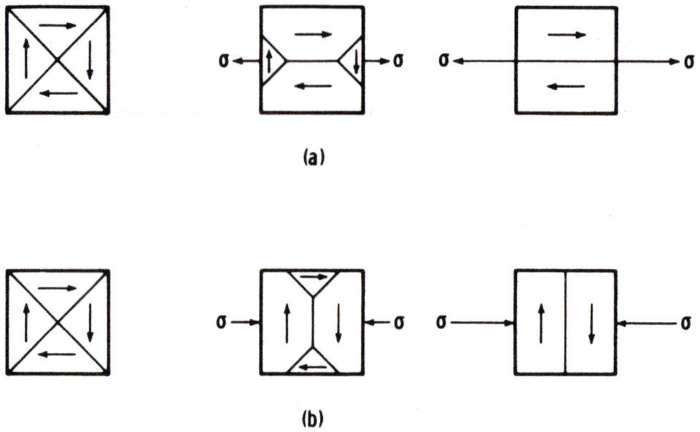

Figure 1: Uniaxial stress-induced magnetic domain alignments in iron-like BCC ferromagnetic materials having positive spontaneous magnetostriction. Arrows inside square blocks indicate the domain magnetization vectors.

negligible at 180-degree walls. This suggests that 90-degree walls can be pinned easily by lattice defects, especially during the stress-induced wall motion. Therefore, in a material like steel, there always exists a fraction of residual domains which are energetically unfavored by uniaxial stress. These residual domains, however, become seed domains when the magnetic field is applied and play an important role in the stress measurement.

3. Domain-Dependent Elastic Modulus of a Ferromagnet

The elastic modulus of a non-linear solid is defined as the ratio between applied infinitesimal uniaxial stress and resultant strain at a given condition of the solid. The uniaxial stress applied to a ferromagnet not only deforms the crystal but also alters the domain structure. This can be described by the following expression:

$$E = \frac{\Delta\sigma}{\Delta\varepsilon_{e1} + \Delta\varepsilon_{em}}$$

where $\Delta\varepsilon_{e1}$ is the direct elastic strain and $\Delta\varepsilon_{em}$ is the magnetoelastic strain. The magnetoelastic strain is produced by rearranging the fraction of unit cells having longer edges, for the case of iron, in certain orientations with respect to the stress axis. The sign of magnetoelastic strain in iron apparently follows the sign of applied stress but it can also be shown to be a general behavior common to any ferromagnetic material [4].

In iron and iron-like materials, the magnetoelastic strain is solely caused by the motion of 90-degree walls and is approximately proportional to the total area of these walls at the moment the modulus is measured. The elastic modulus should increase as the total area of 90-degree walls in the material decreases. Figure 2 shows the contribution of magnetoelastic strain to the total stress-induced strain in a macroscopically unstressed material when the area of 90-degree walls is controlled by an external magnetic field. This variation of elastic modulus can be measured by the acoustic wave velocity changes, i.e., without applying infinitesimal uniaxial stress as shown in figure 2, during the application of magnetic field.

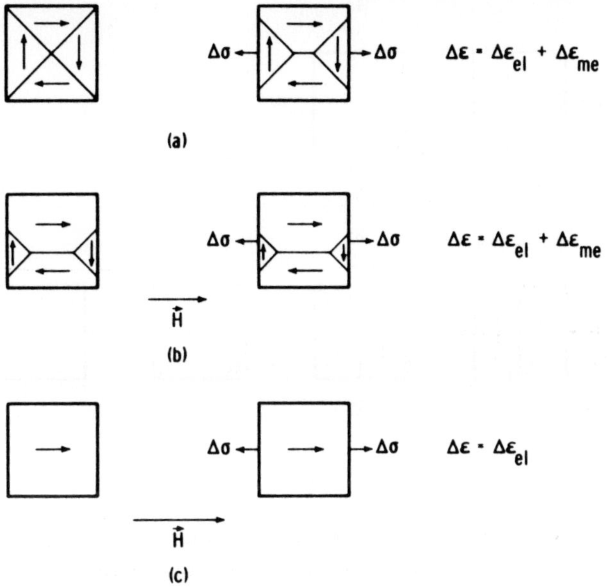

Figure 2: *Domain structure dependence of tensile stress-induced strain in iron-like ferromagnetic materials. Here, H is the magnetic field inside the material. Having a larger area of 90-degree domain walls, the magnetoelastic strain is larger in (a) than in (b). Hence, the elastic modulus should be higher in (b) than in (a). When saturated as in (c), there is no magnetoelastic strain and the modulus is highest. The role of 90-degree domain walls is, therefore, to reduce the elastic modulus of these types of materials.*

4. Effects of Uniaxial Stress on the Domain-Dependent Elastic Modulus

In an impure polycrystalline material like steel, there always exists a fraction of energetically unfavored residual domains and associated 90-degree walls even under maximum possible uniaxial stress. Upon the application of magnetic field, these domains either expand or further reduce their volume depending on their orientations with respect to the magnetic field. Figure 3 shows schematically the domain structure changes during the magnetization of iron-like materials when unstressed, under tension and under compression.

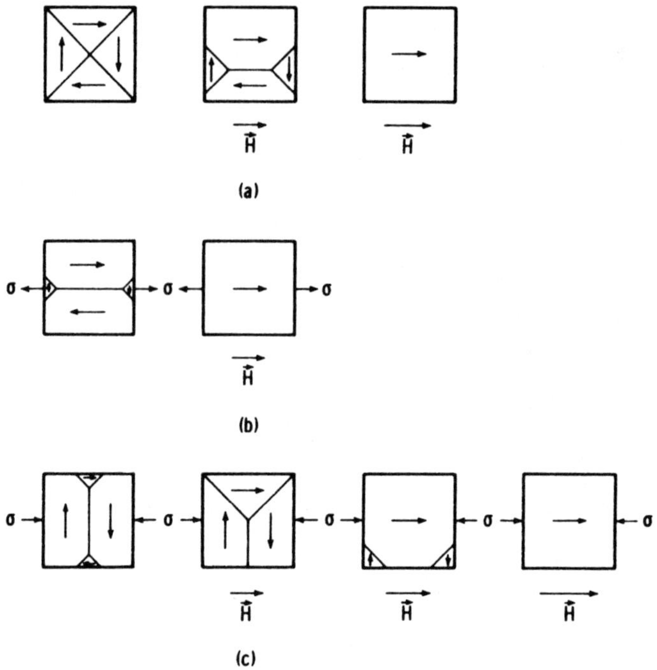

Figure 3: Uniaxial stress effects on the field-induced domain wall motion. The field strength to complete domain wall motion is smaller under tension and is larger under compression than without stress. Under compression, the total area of 90-degree walls increases initially causing a decrease in the elastic modulus.

When magnetized in the tensile stress axis, the increase in elastic modulus is due to the removal of residual 90-degree walls. The total positive change of modulus under tension is thus expected to be smaller than that without stress for a given field strength. Under compression, however, the residual domains with low Zeeman energy density, i.e., oriented close to the field axis, will expand their volume. At the same time the total area of 90-degree walls increases, causing the elastic modulus to decrease. The elastic modulus continues to decrease until it reaches its minimum, which corresponds to the maximum area of 90-degree walls. Beyond this point, as the magnetization process progresses, the modulus begins to increase due to domain rotation. Figure 4 illustrates the expected functional form of domain-dependent elastic modulus for the situation described in figure 3.

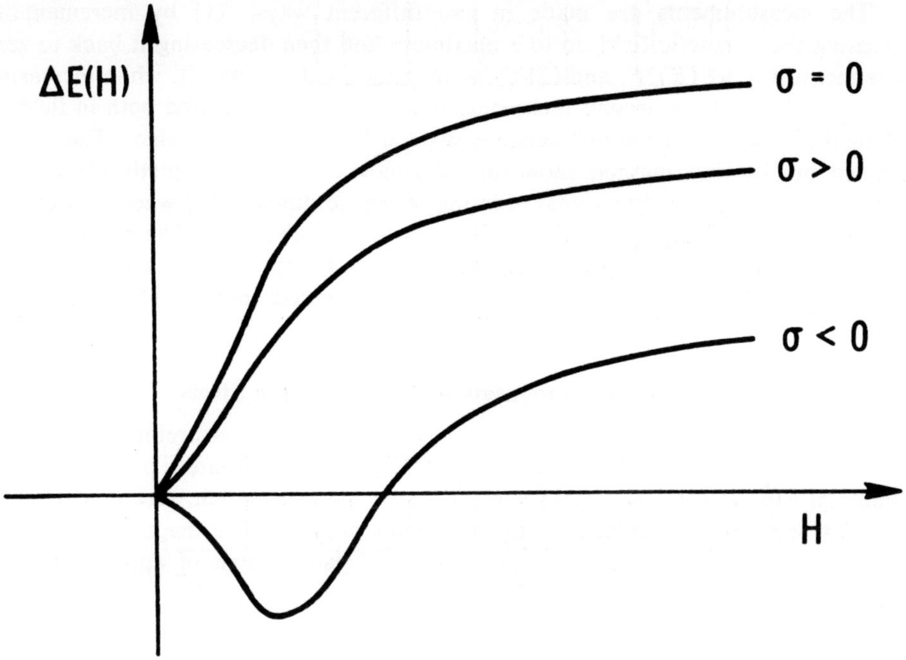

Figure 4: Functional form of elastic modulus changes during magnetization when a magnetic field is applied parallel to the uniaxial stress axis.

5. Experiments

A typical experimental setup consists of a pair of water-cooled electromagnets surrounding low-carbon steel cores, which are in mechanical contact with a steel sample to form closed magnetic circuits. Due to difficulties in extracting the magnitude of interal magnetic field, magnetic induction is measured using a pick-up coil surrounding the sample and an integrating fluxmeter. The fluxmeter reading is calibrated against the values obtained separately when the same coil was exposed to a known uniform magnetic field. For all the experiments described in this paper, magnetic field is applied in the uniaxial stress axis.

The pulsed-phase-locked-loop technique [12] is used to measure the fractional changes in acoustic wave frequency, which are equivalent to those of acoustic natural velocity. Under the phase-locked condition, the fractional change in acoustic wave frequency is given by:

$$\Delta F(B)/F = \Delta V(B)/V - \Delta L(B)/L$$

where V is the actual acoustic phase velocity, L is the acoustic path length and B is magnetic induction. The second term in the right-hand-side of the above equation is due to magnetostriction and its contribution is negligible.

The measurements are made in two different ways: (1) by incrementally increasing the magnetic field up to a maximum and then decreasing it back to zero while measuring $\Delta F(B)/F$, and (2) by turning the field on and off, while increasing its magnitude. For the second technique, frequencies are measured both in field-on and field-off states and the difference is divided by the field-off value. The results obtained by the first method show time dependence due to magnetic after-effect [13] and stress-induced directional ordering of elastic dipoles [14] when a sample is first exposed to a magnetic field or the stress level is changed. The results of the second method, on the other hand, are very much reproducible under any circumstances. Hence, the experimental results presented in this paper are those obtained by the second method.

6. Rayleigh Surface Wave Measurements with Plain Carbon Steels

Figure 5 illustrates the experimental arrangement for this measurement. Rayleigh surface waves with the frequency of 5 MHz were propagated parallel to the cylindrical axes of the 1020, 1045 and 1095 steel samples which were machined to form flat and parallel surfaces along the center portion. The magnetic field was applied in the wave propagation axis, which was also the axis of applied uniaxial stress.

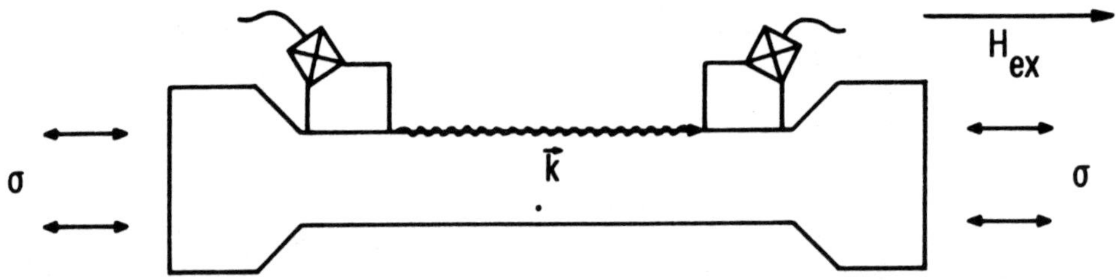

Figure 5: Experimental arrangement for Rayleigh surface wave measurements with applied uniaxial stress. Here $\vec{k}$ is the wave vector.

The results obtained with 1020 steel are shown in figure 6. As was predicted in figure 4, the effect of tension is seen to reduce the total positive shift in frequency. Since a complete domain alignment is not achieved by unaxial stress alone, the slope of the curve should remain positive under any tensile stress level. The curves under low compression are seen to overlap with those under high tension. Nevertheless, at -120 MPa, the curve shows a negative initial slope of $\Delta F(B)/F$ and increased magnitude of compression causes this negative initial slope to become steeper. This is certainly consistent with the model predictions.

Figures 7 and 8 show the results obtained with 1045 and 1095 steel samples, respectively, and these are also consistent with the model. The tension curves in these samples fall between the unstressed curve and the maximum tension curve. The effect of compression is similar to that seen in the previous figure except that the high compression curves in the 1045 steel sample do not show clear inflection

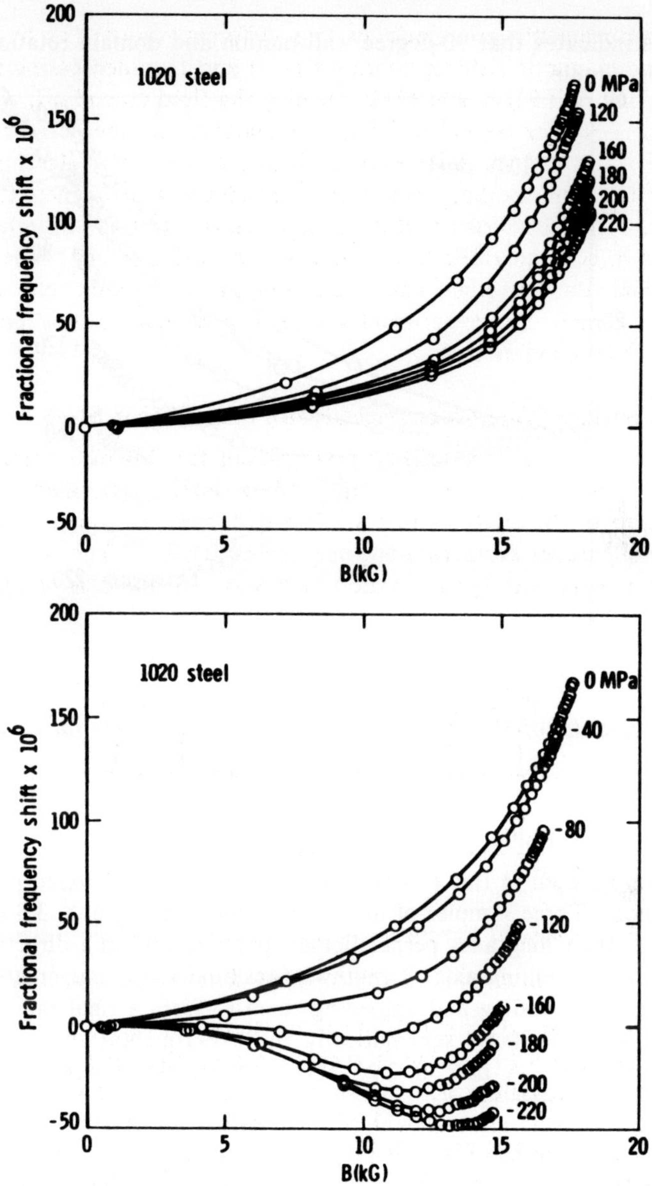

Figure 6: Results of Rayleigh surface wave measurements for a 1020 steel sample: (a) under tension and (b) under compression.

points. This indicates that 90-degree wall motion and domain rotation are not very active in this sample.

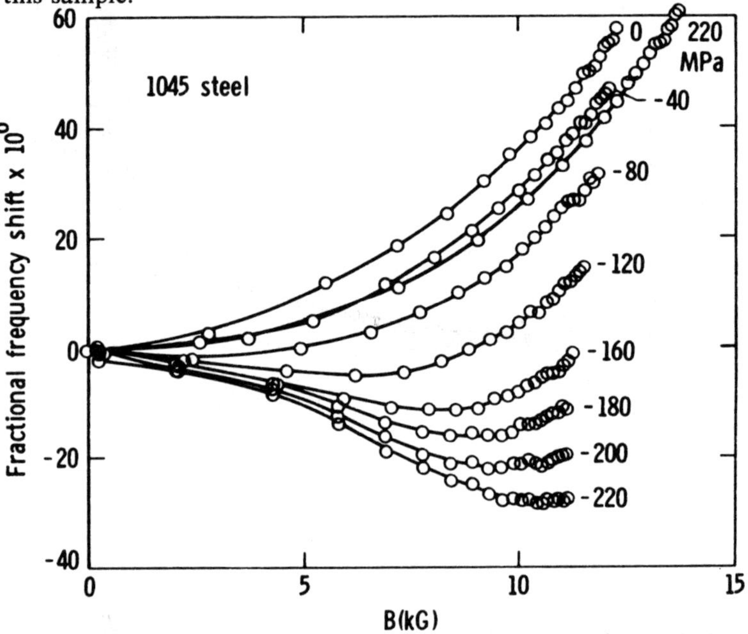

Figure 7: Results of Rayleigh surface wave measurements for a 1045 steel sample.

7. Effect of Severe Rolling Texture

The steel sample used in this experiment was a heavily cold rolled plate having .22 wt.% of carbon. Three samples of identical geometry (40 cm × 3.8 cm × .75 cm) were cut with their long axes perpendicular, parallel, and in a direction 40 degrees with respect to the rolling axis. Lengthwise residual stress was created in the samples by plastically bending and unloading them to form a final radius of curvature of 25 cm. Figure 9 shows schematically the experimental arrangement used for these measurements. A pair of 2.25 MHz compressional wave transducers were used with variable angle blocks.

To avoid a systematic experimental error due to any localized material inhomogeneity, magnetic field non-uniformity and so on, measurements were made by varying the distance between transducers. The variable angle blocks acted as delay lines and, under this particular situation, the acoustic path length could not be defined. Even so, an overall trend is obtained by considering an idealized situation shown in figure 10.

Under such a condition, the frequency change of the phase-locked acoustic wave can be written as follows:

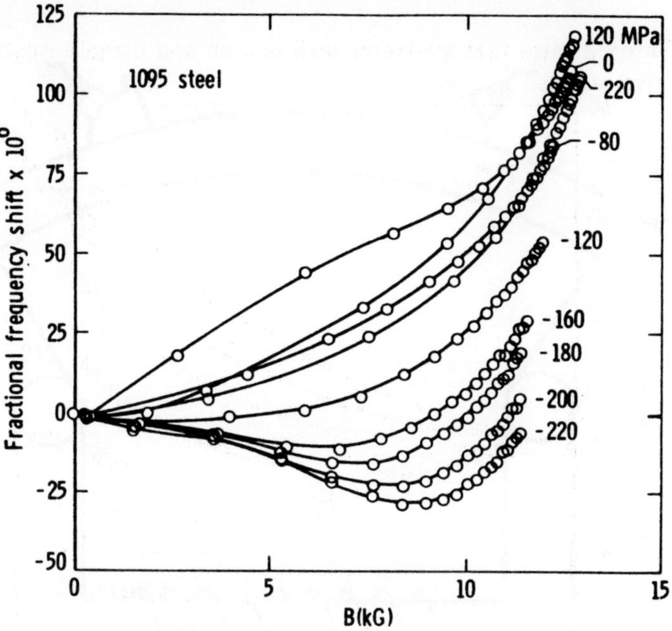

Figure 8: Results of Rayleigh surface wave measurements for a 1095 steel sample.

$$\Delta F(B)/F = \frac{\beta(B)}{1 + \dfrac{2L_d V_{mo}}{L_{mo} V_d}}$$

$$\sim \frac{\beta(B)}{1 + 2(L_d/L_{mo})}$$

where β is an appropriate field-dependent proportionality constant, V_d is the compressional wave velocity in the delay line material and V_{mo} is the Rayleigh surface wave velocity in steel before field application. V_d is found to be about 2700 m/s for lucite, and V_{mo} is estimated to be about 300 m/s.

Figure 11 shows the results obtained in the sample which was cut with its long axis perpendicular to the rolling direction. The label at the end of each curve indicates the separation distance measured between the inner bottom corners of the variable angle blocks. The residual compression curves (from surface A) show slight initial negative slope. As expected, the frequency shift is enhanced with increased path length in the material.

Figures 12 and 13 show the results of measurements with the parallel and 40-degree samples, respectively. The maximum depth of $\Delta F(B)/F$ under compression is almost the same in these samples and is about twice that found previously. According to the above equation, all the compression curves should cross each other at a point where $\Delta F(B)/F$ is zero. This certainly happened for the three samples, indicating that no systematic errors were involved in the measurement.

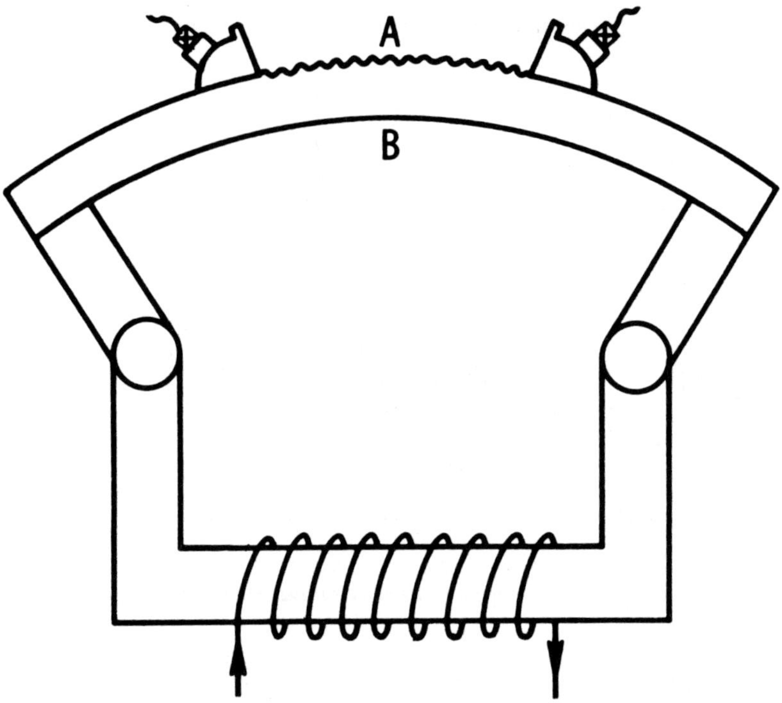

Figure 9: *Experimental arrangement for Rayleigh surface wave measurements with permanently bent samples. Surface A is under lengthwise residual compression and surface B is under residual tension.*

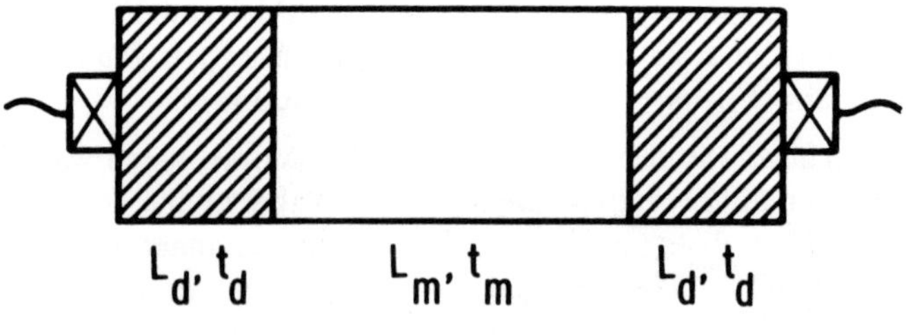

Figure 10: *Sample and delay lines for a simple analysis. L is the acoustic path length and T is the acoustic wave transit time. Subscripts m and d indicate sample and delay line, respectively.*

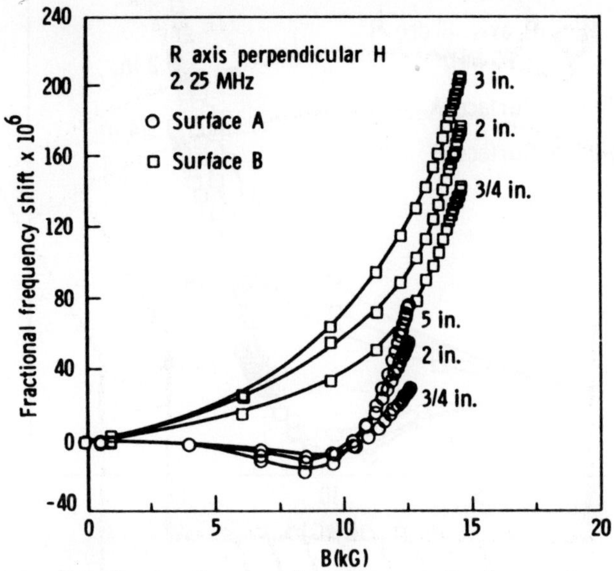

Figure 11: Results for the sample cut perpendicular to the rolling axis.

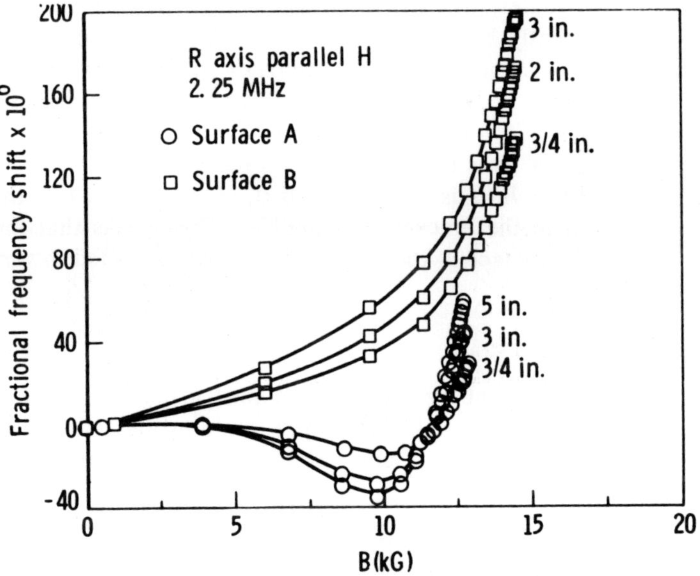

Figure 12: Results for the sample cut parallel to the rolling axis.

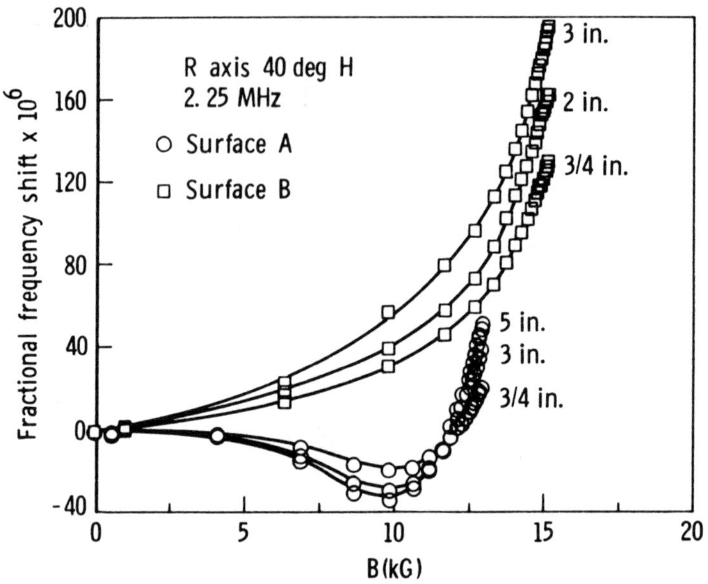

Figure 13: Results for the sample cut in a direction 40 degrees to the rolling axis.

Assuming that the material is non-strain hardening, the residual stress σ_r at the surface of a permanently bent plate, with a finite radius of curvature, can be expressed as follows [15]:

$$\sigma_r = \sigma_y \, [1/2 - 2(\sigma_y R/dE)^2]$$

where σ_y is the yield stress, R is the radius of curvature, d is the thickness and E is Young's modulus. Since R/d is about 30 and σ_y/E is about 10^{-3}, the magnitude of the second term in the bracket is negligible. This means that the lengthwise residual stress at the surface is about 15 ksi and is not expected to vary noticably between these samples.

8. Measurements in Railroad Steel

Samples used in this experiment were cylindrical rods machined in the center to obtain flat and parallel surfaces. 10 MHz bulk compressional waves were propagated perpendicular to the applied uniaxial stress and magnetic field axes as shown in figure 14.

Figure 15 shows the results obtained with a sample cut from a head section of railroad rail. In this figure the tension and compression curves are shown to be separated by the unstressed curve. This behavior of $\Delta F(B)/F$ curves under tension is clearly different from what was shown in figure 4. It was assumed that this effect is caused by a high degree of local lattice strain in this particular sample. There are, actually, two competing effects when a material is under tension:

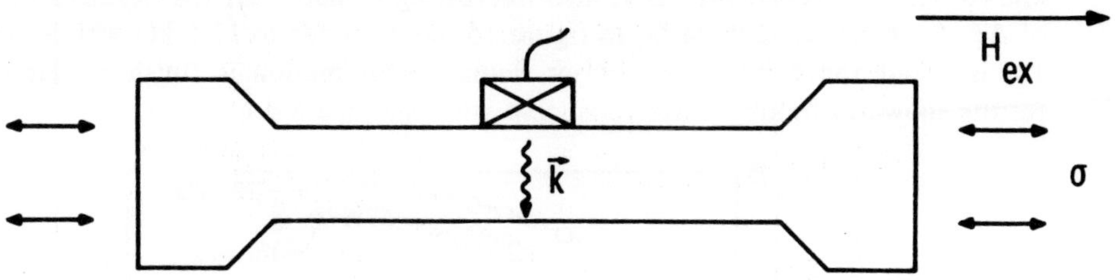

Figure 14: Experimental arrangement for bulk compressional wave measurements with applied uniaxial stress.

(1) The tensile stress-induced domain alignment reduces the total positive field-induced frequency shift compared to the unstressed state as shown in figure 4.

(2) If domain wall motion is difficult in a material due to severe local lattice strain, the effect of tension is merely to help the magnetic field, in the stress axis, remove 90-degree walls. Consequently, domain rotation begins earlier under tension. The combination of these two effects will raise tension curves above the unstressed curve.

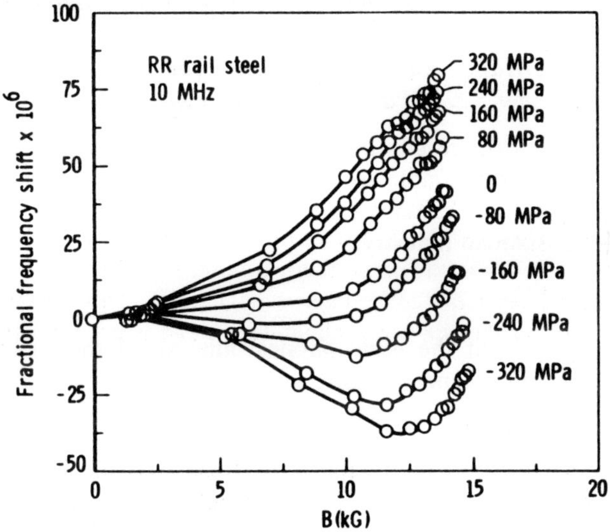

Figure 15: Results for a steel sample cut from a head section of railroad rail.

Figure 16 shows the magnetization curves when unstressed and under ±320 MPa. In the early stage of magnetization, the curve is higher under tension than without stress. This is interpreted as the tensile stress-assisted removal of 90-degree walls by magnetic field which enhances the positive frequency shift. The early occurrence of domain rotation under tension is not very clear in the figure because the domain wall motion and domain rotation regions in a polycrystalline sample overlap. The crossing between the magnetization curves in this figure is the well known Villari reversal [4]. It is also interesting to note that the negative minimum of $\Delta F(B)/F$ under -320 MPa, in figure 15, corresponds to 12.5 kG which, in figure 16, is about the point where major domain wall motion is finished. Hence, the results shown in figure 16 are consistent with the model prediction.

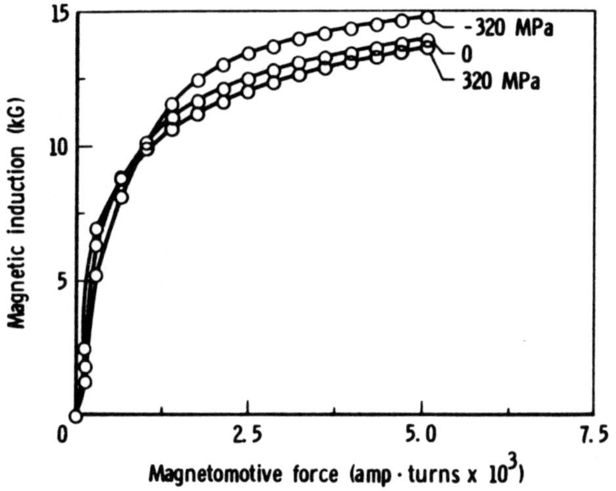

Figure 16: Magnetization curves for the railroad rail sample when unstressed, and under 320 MPa of tension and compression.

Figure 17 shows the results for a sample cut from a railroad wheel. The general stress effects in this sample are the same as shown in figure 15. These results strongly suggest that applications exist for using this technique as a field tool for NDE work in the railroad industry.

9. Conclusion

In this paper we present the basic physical analysis of the uniaxial stress dependence of the low-field magnetoacoustic interaction in iron-base alloys with the BCC crystal structure. A review is made of results obtained under various experimental conditions. The results show that the uniaxial stress effects are consistent with the model prediction. Because of the negative initial slope of $\Delta F(B)/F$ curves under compression, the sign of residual stress can be determined unambiguously without a calibration standard. A further study is currently in progress to obtain quantitative residual stress measurements and to allow practical application of the present technique, with particular emphasis on the measurement of residual stress in railroad wheels.

Acknowledgment

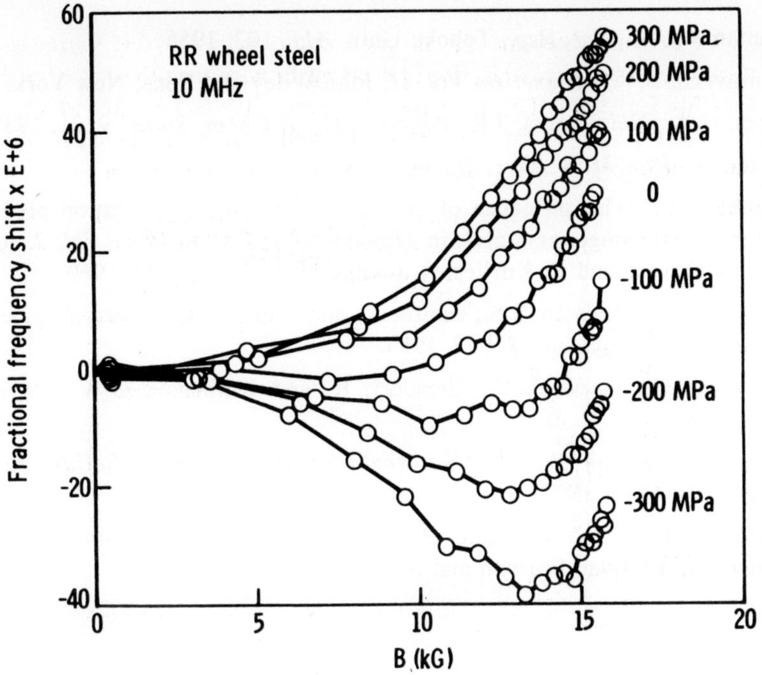

Figure 17: Results for a steel sample cut from the rim of a railroad wheel.

The authors are grateful to the Federal Railroad Administration for support, W.T. Yost and J.H. Cantrell, Jr. for helpful discussions and F.D. Stone, P.W. Kushnick and C.G. Clendenin for their assistance.

References

[1] Namkung, M., Utrata, D., Allison, S.G. and Heyman, J.S., "Magnetoacoustic stress measurement in railroad rail steel", Submitted to the *Review of Progress in Quantitative NDE*, Williamsburg, VA, June, 1985.

[2] Namkung, M., Utrata, D., Allison, S.G. and Heyman, J.S., "Effect of texture on magnetoacoustic stress measurement in steel", Submitted to the *Review of Progress in Quantitative NDE*, Williamsburg, VA, June, 1985.

[3] Namkung, M., Utrata, D., Allison, S.G. and Heyman, J.S., "Magnetoacoustic stress measurements in steel", Submitted to *IEEE Ultrasonics Symposium*, San Francisco, CA, October, 1985.

[4] Cullity, B.D., *Introduction to Magnetic Materials*, Addison-Wesley, Menlo Park, 1972.

[5] Bozorth, D.M., *Ferromagnetism*, Van Nostrand, New York, 1972.

[6] Brown, W.F., Jr., *Phys. Rev. 75*, 147, 1949.

[7] Yamamoto, M., *Sci. Rep.*, Tohoku Univ. *A11*, 102, 1955.

[8] Vonsovskii, S.V., *Magnetism Vol. II*, John Wiley and Sons, New York, 1974.

[9] Alers, G.A., Neighbours, J.R. and Sata, H., *J. Chem. Solids 9*, 21, 1958.

[10] Kittel, C., *Rev. Mod. Phys. 21*, 541, 1949.

[11] Trauble, H., "The influence of crystal defects on magnetization processes in ferromagnetic single crystals", in *Magnetism and Metallurgy Vol. 2*, edited by A.E. Berkowitz and E. Kneller, Academic Press, New York, 1969.

[12] Allison, S.G., Heyman, J.S., Smith, K. and Salama, K., *Proceedings of IEEE Ultrasonics Symposium 2*, 997, 1984.

[13] De Vries, G., Geest, D.V., Gersdorf, R. and Rathenau, G.W., *Physica 25*, 1131, 1959.

[14] Nowick, A.S. and Berry, B.S., *Anelastic Relaxation in Solids*, Academic Press, New York, 1972.

[15] Warnock, F.V. and Benham, P.P., *Mechanics of Solids and Strength of Materials*, Sir Issac Pittman and Sons, London, 1965.

ULTRASONIC DETERMINATION OF TEXTURE AND RESIDUAL STRESS IN POLYCRYSTALLINE METALS

C.M. Sayers†
Materials Physics and Metallurgy Division
AERE Harwell
OX11 0RA, UK

1. Introduction

In a polycrystalline aggregate the elastic constants in the specimen reference frame vary from grain to grain due to the random orientation of the grains. Polycrystalline metals are therefore elastically inhomogeneous, and the elastic constant mismatch at the grain boundaries leads to scattering of the ultrasonic wave. In the long wavelength limit, however, the metal can be modelled as an elastic continuum with elastic constants determined by the elastic constants of the grains and the crystallite orientation distribution function (CODF). This function gives the probability of a crystallite having a given orientation with respect to the specimen frame, and gives a quantitative description of the texture, or crystallographic alignment, of the material. In a strongly textured metal the yield stress varies as a function of direction and this can lead to non-uniform flow in deep drawing for example. As a result there is a need for a non-destructive measurement of texture in process control, and there is considerable interest in the use of ultrasonics for this purpose. In section 2 it is shown how information on the CODF can be obtained from ultrasonic velocity measurements.

---

† Present address: Koninklijke/Shell Exploratie en Produktie Laboratorium, GD Rijswijk ZH, The Netherlands

Also of interest in process control are internal stresses arising from the manufacturing process which must be added to the external stress when determining the response of the structure to an applied load. In the presence of a stress the ultrasonic velocity in the material depends on the propagation and polarisation directions. Figure 1a shows the fractional change in ultrasonic velocity for the three modes of propagation along the plate normal direction in a mild steel specimen subjected to a compressive stress applied along the rolling direction [1]. The fractional change in velocity for the two shear waves when the stress is applied along the transverse direction is shown in Figure 1b. It is seen that the change in velocity is different for the shear waves polarised parallel and perpendicular to the stress axis, and this difference may be used in principle to determine the stress present. This is the basis of the shear wave birefringence technique [2] which involves the measurement of the relative velocities of two orthogonally polarised shear waves travelling in the same direction.

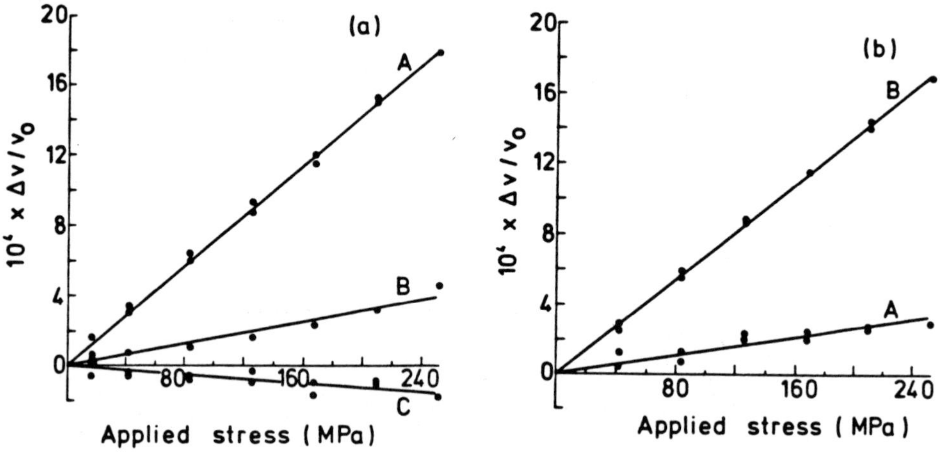

Figure 1: *Fractional change in ultrasonic velocity for propagation along the normal direction of a mild steel plate for an applied uniaxial compressive stress acting (a) parallel to rolling direction and (b) parallel to the transverse direction [1]. Curves: A, shear wave polarised parallel to the rolling direction; B, shear wave polarised perpendicular to the rolling direction; C, longitudinal wave.*

In metals with a texture or non-random distribution of crystallite orientations, the ultrasonic velocity depends on the orientation of the propagation and polarisation directions with respect to the principal texture axes, even in the absence of stress. In general, therefore, a measurement of the shear wave birefringence alone is not sufficient for determining the stress, except in samples in which the texture does not vary and in which only inhomogeneous or localised stresses are present.

For such samples, a stress-free region can be identified allowing the measurement of the stress-free birefringence and hence the correction for the effect of texture to be made in the region of interest. For samples in which the texture varies over a length scale similar to that of the variation in stress, or for samples in which neither the texture nor the stress vary over the accessible area of measurement, shear wave birefringence does not give a unique determination of the stress. In section 3 methods for the separate measurement of both stress and texture are discussed. These rely on the theory of ultrasonic propagation in textured polycrystalline solids presented in section 2.

2. Ultrasonic Propagation in Polycrystalline Aggregates in the Absence of Stress

Let $0x_1x_2x_3$ be an orthogonal set of reference axes fixed in the sample. For a rolled plate these axes could be chosen as the rolling, transverse and normal directions as shown in Figure 2a. It is assumed that the sample has orthorhombic symmetry, i.e. it is assumed to possess three orthogonal mirror planes given by the planes x_1x_2, x_2x_3 and x_3x_1 in Figure 2a. This is the symmetry of rolled plate. It will be assumed, in addition, that the sample is an aggregate of crystallites of cubic crystallographic symmetry, as is the case for steel and aluminum. The crystallographic alignment, or texture, of the plate is most conveniently described by the crystallite orientation distribution function (CODF). This method of analysis has been developed extensively by Roe [3-5] and by Bunge [6,7]. In this paper the notation of Roe is used.

Let $0X_1X_2X_3$ be an orthogonal set of axes for a crystallite given by the (100), (010) and (001) crystallographic directions. The orientation of a given crystallite with respect to the sample axes $0x_1x_2x_3$ can be specified by the three Euler angles ψ, θ and ϕ shown in Figure 2b. The CODF is denoted by $W(\xi,\psi,\phi)$ with $\xi=\cos\theta$. $W(\xi,\psi,\phi)\, d\xi d\psi d\phi$ gives the fraction of crystallites with orientations between ξ and $\xi+d\xi,\psi$ and $\psi+d\psi$, and ϕ and $\phi+d\phi$. It is convenient to expand $W(\xi,\psi,\phi)$ as

$$W(\xi,\psi,\phi)=\sum_{\ell=0}^{\infty}\sum_{m=-\ell}^{\ell}\sum_{n=-\ell}^{\ell} W_{\ell mn}\, Z_{\ell mn}(\xi)\exp(-im\psi)\exp(-in\phi) \quad (1)$$

where the $Z_{\ell mn}$ are the generalised Legendre functions defined by Roe [5]. For an orthorhombic aggregate of crystallites of cubic crystallographic symmetry, the elastic constant tensor depends only on the coefficients W_{400}, W_{420} and W_{440} in Equation (1). Explicit expressions for the elastic constants in terms of these coefficients have been given elsewhere [8-10].

Using these expressions for the elastic constants, the ultrasonic wave velocities in aggregates with small anisotropy can be calculated using Hamilton's principle [11]. For orthorhombic aggregates of cubic crystallites, the change $\delta v_R(\theta)$ in the Rayleigh wave velocity for propagation on the x_1x_2 surface from its isotropic value v_R^o is given by

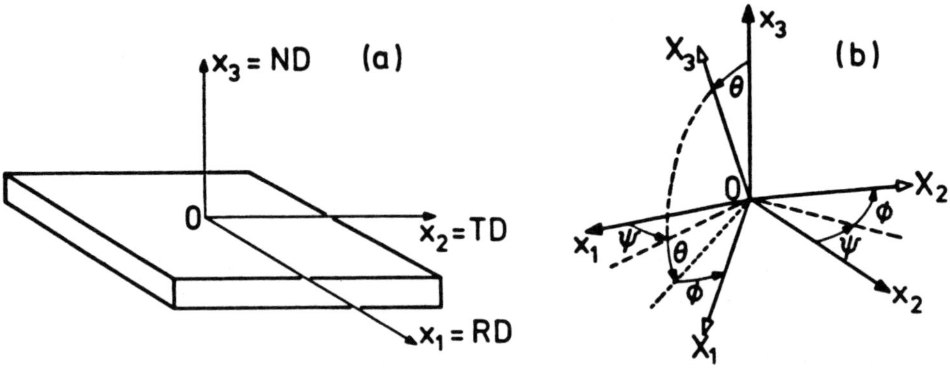

Figure 2: *(a) Choice of specimen axes for a rolled sheet, with $0x_1$ along the rolling direction (RD), $0x_2$ along the transverse direction (TD) and Ox_3 along the normal direction (ND). (b) The orientation of the crystallite coordinate system $0X_1X_2X_3$ with respect to the sample coordinate system $0x_1x_2x_3$ specified by the Euler angles ψ, θ and ϕ [5].*

$$\delta v_R(\theta) = (R_1 + R_2\cos 2\theta + R_4\cos 4\theta)/2v_R^o \tag{2}$$

in the absence of stress, where θ is the angle between the propagation and rolling directions [11]. The amplitudes R_1, R_2 and R_4 are functions of Poisson's ratio ν and are proportional to the coefficients W_{400}, W_{420} and W_{440} respectively. These coefficients may therefore be obtained from a fit of the measured angular variation of the Rayleigh wave velocity to Equation (2) and used to plot ultrasonic pole figures as described in reference [9].

The texture coefficients W_{400}, W_{420} and W_{440} may also be determined from bulk wave velocity measurements. For propagation in the plane of the plate

$$\delta v_{SH}(\theta) \equiv v_{SH}(\theta) - v_s^o = \frac{2\sqrt{2}\pi^2 c}{35\rho v_s^o}(W_{400} - \sqrt{70}\,W_{440}\cos 4\theta) \tag{3}$$

$$\delta v_{SV}(\theta) \equiv v_{SV}(\theta) - v_s^o = \frac{-8\sqrt{2}\pi^2 c}{35\rho v_s^o}(W_{400} - \sqrt{5/2}\,W_{420}\cos 2\theta) \tag{4}$$

$$\delta v_\ell(\theta) \equiv v_\ell(\theta) - v_\ell^o = \frac{6\sqrt{2}\pi^2 c}{35\rho v_\ell^o}\Big(W_{400} - \frac{2\sqrt{10}}{3}W_{420}\cos 2\theta + \frac{\sqrt{70}}{3}W_{440}\cos 4\theta\Big) \tag{5}$$

where $c = c_{11} - c_{12} - 2c_{44}$, the c_{ij} being the single crystal elastic constants. $v_{SH}(\theta)$, $v_{SV}(\theta)$ and $v_\ell(\theta)$ are the velocities of the shear horizontal, shear vertical and longitudinal waves propagating in the plane of the plate at an angle θ to the rolling direction. v_s^o and v_ℓ^o are the isotropic shear and longitudinal velocities.

For propagation in the through thickness direction $0x_3$

$$\delta v_{33} \equiv v_{33} - v_\ell^o = \frac{16\sqrt{2}\pi^2 c}{35\rho v_\ell^o} W_{400} \tag{6}$$

$$\delta v_{31} \equiv v_{31} - v_s^o = \frac{-8\sqrt{2}\pi^2 c}{35\rho v_s^o} (W_{400} - \sqrt{5/2}\, W_{420}) \tag{7}$$

$$\delta v_{32} \equiv v_{32} - v_s^o = \frac{-8\sqrt{2}\pi^2 c}{35\rho v_s^o} (W_{400} + \sqrt{5/2}\, W_{420}) \tag{8}$$

where v_{ij} is the velocity of the wave propagating in the direction $0x_i$ with polarisation in the $0x_j$ direction.

3. Measurement of Stress in the Presence of Texture

3.1. *The Velocity Combinations Approach*

Let λ and μ be the second order and ℓ, m and n be the third order elastic constants of the aggregate in the absence of texture. The shear wave birefringence $\Delta v_s / v_s^o$ is given in the presence of texture and stress by

$$\frac{\Delta v_s}{v_s^o} \equiv \frac{v_{31} - v_{32}}{v_s^o} = \left.\frac{(v_{31} - v_{32})}{v_s^o}\right|_{\text{texture}} + \left.\frac{(v_{31} - v_{32})}{v_s^o}\right|_{\text{stress}} \tag{9}$$

where v_s^o is the shear wave velocity in the absence of texture and stress and v_{ij} is the velocity of a wave propagating in the direction x_i and polarised in the direction x_j. From Equations (7) and (8)

$$\left.\frac{v_{31} - v_{32}}{v_s^o}\right|_{\text{texture}} = \frac{16\pi^2 c W_{420}}{7\sqrt{5}\mu} \tag{10}$$

the stress dependent term being given [2] by

$$\left.\frac{v_{31} - v_{32}}{v_s^o}\right|_{\text{stress}} = \left(1 + \frac{n}{4\mu}\right)(\varepsilon_{11} - \varepsilon_{22}) \ . \tag{11}$$

Here ε_{ij} are components of the infinitesimal strain tensor, $(\varepsilon_{11} - \varepsilon_{22})$ being the difference in principal strains in the plane of the plate.

It is therefore not possible to obtain the difference in the principal strains $(\varepsilon_{11} - \varepsilon_{22})$ in the plane of the plate without a separate determination of the texture parameter W_{420}. The aim of the velocity combination approach is to use the additional information provided by the compressional phase delay to separate the effects of texture and stress in the birefringence measurement [12].

If the sample thickness can be measured, the three velocities v_{31}, v_{32} and v_{33} can be calculated from the measured times of flight. To first order in the texture and stress

$$\rho_o \Sigma v^2 = \rho_o (v_{31}^2 + v_{32}^2 + v_{33}^2) = (\lambda + 4\mu) + P(T_1 + T_2)/3K_o \tag{12}$$

Here $K_o = \lambda + 2\mu/3$ is the bulk modulus and ρ_o is the density of the material in the unstressed state, and

$$P = (2\ell + 2m + 2\mu - 5\lambda) - (\lambda n + 8\lambda m + 8\lambda^2 + 2\mu n)/4\mu \tag{13}$$

It is assumed that the principal stress directions coincide with the principal texture axes $0x_1x_2x_3$, and that the stress component in the x_3 direction is zero, the components along $0x_1$ and $0x_2$ being denoted by T_1 and T_2. Equation (12) is independent of the texture or crystallographic alignment and may therefore be used for a texture independent determination of the sum of the principal stresses $(T_1 + T_2)$. Using the elastic constants of mild steel measured by Sayers and Allen [1], Equation (13) gives $P = -55.38 \times 10^{10}$ Pa. For an applied stress $(T_1 + T_2) = 100$ MPa Equation (12) gives a change in $\rho_o \Sigma v^2$ of 0.0246%.

Although Equation (12) is independent of texture, it is often difficult to obtain an accurate measurement of the path length in practice. A second texture independent combination of the ultrasonic velocities would allow the thickness to be eliminated, but depends on some knowledge of the texture. For example if $W_{400} = 0$, $\rho_o v_{33}^2$ and $\rho_o (v_{31}^2 + v_{32}^2)$ are independent of texture and give the sum of principal stresses $(T_1 + T_2)$. For fibre symmetry about $0x_1$, $W_{420} = -\sqrt{10}\, W_{400}/3$, independently of the magnitude of the texture. $\rho_o (v_{31}^2 + 4v_{32}^2)$ is therefore independent of texture to first order. If $W_{420} = A\, W_{400}$ so that the texture coefficients W_{420} and W_{400} vary in a proportionate manner then $v_{33}^2/\Sigma v^2$ will vary linearly with $(v_{31}^2 - v_{32}^2)/\Sigma v^2$:

$$\frac{v_{33}^2}{\Sigma v^2} = \frac{(K + 4\mu/3)}{(K + 10\mu/3)} + \frac{1}{A}\sqrt{2/5}\, \frac{(v_{31}^2 - v_{32}^2)}{\Sigma v^2} \tag{14}$$

in the absence of stress [12]. Deviations from this line can then be used to evaluate the level of stress present. Figure 3a shows a plot of Equation (14) for the case of steel with fibre texture, the propagation direction being perpendicular to the fibre axis $0x_1$. The effect of an applied stress of 300 MPa is also shown for a point with texture induced birefringence of 0.2%. Measurements of the three times of flight on a large number of steel samples [12] showed the ratio W_{420}/W_{400} to vary from sample to sample. This method is therefore most suitable for determining the variation of stress across a sample or between specimens made by the same process for which the assumption $W_{420} = A\, W_{400}$ is valid. Figure 3b shows a velocity combinations plot for a scan near the crack in a 19 mm thick compact tension specimen of 22 NiMoCr 3.7 steel with small anisotropy [12]. Also shown in the figure are the displacements associated with different types of stress of 3×10^8 Pa in magnitude, calculated from the measured elastic constants *lamda*,μ,ℓ,m, and n [1]. This provides a calibration for the determination of the stress present. Figure 4a shows a schematic diagram of the sample. Taking account of the measurement error shown, the distribution of data points suggests that the main component of stress lies in a direction perpendicular to the line of the crack. This information is unobtainable with the birefringence measurement alone, since the birefringence can only give the difference in principal stresses. The velocity combinations plot can be considered to provide a two-dimensional map of the stress and texture states in the sample, whereas the birefringence is a projection of this map onto one dimension with a resultant loss of information. The calculated stress distribution is shown in Figure 4b and is in reasonable agreement with theory [13].

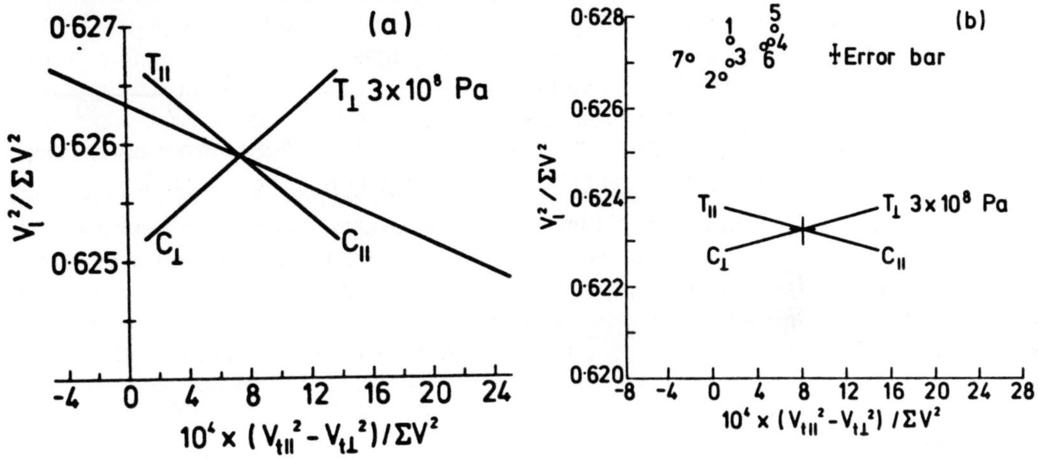

Figure 3: *(a) Velocity combinations plot for steel with fibre texture, the propagation direction being perpendicular to the fibre direction* $0x_1$. *The effect of an applied stress of 300 MPa is shown for a point with a stress free birefringence of 0.2% T - tensile stress; C - compressive stress; // - parallel to rolling to rolling direction;* ⊥ *- perpendicular to the rolling direction. (b) Velocity combinations plot for the seven measurement positions on the compact tension specimen shown in Figure 4a [12]. The effect of an applied stress of 300 MPa on a point + is indicated.*

3.2. *The SH Wave Method*

In the absence of stress it is seen from Equation (3) that an ultrasonic *SH* wave propagating in the plane of the plate has the same velocity for propagation parallel ($\theta=0°$) and perpendicular ($\theta=90°$) to the rolling direction. This corresponds to an interchange of the propagation and polarisation directions of the wave. For an orthorhombic material in the presence of stress with components σ_{ij}

$$\rho(v_{ij}^2 - v_{ji}^2) = \sigma_{ii} - \sigma_{jj} \tag{15}$$

Here ρ is the density and v_{ij} is the phase velocity of a shear wave propagating in the direction x_i and polarised in the direction x_j, x_i and x_j being the principle axes of the elastic constant tensor [14-16]. It is assumed that the principal stresses are

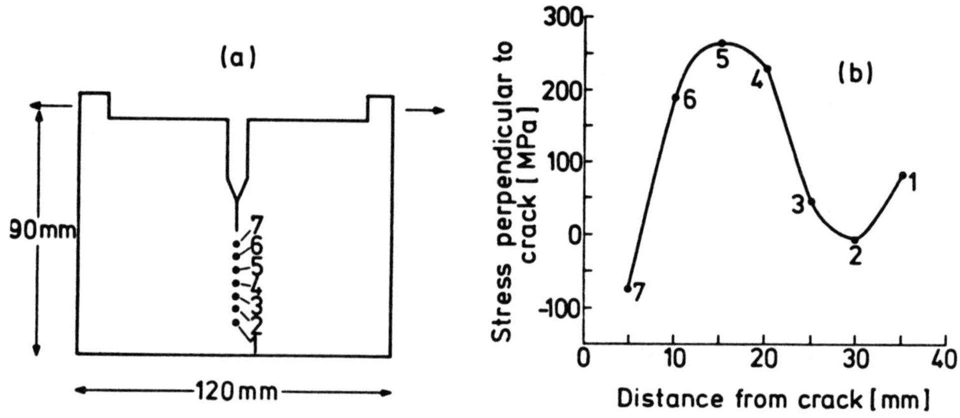

Figure 4: *(a) Schematic diagram of the compact tension specimen of 22 NiMoCr 3.7 steel [12,13] with the seven measurement points indicated. (b) Plot of residual stress as a function of distance from the crack tip for the compact tension specimen deduced from the velocity combinations plot [12].*

parallel to the principal texture axes. This relationship is independent of the magnitude of the texture of the material. Its use for separating texture and stress was suggested by MacDonald [14] following earlier work of Biot [15] and Thurston [16].

A disadvantage of the method is that propagation of waves along the stress axis is necessary. However Thomson et al [17-19] have overcome this problem for thin plates by using electromagnetic acoustic transducers (EMATs) to excite the fundamental horizontally polarised shear (*SH*) wave of the plate travelling in the plane of the plate. The velocity of this mode is frequency independent, and is identical to the *SH* wave velocity in an unbounded medium with the same texture and stress distribution. Measurements on 6061-T6 aluminum, 304 stainless steel and commercially pure copper plate gave essentially equal velocities for propagation along the rolling and transverse directions as predicted by theory. The stress predicted using Equation (15) was found to be in good agreement with the uniaxial tensile stress applied. Measurements on a rolled titanium sample did not show the expected behaviour, and this may be due to a failure of the assumption of orthorhombic symmetry for this sample.

Allen and Langman [20,21] generated the zero-order *SH* plate wave using a conventional piezoelectric shear wave transducer aligned to produce horizontally polarised waves. Two EMATs mounted in a rigid base were used as receivers. The three transducers lie along the same axis and the phase delay associated with the propagation of the elastic wave between the two EMATs arising from a stimulus of the piezoelectric transducer is measured. Their results are in good agreement with theory at zero stress as may be seen in Table 1 which summarises the results obtained on all the samples measured. The angular dependence of the *SH* wave velocity measured on four of the plates is shown in Figure 5. Deviations from the theoretical curves shown in Figure 5 occur because in anisotropic media the phase and group velocities are not parallel, this phenomenon being referred to as beam skewing. The effect of this on the angular variation of the *SH* wave velocity has been discussed elsewhere [22].

Table 1: Comparisons between ultrasonic phase delays for various samples of thin sheet measured by Allen and Langman [13,14]. t_θ *is the phase delay for propagation of SH waves at an angle* $\theta°$ *to the rolling direction, and* t_{av} *is the average of* t_o *and* t_{90}.

Material	Sample number	Thickness/mm	$\frac{t_o - t_{90}}{t_{av}} \times 10^4$	$\frac{t_{45} - t_{av}}{t_{av}} \times 10^4$
Aluminum	1	1.60	1.8	374
	2	1.20	-1.3	-219
	3	1.00	-2.7	-335
	4	1.00	0.9	-231
	5	0.70	0.0	122
	6	0.50	0.4	-154
Stainless	1	1.19	-4.9	-331
Steel	2	0.91	-1.3	-537
	3	0.56	0.0	-733
Copper	1	1.17	-1.3	595
	2	0.98	-4.4	870
	3	0.60	-1.4	1684
Brass	1	1.00	0.6	-369

3.3. *Rayleigh Wave Method*

Let v_{Ri} denote the velocity of the Rayleigh wave propagating on the x_1x_2 surface in the direction x_i. In the presence of texture and stress

$$\frac{\Delta v_R}{v_R^o} \equiv \frac{v_{R1} - v_{R2}}{v_R^o} = \left.\frac{(v_{R1} - v_{R2})}{v_R^o}\right|_{\text{texture}} + \left.\frac{(v_{R1} - v_{R2})}{v_R^o}\right|_{\text{stress}} \tag{16}$$

where, from Equation (2)

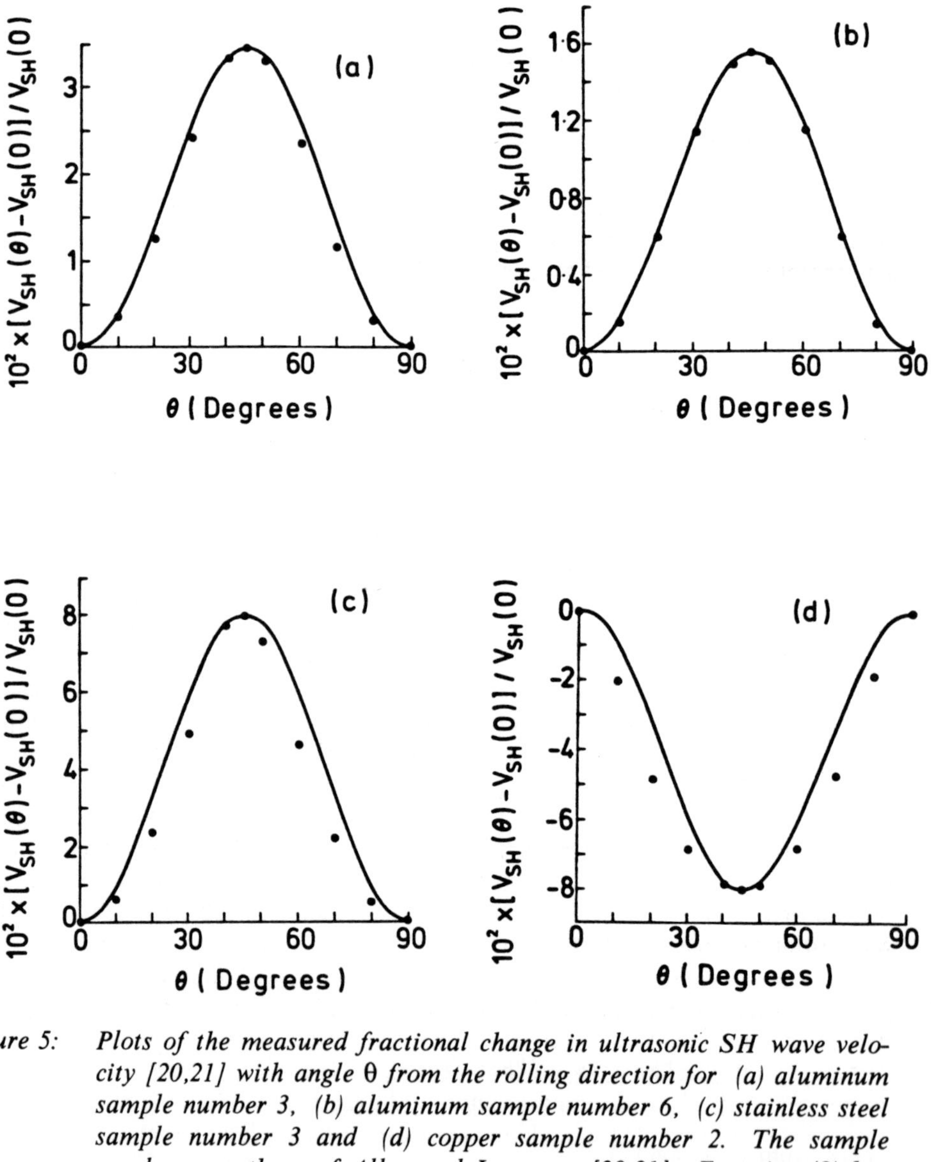

Figure 5: *Plots of the measured fractional change in ultrasonic SH wave velocity [20,21] with angle* θ *from the rolling direction for (a) aluminum sample number 3, (b) aluminum sample number 6, (c) stainless steel sample number 3 and (d) copper sample number 2. The sample numbers are those of Allen and Langman [20,21]. Equation (3) has been plotted for comparison by fitting to the experimental values at* $\theta=0°$, $45°$ *and* $90°$.

$$\left.\frac{v_{R1}-v_{R2}}{v_R^o}\right|_{\text{texture}} = \frac{16\pi^2 c\, W_{420}\beta}{7\sqrt{5}\mu} \tag{17}$$

β is a function of Poisson's ratio and is given in reference [11].

The stress dependent is given by

$$\left.\frac{v_{R1}-v_{R2}}{v_R^o}\right|_{\text{stress}} = \frac{\alpha_1}{2\alpha_o}(\varepsilon_{11}-\varepsilon_{22}) \tag{18}$$

α_o and α_1 being functions of λ, μ and the third elastic constants ℓ, m and n [23,24].

It is seen from Equations (10), (11), (17) and (18) that Equations (9) and (16) depend only on two unknowns W_{420} and $(\varepsilon_{11}-\varepsilon_{22})$. The texture parameter W_{420} may therefore be eliminated to give the difference in principal strains $(\varepsilon_{11}-\varepsilon_{22})$:

$$\varepsilon_{11}-\varepsilon_{22} = \frac{\Delta v_R/v_R^o - \beta\Delta v_s/v_s^o}{\alpha_1/2\alpha_o-(1+n/4\mu)\beta} \tag{19}$$

Equation (19) depends, for its validity, on the accuracy of Equation (2). This has been checked by determining the parameters W_{420} and W_{440} which determine the angular dependence of the Rayleigh wave velocity using bulk waves [25] and comparing the predicted variation with that measured. For a stress free plate W_{440} can be determined from the angular dependence of the *SH* wave velocity (Equation (3)), whilst the measured shear wave birefringence gives W_{420} (Equation (10)). Two samples were used. The first was a 99.5% pure aluminum plate (BS 1470/1050A) and the second an aluminum 4.5% Mg alloy plate (BS 1470/5083/0). Both plates were 50 mm thick. Figure 6 shows a fit of Equation (3) to the data for the aluminum and aluminum alloy plates giving $W_{440}=-0.00256$ for the pure aluminum plate and $W_{440}=-0.00306$ for the alloy. The measured shear wave birefringence for the two plates were $\Delta v_s/v_s^o=92.62\times10^{-4}$ and $\Delta v_s/v_s^o=70.32\times10^{-4}$ respectively giving $W_{420}=-0.002267$ for the pure aluminum plate and $W_{420}=-0.001721$ for the alloy. Figure 7 shows the prediction of Equation (2) using these values compared with the measured Rayleigh wave velocity made at several different frequencies [25]. The main angular dependent term is seen to arise from the term R_2 involving W_{420}, which was obtained from the shear wave birefringence. Since this represents a through thickness average of the texture of the plate, the low frequency results should be in better agreement with the prediction of Equation (2), and this is verified by the results shown in Figure 7. Assuming the plates to contain zero stress, the results show that both plates have a surface texture different to the bulk average. However, in contrast to the pure aluminum plate which contains a texture gradient over a thin layer only, the aluminum alloy plate has a texture which varies over a longer length scale.

The theoretical model presented in section 2 was also checked by comparing the values of W_{420} and W_{440} obtained from ultrasonic bulk wave velocity measurements with values obtained by a neutron diffraction measurement of the texture in

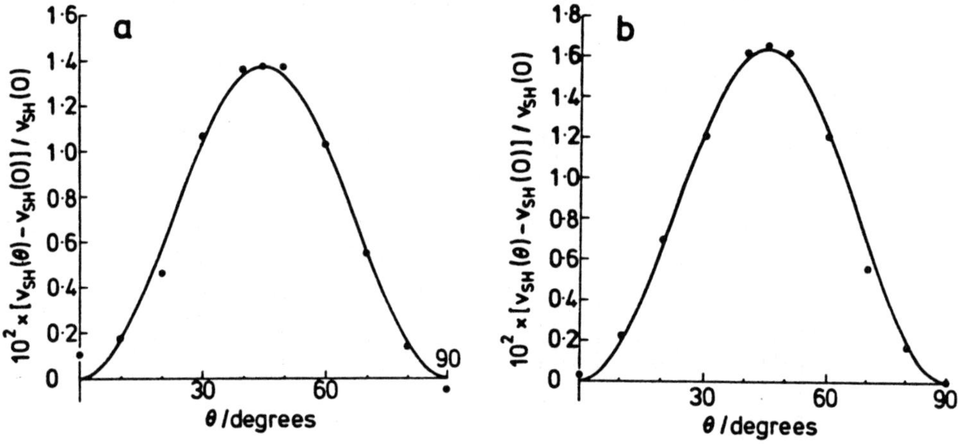

Figure 6: *Variations of* $(v_{SH}(\theta) - v_{SH}(0))/v_{SH}(0)$ *with angle* θ *from the rolling direction for (a) a 99.5% pure aluminum plate and (b) an aluminum 4.5% magnesium alloy [25].*

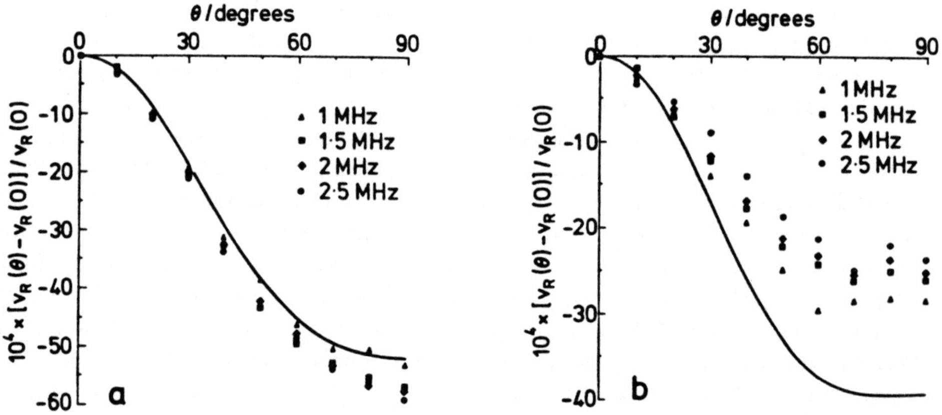

Figure 7: *Variation of* $(v_R(\theta) - v_R(0))/v_R(0)$ *with angle* θ *from the rolling direction for (a) a 99.5% pure aluminum plate and*
(b) an aluminum 4.5% magnesium alloy at several different frequencies [25]. The curve shows the prediction of Equation (2).

the sample. The values obtained from the (200) and (111) pole figures are in reasonable agreement with each other and with the values determined using ultrasonics as seen in Table 2. This confirms that the main effect of the texture on the ultrasonic velocities is due to the crystallographic alignment of the grains and not due to other factors such as any grain shape anisotropy, etc.

Table 2: Comparison of values of W_{400}, W_{420} and W_{440} determined by ultrasonics with those determined by neutron diffraction.

99.5% pure Aℓ	W_{400}	W_{420}	W_{440}
Ultrasonics	-	-0.002267	-0.002560
(200)	0.005671	-0.0025925	-0.0027206
(111)	0.0060636	-0.0024745	-0.0026664
Aℓ 4.5% Mg	W_{400}	W_{420}	W_{440}
Ultrasonics	-	-0.001721	-0.003060
(200)	0.0073897	-0.0015506	-0.002420
(111)	0.0072983	-0.0016496	-0.0024196

4. Conclusion

In this paper the theoretical basis for treating the propagation of ultrasound in textured polycrystalline metals has been outlined. It has been demonstrated how quantitative information concerning the crystallite orientation distribution function can be obtained from the angular variation of ultrasonic velocities. Three methods for the texture independent measurement of stress in anisotropic polycrystalline metals have been described. Of these the Rayleigh wave method is capable of determining the variation of stress with depth by using the frequency dependence of the Rayleigh wave velocity. However, in the presence of a depth dependent texture, Equation (19) is only approximately texture independent. As a result the combination of the Rayleigh wave velocity with that of either a skimming SV or longitudinal wave would be more satisfactory (Equations (4) and (5)). The velocities of these waves need to be measured at a frequency at which the waves average over the depth in the same way as the Rayleigh wave. This should allow the depth dependence of both texture and stress to be obtained from the frequency dependence of the Rayleigh wave velocity.

5. References

[1] Sayers, C.M. and Allen, D.R., *J. Phys. D, 17*, 1984, pp. 1399.

[2] Hsu, N.N., *Exp. Mech., 14*, 1974, pp. 169.

[3] Roe, R.J. and Krigbaum, W.R., *J. Chem. Phys., 40*, 1964, pp. 2608.

[4] Roe, R.J., *J. Appl. Phys., 36*, 1965, pp. 2024.

[5] Roe, R.J., *J. Appl. Phys., 37*, 1966, pp. 2069.

[6] Bunge, H.J., *Z. Metallk., 56*, 1965, pp. 872.

[7] Bunge, H.J., *Krist. Tech., 3*, 1968, pp. 431.

[8] Morris, P.R., *J. Appl. Phys., 40*, 1969, pp. 447.

[9] Sayers, C.M., *J. Phys. D., 15*, 1982, pp. 2157.

[10] Allen, A.J., Hutchings, M.T., Sayers, C.M., Allen, D.R. and Smith, R.L., *J. Appl. Phys., 54*, 1983, pp. 555.

[11] Sayers, C.M., *Proc. Roy. Soc. A, 400*, 1985, pp. 175.

[12] Allen, D.R. and Sayers, C.M., *Ultrasonics, 22*, 1984, pp. 179.

[13] Schneider, E. and Goebbels, K., Periodic Inspection of Pressurised Components. *Conference at Inst. Mech. Engs. London*, 1982.

[14] MacDonald, D.E., *IEEE Trans. on Sonics and Ultrasonics, SU-28*, 1981, pp. 75.

[15] Biot, M.A., *J. Appl. Phys., 11*, 1940, pp. 522.

[16] Thurston, R.N., *J. Acoust. Soc. Am., 37*, 1965, pp. 348.

[17] Thompson, R.B., Smith, J.F. and Lee, S.S., in *'Review of Progress in Quantitative NDE 2'*, ed. D.O. Thompson and D.E. Chimenti, Plenum Press, New York, 1983.

[18] Thompson, R.B., Lee, S.S. and Smith, J.F., in *'Review of Progress in Quantitative NDE 3'*, ed. D.O. Thompson and D.E. Chimenti, Plenum Press, New York, 1984.

[19] Lee, S.S., Smith, J.F. and Thompson, R.B., in *'Review of Progress in Quantitative NDE 4'*, ed. D.O. Thompson and D.E. Chimenti, Plenum Press, New York, 1985.

[20] Allen, D.R. and Langman, R., *AERE-R11573, 1985.*

[21] Langman, R. and Allen, D.R., *AERE-R11598, 1985.*

[22] Allen, D.R., Langman, R. and Sayers, C.M., *Ultrasonics, 23*, 1985, pp. 215.

[23] Hayes, M. and Rivlin, R.S., *Arch. Rat. Mech. Anal., 8*, 1961, pp. 358.

[24] Hirao, M., Fukuoka, H. and Hori, K., *J. Appl. Mech., 48*, 1981, pp. 119.

[25] Sayers, C.M., Allen, D.R., Haines, G.E. and Proudfoot, G.G., *Phil. Trans. Roy. Soc.*, 1986.

CONTRIBUTED PAPERS

ANALYTICAL CONSIDERATIONS FOR INELASTIC ZONE AND CRACK GROWTH CHARACTERIZATIONS

L.M. Brock
Engineering Mechanics
University of Kentucky
Lexington, Kentucky

Abstract

To illustrate how the characteristics of cracks and crack edge inelastic zones are manifested in the dynamic signals radiated during fracture, results for two idealized situations are presented. In the first situation, the inelastic zone is viewed as distributions of screw dislocations, and one such distribution growing from a crack edge prior to crack propagation is considered. In the second, the zone is viewed as a screw dislocation array, and a single dislocation moving near a running crack is treated.

The results for both situations indicate that the crack/zone characteristic representations are simple in form. The results also indicate that, in terms of amplitude, the crack/zone intraction effects probably constitute an important part of the fracture process signal.

1. Introduction

The use of dynamic field measurements to study fracture is important both for structural integrity evaluation [1] and for the development of fracture theory itself [2]. Interpretation of these measurements is generally, e.g. [3], based on the assumption that the cracked material is linearly elastic. However, observations indicate that the material in zones around the crack edges themselves is inelastic. Moreover, it is known that these zones can be an important feature of the fracture process [4,5]. It is worthwhile, therefore, to determine the extent to which inelastic zone characteristics can be discerned in the dynamic signals radiated by the fracture process.

In making this determination, however, expressions for the signals themselves must be obtained, and studied to see just how crack/zone details are manifested. As a first-step contribution to this endeavor, we here present results for two idealized situations involving cracks with inelastic edge zones in unbounded solids. The first situation considers zone-crack interaction prior to crack propagation, the second involves interaction after the crack begins to run. In both situations, the solids are linearly elastic, isotropic and homogeneous and, following [6-8], dislocations are used to represent the zones.

2. Dislocation Distribution Growth from a Stationary Crack

From Taylor's model [6] of plastic flow, Bilby et al. [7] argue that an inelastic zone near a stationary crack be represented by a distribution of dislocations growing from the edge. As a simple example of this idea, we here consider the situation illustrated in Figure 1: A semi-infinite crack defined by the Cartesian coordinates (x_1,x_2) as $x_2=0$, $x_1<0$ is initially at rest in an unbounded plane. For $s>0$, where $s=(\text{time})\times(\text{rotational wave speed})$, a distribution of screw dislocations grows out from the crack edge along a continuous contour $x_2=C(x_1)$.

As seen in Figure 1, $\mathbf{x}=(x_1,x_2)$ is the position vector of any point in the plane, and $\mathbf{X}(S)=(X_1,X_2)$ is the position vector of those points on the contour, where S is the contour arc length measured from the crack edge. In terms of this variable, $S=h(s)$ defines the distribution length, and $W(S,s)$ defines the distribution strength. That is, W is the discontinuity in the out-of-plane displacement as the contour is crossed from its (-) side to its (+) side. In Figure 1, the ($\pm$) sides are the upper and lower sides, respectively.

The distance parameter h is continuous and single-valued, with $h(0)=0$ and $0\leq\dot{h}<1$, $s>0$, where $(\dot{\ })\equiv\partial(\ )/\partial s$. The distribution function W is bounded, continuous and piecewise smooth for all $s\geq0$, $0\leq S\leq h(s)$. Other than those physically-motivated restrictions, W and h are arbitrary.

The governing equations for this situation are not trivial, but partial transient solutions have been obtained in [9]. In particular, only the out-of-plane displacement $w(\mathbf{x},s)$ exits and has the form

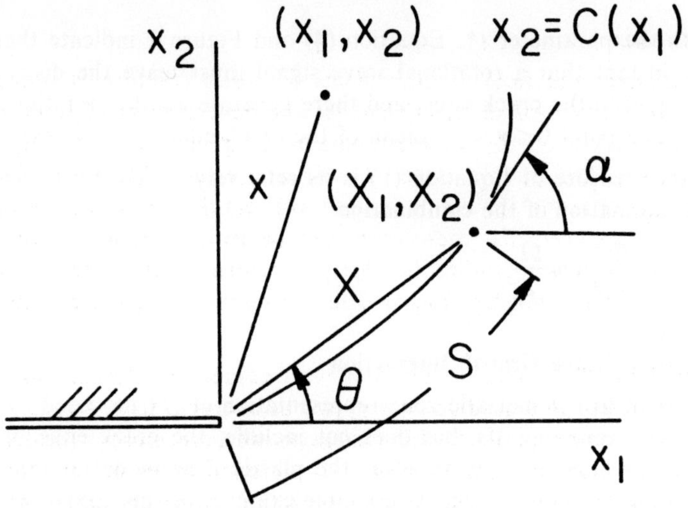

Figure 1: Dislocation distribution and crack geometry

$$2\pi w = \int_x^s \frac{du}{\sqrt{(u-x_1)}} \left[\frac{W^*}{\sqrt{(2R)^*}} \cos\frac{\Omega^*}{2} + \int_0^{h^*} W_s(S, s-u-X) \frac{1}{\sqrt{(2R)}} \cos\frac{\Omega}{2} dS \right] \quad (1)$$

where

$$x = |\mathbf{x}|, \; X = |\mathbf{X}|, \; W_S = \frac{\partial W}{\partial S} - \dot{W}\cos(\theta-\alpha) \quad (2)$$

$$X_1 - \frac{1}{2}\frac{u^2-x^2}{u-X_1} = R\cos\Omega, \; X_2 = R\sin\Omega \quad (3a,b)$$

and ( )* denotes a quantity evaluated at $S=h(t^*)$, $s=t^*$. The angles (θ,α) are shown in Figure 1, while the parameter t^* is defined by

$$t^* = s - u - X[h(t^*)] \quad (4)$$

Equation (1) is valid if $1+X'(h)\dot{h} \geq 0$, where ( )′ denotes differentiation with respect to the argument. This restriction can only be violated if the edge of the dislocation distribution nears the crack edge at a rate which exceeds the rotational wave speed. Although no restriction on the contour function C precludes this, the restrictions on $\dot{h}$ imposed above do.

Equation (1) shows that the signal radiated by the crack-distribution interaction process contains two components. The first component disappears if $W(h,s)=0$; that is, the displacement discontinuity vanishes continuously at the distribution edge. The second component vanishes if the distribution is uniform; that is, the discontinuity is constant along the contour and does not vary with time.

From a signal characterization standpoint, Equation (1) shows that the period s after the dislocation distribution begins to grow and the distance x from the crack edge to the observation point are the most obvious independent variables, along with the parameter t^*. Equation (4) and Figure 1 indicate that this parameter is the instant that a rotational wave signal must leave the distribution edge in order to travel to the crack edge, and there generate another rotational signal which then reaches a point $0<u<x_1$ ahead of the crack edge at the instant s.

Another feature of Equation (1) is its relatively simple form; essentially it is a weighted summation of the combination

$$\frac{W}{\sqrt{2R}}\cos\frac{\Omega}{2} \tag{5}$$

3. Dislocation-Crack Growth Interaction

An alternative to the inelastic zone representation of [7] proposed by Hart [8] suggests that the zone engulfs, but does not include, the crack edge itself. Thus, the dislocations arising in the zone, after the plastic flow model of [6], exist at finite distances from the crack edge. As a simple example of this model, we here consider the situation illustrated in Figure 2: A semi-infinite crack, again defined as $x_2=0$, $x_1<0$, is initially in equilibrium with a single screw dislocation of strength $b>0$, i.e. the Burgers vector is out-of-plane with magnitude b. This dislocation lies a distance d from the crack edge. As another variation on the previous example, we let this crack grow in its own plane for $s>0$. The rotational wave generated by this growth reaches the dislocation at $s=d$ and we assume that the dislocation begins to move at some subsequent $s=s_o>d$.

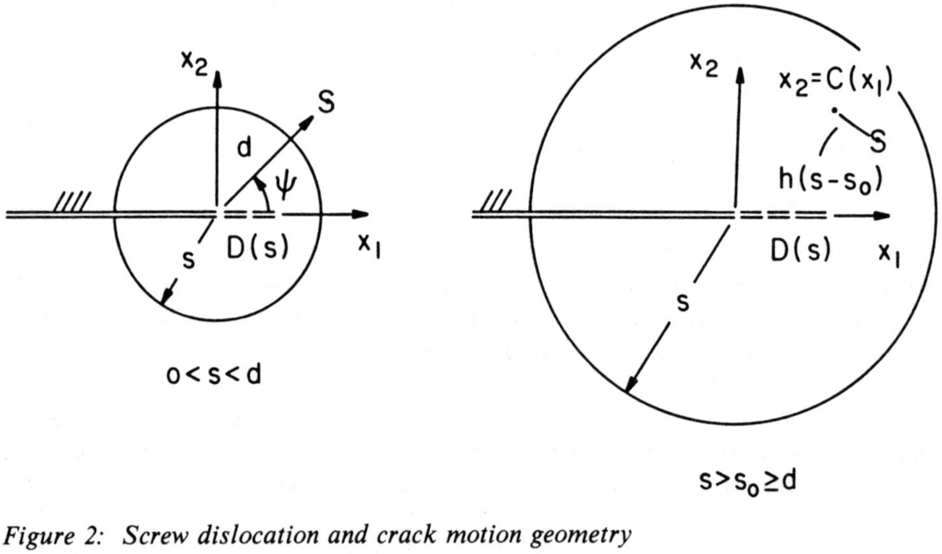

Figure 2: Screw dislocation and crack motion geometry

As shown in Figure 2, the dislocation moves along a continuous contour $x_2=C(x_1)$ a distance $S=h(s-s_o)$. Here h is continuous and single-valued, with $h(s_o)=0$ and $0\leq\dot{h}<1$, $s\geq s_o$. The crack extension function $D(s)$ shown in Figure 2 satisfies analogous requirements $D(0)=0$, $0\leq\dot{D}<1$, $s\geq 0$, and must also be continuous and single-valued.

The non-trivial governing equations for this problem have been treated in [10], and a partial transient solution obtained. This shows that only the out-of-plane displacement exists, and has two components: one component is the equilibrium solution, and is, therefore, a time-invariant background upon which the dislocation-crack growth interaction process is superimposed. The second component is this signal, and has the form

$$2\pi w=b\int_{u_1}^{u_2}\frac{1}{\sqrt{(2R)^*}}\sqrt{(R_x/R_u)^*}\cos\frac{1}{2}(\Omega+\Omega_u-\Omega_x)^*\frac{du}{\sqrt{u-x_1}} \tag{6}$$

where ( )* again denotes a quantity evaluated at $S=h(t^*)$, $s=t^*$ and

$$t^*=s-u+D(t_x)-R_x^*,\quad t_x=s+x_1-D(t_x) \tag{7}$$

The polar coordinats (R,Ω) in Equation (6) are again defined in Equation (3), while (R_x,Ω_x) and (R_u,Ω_u) follow simply by replacing the second term on the left-hand side of Equation (3a) with $D(t_x)$ and $D(t_u)$, respectively, where

$$t_u=s+u-D(t_u) \tag{8}$$

Finally, the integration limits in Equation (6) are obtained from the equations

$$u_2=s-\sqrt{2}w_2,\ \sqrt{2}w_2[d\cos\psi+s_o-\sqrt{2}N(w_2)]=s_o^2-d^2+(d\cos\psi-s_o)\sqrt{2}N(w_2) \tag{9}$$

$$u_1=s-\sqrt{2}w_1,=\sqrt{2}w_1[\sqrt{2}N(w_1)-s-x_1]=x^2-s^2+(s-x_1)\sqrt{2}N(w_1) \tag{10}$$

where the function N follows from the implicit relation

$$N(u)=u+\sqrt{2}D\left[\frac{N(u)+u}{\sqrt{2}}\right] \tag{11}$$

As in the previous example, the restriction $1+X'(h)\dot{h}\geq 0$ must hold.

Equation (6) shows that the signal received from the dislocation-crack growth interaction process depends on process parameters in a more subtle way than in the previous example, chiefly through the polar coordinate sets (R,Ω), (R_u,Ω_u), (R_x,Ω_x). Moreover, it is noted that three time parameters (T^*,t_x,t_u) appear, as opposed to only one in the previous example. Nevertheless, the essential form of the signal is again simple, and not too different from that given in Equation (5).

4. Signal Amplitude Considerations

Both examples considered here have solutions which are directly proportional to the magnitude of the dislocation or dislocation distribution. Observations [11] suggest that these are generally on the order of 2.5 angstroms; that is,

$$b, |W| \sim 2.5 \times 10^{-10} m \tag{12}$$

For purposes of comparison, consider the diffraction of a plane step-stress SH wave by the cracks in the two examples: Achenbach [12] has shown that, if the wavefront is parallel to the crack, then the out-of-plane displacement w in both cases is proportional to τ/μ, where μ is the shear modulus, and τ is the step-stress magnitude. For steels, a nominal value of this dimensionless ratio is

$$\tau/\mu \sim 10^{-4} \tag{13}$$

The orders-of-magnitude disparity between the quantities in Equations (12) and (13) suggest that the inelastic zone contribution to a signal received from a fracture process will be impossible to detect. However, the solutions presented here are for single dislocations and dislocation distributions. Since dislocation densities per grain in many crystalline solids can be on the order of 10^{10}, the numbers given above actually suggest that only a fraction of the dislocations in an inelastic zone need to move before their effect on a signal received from a fracture process becomes important.

5. Discussion

In the idealized examples considered here, the signals received from the crack-inelastic zone interaction component of a fracture process – both before and after crack extension begins – have simple forms. This suggests that the development of tractable analytical or numerical signal representations for more realistic cases may be feasible. Moreover, the results presented here indicate that the magnitudes of the contributions of the interaction component can be an important part of the total signal. These observations suggests that the inelasticity of fracture processes could be discerned by comparing measured signal responses with analytical/numerical signal models which, respectively, ignore, as in [3], inelastic zones, and models which, based on expressions such as those presented here, include crack-zone interaction effects.

Acknowledgement

This work was sponsored by NSF Grant MEA 8319605, and was conducted while the author was an ASEE/Navy Summer Faculty Research Associate at the Naval Research Laboratory in Washington, D.C.

Reference

[1] Achenbach, J.D., Pao. Y.H. and Tiersten, H.F., "Report on the Workshop on Application of Elastic Waves in Electrical Devices, Non-destructive Testing and Seismology", Northwestern University, Evanston, Illinois, 1976.

[2] Knauss, W.G., Ravi-Chandar, K. and Rosakis, A.J., "Workshop on Dynamic Fracture", California Institute of Technology, Pasadena, California, 1983.

[3] Angel, Y.C. and Achenbach, J.D., "Reflection and Transmission of Elastic Waves by a Periodic Array of Cracks", *ASME Journal of Applied Mechanics, 52*, 1985, 33-41 pp.

[4] Fu, L.S., et al., "Fundamentals of Microcrack Nucleation Mechanics", NASA Contractor Report 3851, January, 1985.

[5] Rosakis, A.J. and Freund, L.B., "Optimal Measurements of the Plastic Strain Concentration at the Crack Tip in a Ductile Steel Plate", *ASME Journal of Engineering Materials and Technology, 104*, 1982, 115-120 pp.

[6] Taylor, G.I., "The Mechanism of Plastic Deformation of Crystals", *Proceedings of the Royal Society, A145*, London, 1934, 362-415 pp.

[7] Bilby, B.A., Cottrell, A.H. and Swinden, K.H., "The Spread of Plastic Yield from a Notch", *Proceedings of the Royal Society, A272*, London, 1963, 304-314 pp.

[8] Hart, E.W., "A Theory for Stable Crack Growth Extension Rates in Ductile Materials", *International Journal of Solids and Structures, 16*, 1980, 807-823 pp.

[9] Brock, L.M., "Transient Solutions for Arbitrary Growth of Screw Dislocation Near a Crack", University of Kentucky College of Engineering Technical Report, October, 1985.

[10] Brock, L.M., "Transient Solution for Arbitrary Screw Disloction Motion Near a Running Crack," University of Kentucky College of Engineering Technical Report, December, 1985.

[11] Guy, A.G., *Elements of Physical Metallurgy*, (2nd Ed.), Addison-Wesley, Reading, Maryland, 1962.

[12] Achenbach, J.D., "Extension of a Crack by a Shear Wave", *Zeitschrift für angewandte Mathematik und Physik, 21*, 1970, 887-900 pp.

A COMPARISON OF SEVERAL ULTRASONIC TECHNIQUES FOR ABSOLUTE STRESS DETERMINATION IN THE PRESENCE OF TEXTURE*

A.V. Clark and **J.C. Moulder**
Fracture and Deformation Division
National Bureau of Standards
Boulder, Colorado

R.B. Mignogna and **P.P. DelSanto**
Naval Research Laboratory
Washington, D.C.

Abstract

Most ultrasonic stress measurement techniques measure some combination of stresses, such as the sum or difference of principal stresses, relative to some (possibly unknown) stress state. In order to apply fracture mechanics methods to failure/life assessment of flawed structures, stresses must be determined absolutely.

Ultrasonic stress measurement methods rely on the acoustoelastic effect: the stress-induced change in phase velocity. Unfortunately, texture also has a comparable effect. This has limited most applications to measurement of applied (as opposed to residual) stress.

We have considered several possible methods of absolute stress determination by ultrasonic means. In some instances, combinations of stresses can be measured and then used in conjunction with other information (e.g. stress equilibrium equations) for absolute stress determination. Some of these techniques are used in photoelasticity; their extension to acoustoelasticity is made more difficult by the influence of texture. Elimination of texture places resolution requirements on velocity measurements which vary from technique to technique. We consider several methods of eliminating texture and develop criteria which must be met to achieve a desired (absolute) stress resolution. Illustrations are given for several techniques by using data from acoustoelastic experiments.

1. Introduction

We consider several ultrasonic methods of making absolute determination of components of plane stress in the presence of texture. These methods are based on the acoustoelastic effect: the presence of stress induces small changes in phase velocity.

For most engineering applications, it is sufficient to consider structural elements such as rolled plates of common aluminum and steel alloys as isotropic. However, the acoustoelastic effect is quite small, requiring second-order effects such as texture to be considered.

Consider for instance, the case of a rolled plate of 2024-T351 aluminum. The phase velocity of a shear-horizontal wave (at normal incidence) in the unstressed state is greatest when polarized along the rolling direction and least along the transverse direction. The acoustic birefringence (normalized difference in phase velocity of orthogonally polarized shear-horizontal waves) for a 9-mm thick rolled plate of 2024-T351 aluminum is about 3.5×10^{-3} [1]. This unstressed birefringence is due to a slight degree of preferential orientation of the crystallites that make up the plate. The crystals have preferred orientation due to the rolling process. The corresponding stress-acoustic constant, C_A, is about 3.5×10^{-5}/MPa. (The stress-acoustic constant is a measure of the material's sensitivity to the acoustoelastic effect.) If the unstressed birefringence, B_o, is ignored, an error in stress of B_o / C_A results; for this example, the stress error is about 100 MPa. This clearly illustrates the need to account for texture when making ultrasonic stress measurements.

Several methods of separating the effects of texture and stress have been proposed in the literature. We will first review the theory for each method and then compare the stress resolution for each method. These values are based either on our own data or on the published results of others. We do not espouse any particular method, nor do we claim that we have considered every possible method. In fact, the method of choice may depend upon the particular application, the transducers available, the stress resolution desired, and the available time or resources for making the measurements. Our purpose here is to point out, in a quantitative fashion, the resolution possible at present with each technique.

2. Theory

It is assumed that the stressed material can be idealized as slightly orthotropic; that is, any volume element has three orthogonal axes of two-fold symmetry. We call these axes the "material symmetry axes" and refer all stresses to a coordinate system oriented along these axes.

For a shear-horizontal (SH) wave propagating in the $x_2 - x_3$ material symmetry plane at an angle of incidence ϕ, the difference in arrival times between the stressed, anisotropic state and the unstressed state is given by [2]

$$\frac{t_1^s - t_1^u}{t_o} = \frac{d^s - d^u}{d_o} - [a(\phi)\sigma_1 + b(\phi)\sigma_2]$$
$$- (1/2\mu)(\sin^2\phi(C_{66}^s - C_{66}^u) + \cos^2\phi(C_{55}^s - C_{55}^u)). \quad (1a)$$

The superscripts s and u refer to stressed and unstressed states, respectively. For example, t_1^s is the arrival time for the SH-wave (polarized in the x_1-direction) in the stressed state and d^s is the corresponding acoustic path length. t_o is the average of t_1^s and t_1^u; $a(\phi)$ and $b(\phi)$ are stress-acoustic constants which depend upon the second- and third-order elastic moduli of the material. σ_1 and σ_2 are the principal stresses, referred to the in-plane material symmetry axes x_1 and x_2 (x_3 is in the thickness direction). μ is the isotropic shear modulus, defined as the average of the orthotropic second-order shear moduli: $\mu \equiv 1/2(C_{55} + C_{66})$.

t_1^s may be measured at a different location than t_1^u. Consequently, quantities such as C_{55}^s may differ from C_{55}^u, due to inhomogeneity. In fact, we will later show that in many cases the ability to remove texture depends critically on the degree of homogeneity.

The difference in arrival times for SH-waves polarized in the $x_1 - x_3$ plane of material symmetry is

$$(\Delta t_2/t_o) = \Delta d/d - [a(\phi)\sigma_2 + b(\phi)\sigma_1] - (1/2\mu)[\Delta C_{66}\sin^2\phi - \Delta C_{55}\cos^2\phi], \quad (1b)$$

where we denote $\Delta d/d = (d^s - d^u)/d_o$ and $\Delta C_{66} = C_{66}^s - C_{66}^u$, etc. There are 6 unknowns in Equations (1a) and (1b): σ_1, σ_2, $\Delta d/d$, $\Delta C_{ii}/2\mu$ where $i = 4,5,6$.

We now consider some special cases. In the first case, we can measure t_1 and t_2 before stress is applied. If no change in microstructure occurs, then $\Delta C_{ii} = 0$. If stresses are in the elastic range, then $\Delta d/d = -(\nu/E)(\sigma_1 + \sigma_2)$. Let $\bar{a} = a + \nu/E$, $\bar{b} = b + \nu/E$ where ν and E are Poisson's ratio and Young's modulus. From Equations (1a) and (1b),

$$\Delta t_1/t_o = -[\bar{a}\sigma_1 + \bar{b}\sigma_2]; \; \Delta t_2/t_o = -[\bar{a}\sigma_2 + \bar{b}\sigma_1], \quad (2)$$

which can be solved for σ_1 and σ_2.

In the second case, we must measure the stress in-situ, but we have a homogeneous material ($C_{ii}^s = C_{ii}^u$). Then $\Delta d/d$ is an unknown, and Equations (1a) and (1b) give

$$\Delta t_1/t_o = \Delta d/d - [a\sigma_1 + b\sigma_2];\ \Delta t_2/t_o = \Delta d/d - [\bar{a}\sigma_2 + \bar{b}\sigma_1]. \tag{3}$$

If the longitudinal wave speed is measured at a stressed and an unstressed location, we have the additional information:

$$\Delta t_p/t_p = \Delta d/d - M(\sigma_1 + \sigma_2). \tag{4}$$

Here t_p is $1/2(t_p^s + t_p^u)$ and M is the stress-acoustic constant for longitudinal waves. We now have three equations in three unknowns, so we can determine both σ_1 and σ_2.

In the special third case, we have axial symmetry; from stress equilibrium in polar coordinates

$$d\sigma_r/dr = (\sigma_\theta - \sigma_r)/r. \tag{5}$$

σ_r and σ_θ are radial and hoop stresses, respectively, and r is the radial coordinate. We obtain $\sigma_\theta - \sigma_r$ from the difference of Equations (1a) and (1b):

$$(t_r - t_\theta)/t_o = B_o \cos^2\phi + (\bar{a} - \bar{b})(\sigma_\theta - \sigma_r), \tag{6}$$

where the unstressed birefringence, B_o, equals $(t_r^u - t_\theta^u)/t_o$. If B_o can be determined, then $\sigma_\theta - \sigma_r$ can be found and substituted into stress-Equilibrium Equation (5) to determine σ_r (by integration); σ_r can then be added to the measured value of $\sigma_\theta - \sigma_r$ to obtain σ_θ.

There are (at least) two possible methods to determine B_o. The first, due to King and Fortunko [3], uses off-axis SH-waves at two different angles of incidence. Since the acoustoelastic constant $\bar{a} - \bar{b} = m$ does not depend upon $\cos^2\phi$ (whereas the texture term $B_o \cos^2\phi$ does), we obtain 2 equations for B_o and $\sigma_\theta - \sigma_r$.

The second method uses the fact that the unstressed birefringence is dispersive. It has been shown that [4]

$$B_o = C_1 + C_2(k\hat{a})^2\ , \tag{7}$$

where k = wave number and $\hat{a}$ = average grain size. Combining Equations (6) and (7) gives

$$B = [C_1 + m(\sigma_\theta - \sigma_r)] + C_2(k\hat{a})^2\ , \tag{8}$$

where the first term (in braces) is independent of frequency. The effect of texture and stress can be separated by measuring B_o as a function of frequency, and obtaining slope C_2 and intercept, C_1.

Another special case occurs when the principal stress direction is not coincident with the material symmetry axes. The pure-mode SH-wave polarization directions (acoustic axes) are at an angle ψ to the material axes; ψ is given by [5]

$$\tan 2\psi = \frac{2m\sigma_{12}}{B_o + m(\sigma_{11} - \sigma_{22})} \,, \tag{9}$$

where σ_{11} and σ_{22} are normal stresses and σ_{12} is the shear stress. The birefringence is now the normalized difference of arrival times of SH-waves polarized along the acoustic axes, and is related to stresses by [5]

$$B^2 = [B_o + m(\sigma_{11} - \sigma_{22})]^2 + (2m\sigma_{12})^2 \ . \tag{10}$$

From the above two equations,

$$\sigma_{12} = \frac{B \sin 2\psi}{2m} \ ; \tag{11}$$

the effect of texture has been removed by combining (9) and (10).

To calculate the normal stresses we use the stress-equilibrium equations

$$\partial\sigma_{11}/\partial x_1 = -\partial\sigma_{12}/\partial x_2 \ ; \quad \partial\sigma_{22}/\partial x_2 = -\partial\sigma_{12}/\partial x_1 \ . \tag{12}$$

The normal stresses can be obtained by measuring the shear stress gradient and integrating (12). Alternatively, we can fit a series of functions to our shear stress data, determine the coefficients which give the best fit, and then substitute the series into Equations (12) and integrate.

The last case we consider is the propagation of Rayleigh waves at an angle α to one of the in-plane material symmetry directions. The difference ΔV_R between velocity V_R^s in the stressed, anisotropic case and velocity V_R^u in the unstressed, isotropic case is [6]

$$(V_R^s - V_R^u)/V_R = q_1(\alpha) + (q_2 + q_3\cos 2\alpha)\sigma_1 + (q_2 - q_3\cos 2\alpha)\sigma_2 \ . \tag{13}$$

Here we assume that the principal stress and material symmetry axes coincide.

Suppose V_R^u is measured on an unstressed plate of the same (anisotropic) material as the stressed plate. Let $\Delta V_R = V_R^S - V_R^u$, $q^+ = q_2 + q_3$, $q^- = q_2 - q_3$. For Rayleigh waves propagating at $\alpha = 0°$, 90°, we have

$$(\Delta V_R/V_R)_{0°} = \Delta q_1(0°) + q^+\sigma_1 + q^-\sigma_2 \ ; \ (\Delta V_R/V_R)_{90°} = \Delta q_1(90°) + q^+\sigma_2 + q^-\sigma_1 \tag{14}$$

which can be solved for σ_1 and σ_2 for a homogeneous material.

3. Stress Resolution

Here we consider the stress resolution achievable with the various techniques outlined above. First, we consider SH-waves propagating at normal incidence with arrival times measured before and after application of stress. Using Equation (2),

$$\sigma_1 = -[\bar{a}(\Delta t_1/t_o) - \bar{b}(\Delta t_2/t_o)]/(\bar{a}^2 - \bar{b}^2) \tag{15}$$

and similarly for σ_2. Suppose we have an uncertainty in Δt_i; denote this uncertainty by δt_i. The resulting uncertainty $\delta\sigma_1$ is given by

$$\delta\sigma_1 = -[\bar{a}(\delta t_1/t_o) - \bar{b}(\delta t_2/t_o)]/(\bar{a}^2 - \bar{b}^2) \ . \tag{16}$$

The worst case occurs when $\bar{a}\delta t_1$ and $\bar{b}\delta t_2$ have opposite signs. Assume $|\delta t_1| = |\delta t_2| = \delta t$; then $\delta\sigma_1 = \delta t/(\bar{a}+\bar{b})$.

Typical values of the stress-acoustic constants $\bar{a}$ and $\bar{b}$ are, for 2024-T351 aluminum: $\bar{a} = -2.4\times10^{-5}$/MPa, $\bar{b} = 1.2\times10^{-5}$/MPa [7]. For this alloy, $\delta\sigma_1 \cong (\delta t/t_o)\times10^5$ MPa. To obtain a stress resolution of 10 MPa, we must therefore resolve arrival times to one part in 10^4 or better ($\delta t/t_o \cong 10^{-4}$).

For the second case, we measure stresses in-situ and assume that the material is homogeneous ($C^s_{66} = C^u_{66}$, etc.). Arrival time changes are measured with SH and longitudinal waves at normal incidence. Equations (3) and (4) show that the sum of principal stresses σ is

$$\sigma = \frac{2(\Delta t_p/t_p) - (\Delta t_1 + \Delta t_2)/t_o}{a + b - 2M} \ . \tag{17}$$

(We assume that $\sigma_1 - \sigma_2$ can be obtained with sufficient accuracy from the birefringence $(\Delta t_1 - \Delta t_2)/t_o$.)

To estimate the worst case error, let $|\Delta t_p/t_p| = |\Delta t_1/t_o| = |\Delta t_2/t_o| = \delta t/t$; the uncertainty in σ is $\delta\sigma = [4\delta t/t_o](a+b-2M)$. Using the values $a = -2.9\times10^{-5}$/MPa, $b = 0.7\times10^{-5}$/MPa for 2024-T351 aluminum [7] and the value $M = 0.8\times10^{-5}$/MPa [8], we find that to have a stress uncertainty of 10 MPa, we must resolve arrival times to about 10^{-4}.

In the next case, we assume that the stress distribution is axisymmetric; the principal stresses σ_θ and σ_r (equivalent to σ_1 and σ_2 where the local polar coordinate system coincides with the material symmetry axes) are related through Equation (5).

In the method of King and Fortunko [3], B_o and $\sigma_\theta - \sigma_r$ are determined by measuring B at angles of incidence ϕ_1 and ϕ_2; this gives

$$\sigma_\theta - \sigma_r = \frac{B(\phi_1)\cos^2\phi_2 - B(\phi_2)\cos^2\phi_1}{m(\phi_1)\cos^2\phi_2 - m(\phi_2)\cos^2\phi_1} \tag{18}$$

Assuming a worst case error $|\delta B(\phi_1)| = |\delta B(\phi_2)| = \delta B$, we find a stress uncertainty $\delta(\sigma_\theta - \sigma_r) = [\delta B(\cos^2\phi_1 + \cos^2\phi_2)]/D$ where D is the denominator of Equation (18). Typical values for 6061-T651 aluminum are [3]: $m(12.6°) = -4.4\times10^{-5}$/MPa, $m(33.8°) = -3.7\times10^{-5}$/MPa. Using these we find that 10 MPa resolution in $\sigma_\theta - \sigma_r$ requires an uncertainty in B of about 3×10^{-5}.

The second technique for determining B_o is the dispersion method [4]. The unstressed birefringence is measured as a function of frequency squared on a reference specimen; the slope C_2 and intercept C_1 are determined and substituted into Equation (8) to separate the effects of texture and stress.

We measured the dispersion of B_o for a 25-mm thick rolled plate of HY-80 steel and a 9.5-mm thick rolled plate of 7075-T6 aluminum with an EMAT having an aperature 1 cm × 1 cm. We measured B_o at various locations on the plates and obtained the average value $<B_o>$ at each frequency; we then fitted a straight line through values of $<B_o>$ to obtain C_1 and $\overline{C}_2=(2\pi\hat{a}/V)^2C_2$ (recall that $\hat{a}$ = average grain dimension). The results are shown in Figures 1a and 1b.

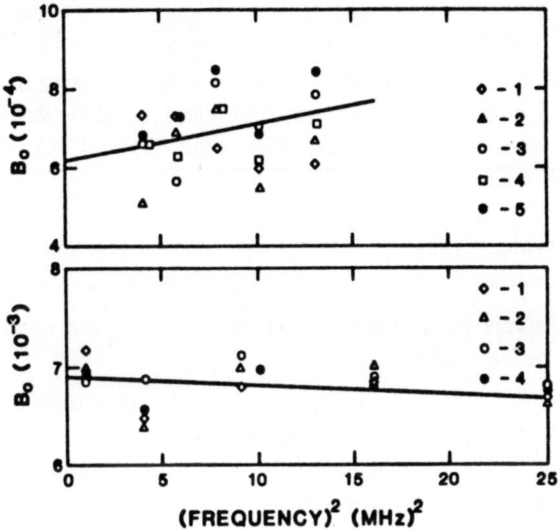

Figure 1: Dispersion of birefringence (a) in HY-80 steel and (b) in 7075-T6 aluminum.

For the steel specimen, we observed a degree of inhomogeneity. If we were to fit straight lines through values of B_o versus frequency squared for each of the locations 1-5 in Figure 1a we would find that some values of $\overline{C}_2$ would be positive and some negative. The intercepts, C_1, would also display scatter. This raises a question; should C_1 and $\overline{C}_2$ be measured for each location and the average values of C_1 and $\overline{C}_2$ be taken, or should we measure $<B_o>$ for each frequency and then fit a straight line to these values?

The results for the aluminum specimen were more encouraging, since the scatter in the data is less and $\overline{C}_2$ is half the value for the steel specimen.

A special case of the dispersion technique is simply to measure $<B_o>$ at a single frequency. Obviously, we cannot then separately determine C_1 and $\overline{C}_2$, but we can subtract $B-<B_o>$ to obtain $\sigma_\theta-\sigma_r$. We can also obtain a measure of the stress resolution by dividing the standard deviation δB_o of our measured values by the stress-acoustic constant: $\delta(\sigma_\theta-\sigma_r)=\delta B_o/m$.

We have measured B_o on three different steel alloys; the results are summarized in Table 1.

Table 1: Measured values of B_o

Alloy	Thickness (mm)	Number of Locations	$<B_o>$ (10^{-4})	δB_o (10^{-4})	$\delta(\sigma_1-\sigma_2)$ (MPa)
HY-80	9.5	20	6.6	1.9	25
HY-80	25	10	12.4	3.1	40
A516	16	14	2.1	2.6	35
304SS	9.5	25	9.4	9.2	115

We have used a typical value of $m \cong 8\times10^{-6}$/MPa [9] and rounded off stress uncertainty to the neares 5 MPa. We conclude that this method gives satisfactory results for ferritic steels (uncertainty $\leqq 40$ MPa), but poor results for austenitic steels (uncertainty $>$ 100 MPa).

In the most general case, the stresses will not be axisymmetric, and the shear stress does not vanish. By determining the orientation of the acoustic axes (finding angle ψ) and measuring B we can eliminate texture and determine σ_{12}; see Equation (11). The remaining (normal) stresses can be determined by integrating Equation (12).

An error analysis of Equation (11) gives

$$\delta\sigma_{12}=\delta B\frac{\sin^2\psi}{2m}+2\delta\psi\frac{B\cos^2\psi}{2m} \quad . \tag{19}$$

Typical values are: $\delta B \cong 10^{-4}$, $\delta\psi \cong 0.05$ rad, $m \cong 4\times10^{-5}$/MPa, $B \cong 5\times\ 10^{-3}$, based on experiments on 9.5-mm thick 2024-T351 aluminum [10]. For these values, errors in measuring ψ are dominant; a worst case estimate of stress uncertainty is $\delta\sigma_{12} \cong 5$ MPa.

These errors may propagate when determining σ_{11} and σ_{22} from Equation (12); in the worst case, errors can accumulate as the integration proceeds. On the other hand, if the errors are random then partial cancellation may occur.

Another possible method that may smooth errors is to fit a series of functions to the σ_{12} data and determine the coefficients in the series. The functional form of σ_{12} can then be substituted into Equation (12) and the integrations performed to obtain σ_{11} and σ_{22}.

Finally, we consider the stress resolution possible with Rayleigh waves (Equation 14). For simplicity, we assume no errors in measuring relative velocity changes and focus on errors $\delta q_1(\alpha)$ due to inhomogeneity. Martin [11] found that for 3-mm thick rolled plates of 2024-T3 aluminum (based on only 2 plates), $|\delta q_1(0°)|=2.2\times10^{-4}$, $|\delta q_1(90°)|=1.1\times10^{-4}$. Performing an error analysis on Equations (14), we have

$$\delta\sigma_1 \cong \frac{q^+[\delta q_1(0°)]-q^-[\delta q_1(90°)]}{(q^+)^2-(q^-)^2}, \tag{20}$$

and similarly for $\delta\sigma_2$. Using values quoted by Jassby and Saltoun [12] for 2024-T351 aluminum, the worst case errors are: $\delta\sigma_1 \cong 60$ MPa, $\delta\sigma_2 \cong 70$ MPa.

For 7075-T6 aluminum, Martin [11] found that (based on data from 2 plates) $\Delta q_1(0°)=1.7\times10^{-4}$, $\Delta q_1(90°)=0.5\times10^{-4}$. Assuming that q^+ and q^- are approximately the same as for 2024-T351 aluminum, we find stress uncertainties of $\delta\sigma_1 \cong 50$ MPa, $\delta\sigma_2 \cong 40$ MPa.

The values above may be conservative, since a larger data base (using more plate specimens) may lead to a smaller value of δq_1. We intend to explore this possibility in the future.

4. Examples

We now illustrate the application of the techniques described above for absolute stress determination in the presence of texture.

In our first example, we measured arrival times with EMATs before and after welding the two 9.5-mm thick 2024-T351 aluminum alloy plates shown in Figure 2. We also measured the principal stresses with strain gages bonded to the plates prior to welding. A comparison of principal stresses σ and σ (acting in directions parallel and perpendicular to the weld seam), determined by EMAT and strain gages, is shown in Figure 3. A fuller account of these measurements is given in Ref. 13.

In our second example, we measured the arrival times in a shrink-fit specimen specimen made of an oversized stainless steel plug inserted into a hole in a rolled plate of 6061-T651 aluminum alloy. The material symmetry axes in the plate are parallel and perpendicular to the rolling direction; consequently, the principal stress directions coincide with material symmetry axes along the radii parallel and perpendicular to the rolling direction.

We measured arrival times along the specimen radius perpendicular to the rolling direction and also in the corners (zero stress state). We took multiple measurements of t_1^u, t_2^u in the corners and averaged to suppress inhomogeneity. We then used $\langle t_1^u\rangle$ and $\langle t_2^u\rangle$ in Equations (3) to determine the radial stress, σ_r. There are three unknowns in these equations (σ_θ, σ_r, and $\Delta d/d$); instead of making P-wave measurements (so that Equation (4) can be used), we used the boundary condition $\sigma_r=0$ at the plate edge. This allows us to eliminate $\Delta d/d$, provided that there is a constant difference between plate thickness in the corners and along the scan line.

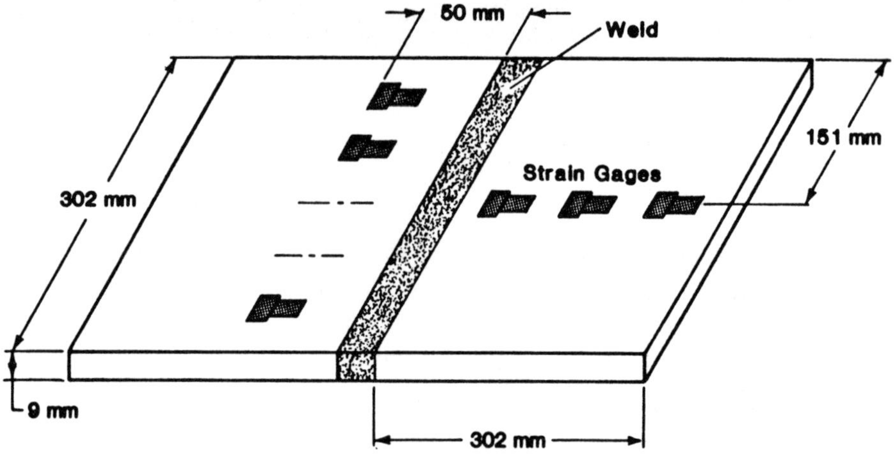

Figure 2: *Location of strain gages on thin welded aluminum alloy plate.*

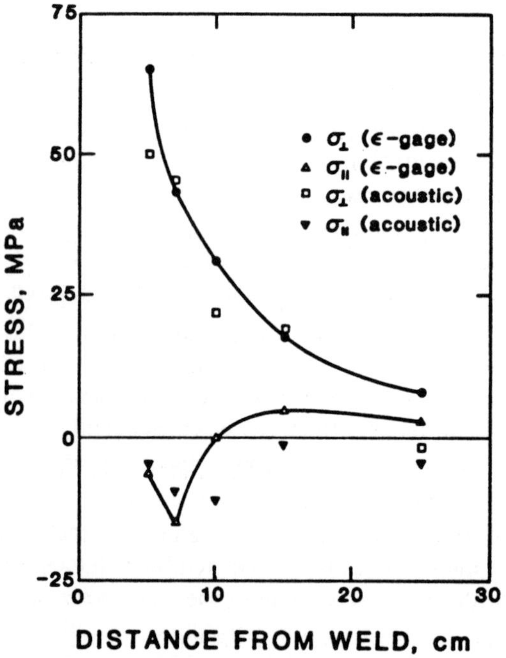

Figure 3: *Comparison of principal stress as measured by strain gages and by EMAT, along plate centerline.*

The resulting values of σ_r are shown as open circles in Figure 4; the solid line is the theoretical value. Reasonably good agreement (within 30 MPa) is obtained. More details are given in Ref. 14.

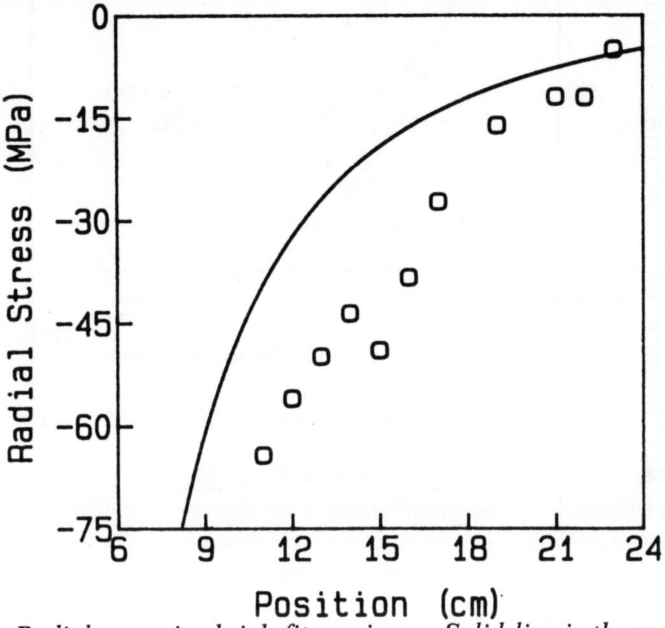

Figure 4: *Radial stress in shrink-fit specimen. Solid line is theory, open circles are EMAT measurements.*

We used the shrink-fit specimen to illustrate the case of axial symmetry. The open circles in Figure 5 are data we obtained by measuring the birefringence along the scanline and subtracting $<B_o>$ obtained from multiple measurements in the plate corners [14]. Values of $\sigma_\theta - \sigma_r$ determined by King and Fortunko [3] using their off-axis technique are shown as stars. The solid line is the theoretical value. Good agreement between theory and experiment was obtained for both the method of King and Fortunko [3] and our multiple measurement technique [14].

When a shear stress exists, we can use Equations (11) and (12) to obtain all components of plane stress. We illustrate this by measuring σ_{12} along the centerline of the welded plates shown in Figure 2, and along parallel lines ± 2 cm from the centerline. If the welding process is symmetric about the centerline, then σ_{12} is antisymmetric, with a gradient of σ_{12} parallel to the weld seam (x_2-direction).

Data measured along the three lines are shown in Figure 6. Measured values of σ_{12} along the centerline are approximately zero (<10 MPa). We calculated $\partial\sigma_{12}/\partial x_2$ from the data, and numerically integrated this gradient along the centerline (x_1-axis), starting from the free edge of the plate ($\sigma_{11}=\sigma_{12}=0$). The resulting normal stress σ_{11} is shown in Figure 7, as well as strain gage data [13]. Good agreement, within 20 MPa, is evident.

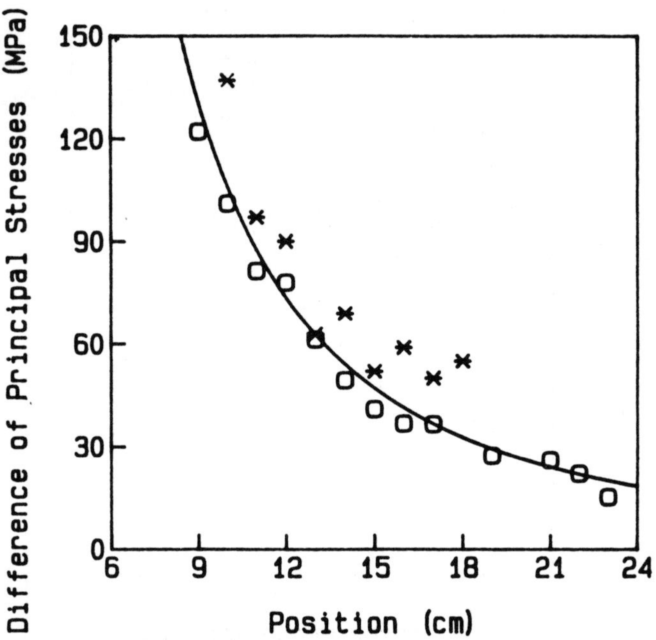

Figure 5: *Difference of principal stresses in shrink-fit specimen. Solid line is theoretical value, open circles are measured with multiple unstressed specimen technique, stars are King and Fortunko data [3].*

In another experiment, we measured σ_{12} in the modified compact tension specimen of Figure 8. We fitted our data to a series of functions having unknown coefficients [15]; the functions are each a solution to stress equilibrium. Using the series of stress functions, we generated contours of shear stress shown in Figure 9 [10]. Theoretical contours (based on photoelastic data [15] are shown in Figure 10. Note the good agreement between both sets of contours.

The stress intensity factor K_I is the coefficient of the leading term in our series representation for shear stress [15]. Our acoustoelastic value of K_I was within a few percent of the theoretical value [10].

5. Conclusions

The problem of absolute stress determination is complicated by the need to separate texture and stress. In many cases, the success of this separation depends not only on the velocity resolution but also upon the degree of material homogeneity. The only in situ stress measurement techniques considered here which appear to avoid this problem of homogeneity are the off-axis technique of King and Fortunko [3] for the axisymmetric case and the measurement of shear stress ($\sigma_{12} \neq 0$). However, in the former case the arrival times must be measured with greater accuracy and, in the latter case, the angle ψ must be measured to within about 0.1 radian, for a stress resolution of 10 MPa. Furthermore, in the shear stress measurement method, errors may accumulate when integrating the stress-

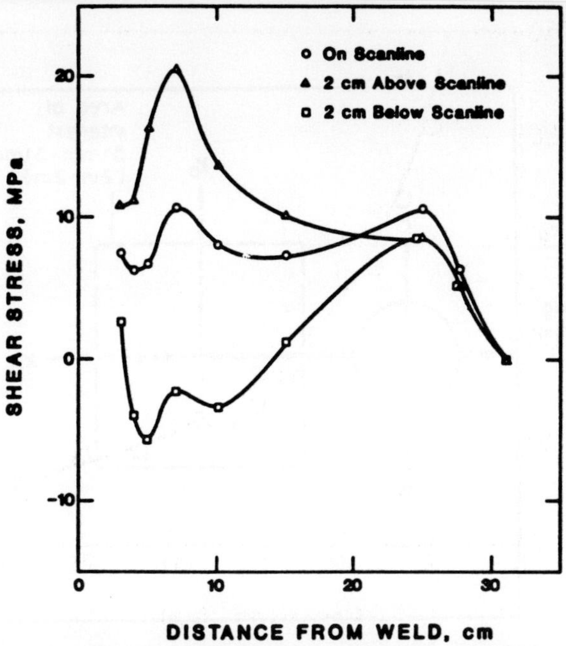

Figure 6: *EMAT measurement of shear stress on scanline (centerline) of thin welded plate (Figure 2) and on lines ±2 cm from centerline.*

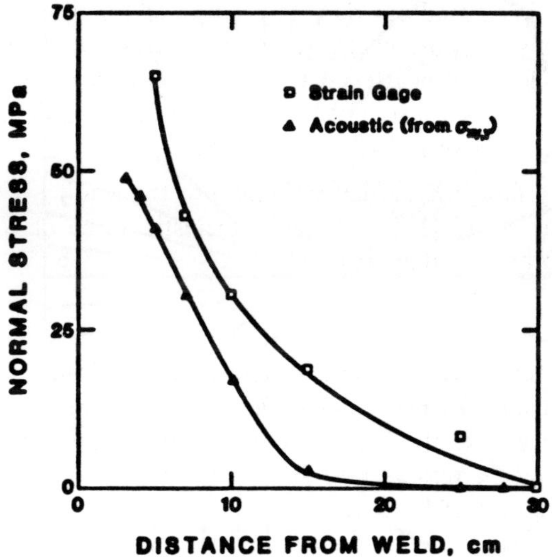

Figure 7: *Normal stress in thin welded plate as determined by strain gages and by integrating shear stress gradient measured with EMAT.*

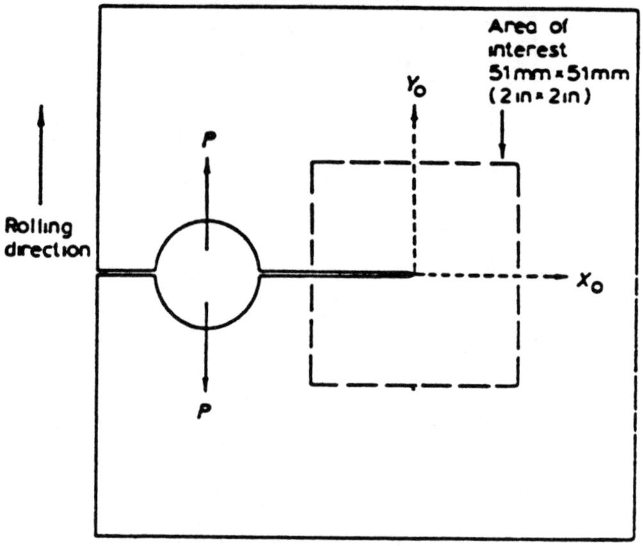

Figure 8: *Modified compact tension specimen. Shear stress measured in "Area of interest".*

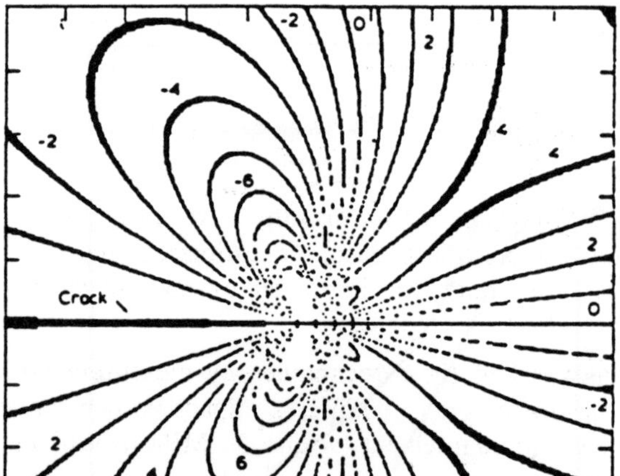

Figure 9: *Ultrasonically determined shear stress contours.*

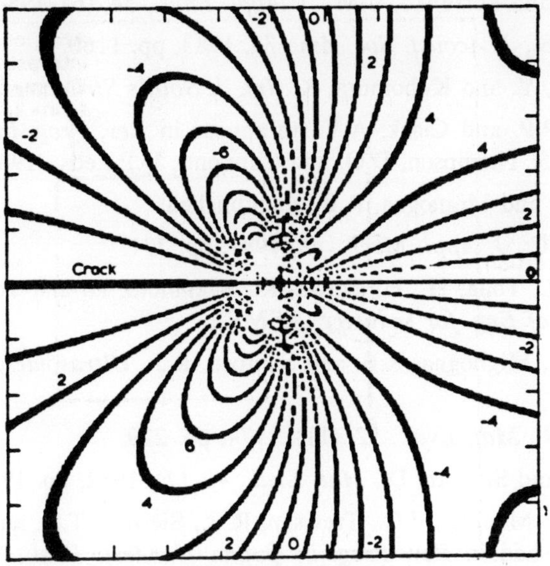

Figure 10: Photoelastic shear stress contours.

equilibrium equation to obtain normal stresses.

For sufficiently homogeneous material, multiple measurements on unstressed reference specimens can be made and used as baseline values to separate texture and stress. For example, we measured $<B_o>$ and δB_o on steel alloys and estimated the resulting stress uncertainty $\delta(\sigma_1 - \sigma_2)$. We found that $\delta(\sigma_1 - \sigma_2)$ was less than 40 MPa for two ferritic steel alloys but more than 100 MPa for a stainless steel alloy.

We also made estimates of the uncertainty in σ_1 and σ_2 for the Rayleigh wave technique; based on a sparse data base, we estimated uncertainties of about 50 MPa. These uncertainties might be improved by making more measurements on reference specimens.

We illustrated several of these techniques by measurements made on welded plates, a shrink-fit specimen, and a modified compact tension specimen, all made of aluminum. We found good agreement (within 30 MPa or less) between theoretical and experimental values of stress.

References

[1] Clark, A.V. and Mignogna, R.B., *Ultrasonics, 21*, (5), 1983, pp. 217.

[2] Clark, A.V. and Moulder, J.C., *Ultrasonics, 23*, (11), 1985, pp. 253.

[3] King, R.B. and Fortunko, C.M., *J. Appl. Phys., 54*, (6), 1983, pp. 3027.

[4] Hirsekorn, S., *J. Acoust. Soc. Am., 73*, 1983, pp. 1160.

[5] Iwashimizu, Y. and Kubomura, K., *Int. J. Solids Structures, 9*, 1973, pp. 99.

[6] DelSanto, P.P. and Clark, A.V., to appear in *Rev. Prog. in Quant. Nondest. Eval.*, Vol. 5, Thompson, D.O. and Chimenti, D.E., eds., 1985.

[7] Clark, A.V. and Mignogna, R.B., unpublished.

[8] Johnson, G.C., *J. Appl. Mech., 48*, 1981, pp. 791.

[9] Fukuoka, H., Toda, H., Hirakawa, K., Sakamoto, H. and Toya, Y., to be published in *J. of Eng. for Industry* (ASME).

[10] Clark, A.V., Mignogna, R.B. and Sanford, R.J., *Ultrasonics, 21*, (2), 1983, pp. 57.

[11] Martin, B.G., *Mat. Eval., 32*, (11), 1974, pp. 229.

[12] Jassby, K. and Saltoun, D., *Mat. Eval., 40*, (2), 1982, pp. 198.

[13] Clark, A.V., Moulder, J.C., Trevisan, R.E., Siewert, T.A. and Mignogna, R.B., to be published in *Rev. Prog. in Quant. Nondest. Eval.*, Vol. 5, Thompson, D.O. and Chimenti, D.E., eds., 1985.

[14] Clark, A.V. and Moulder, J.C., to be published in *Proc. 15th Symp. on Nondest. Eval.*, sponsored by Southwest Res. Inst., 1985.

[15] Irwin, G.R., et al., U.S. Nuclear Reg. Comm. Rep. NUREG/CR-1455, May, 1980.

ULTRASONIC SCATTERING AND NDE OF MATERIALS AND CRACKS

S.K. Datta
Department of Mechanical Engineering and CIRES
University of Colorado
Boulder, Colorado

A.H. Shah
Department of Civil Engineering
University of Manitoba
Winnipeg, Manitoba, Canada

H.M. Ledbetter
Fracture and Deformation Division, NBS
Boulder, Colorado

Abstract

In this paper we have summarized our recent research work on single and multiple scattering of elastic waves. In recent years we have analyzed scattering of longitudinal and shear waves by a single planar or non-planar crack near the free surface of a semi-infinite medium. It has been shown that scattered displacements at the free surface have characteristic features related to the cracks that can be used for quantitative non-destructive evaluation. We have also examined scattering of SH waves in a homogeneous plate of finite thickness. Comparison of theoretical results with experiments performed with EMAT's shows remarkable agreement.

Additionally, using a multiple scattering formulation and making use of long wavelength expansions and quasi-crystalline approximation we have developed equations governing phase velocities and attenuation in composite media with a distribution of voids or reinforcing particles. Theoretical results for particle- and fiber-reinforced materials have been found to agree well with experiments.

1. Introduction

The scattering of elastic waves by near-surface and surface-breaking planar cracks has received considerable attention in recent years. The problem is of interest for quantitative ultrasonic evaluation of cracks in structural parts. Most notable among recent publications are those by Achenbach and his co-workers and these have been described elsewhere in this volume. We have used a combined finite element and eigenfunction expansion approach to solve this class of problems. The technique has the advantage that a non-planar crack as well as more than a single crack can be analyzed. However, it is not very economical to use at high frequencies for which the ray method developed by Achenbach et al. [1] is still well-suited.

Results for scattering of SH waves by planar cracks in a semi-infinite medium were reported in [2-4]. Corresponding problem for a thick plate was studied in [5-8]. Predicted results for SH waves were found to compare well with experiments performed with EMAT's [9-10].

We have used the same technique for analyzing scattering of longitudinal (P) and vertically polarized shear (SV) waves by near-surface and surface-breaking cracks in a half-plane. Numerical results for the scattered surface displacements have been reported in [11,12].

In addition to the single scattering studies we have also investigated wave propagation through a medium containing a random distribution of fibers and particles. Using long wavelength and quasi-crystalline approximations we have derived equations governing phase velocities of coherent plane waves propagating through such a medium. Results for longitudinal and shear waves propagating through fiber-reinforced composite media have been reported in [13-15]. Theoretical results are shown to compare well with experiments. Longitudinal and shear wave propagation through a medium with inclusions has been studied in [16,17]. Details of the calculations for a particular particle-reinforced composite and comparison with experiments have been presented in [18-21]. As a special case of the particle-reinforced composite we also have derived properties of a medium with a distribution of voids [22]. Only phase velocities were considered in all these works. More recently [23] we have studied damping of longitudinal and shear waves in a particle-reinforced composite. Results of a preliminary investigation of the effect of interface properties on damping and phase velocities are in [24].

In the following sections we summarize the important findings of some of our studies referenced above.

2. Scattering by a Crack

2.1. *Half-Space*

Consider a surface-breaking planar crack in a half-space as shown in Figure 1. The crack is oriented at an angle α to the x-axis and is of length D. It is assumed that a plane body wave is incident on the crack. The direction of the wave makes an angle γ with the negative y-axis as shown.

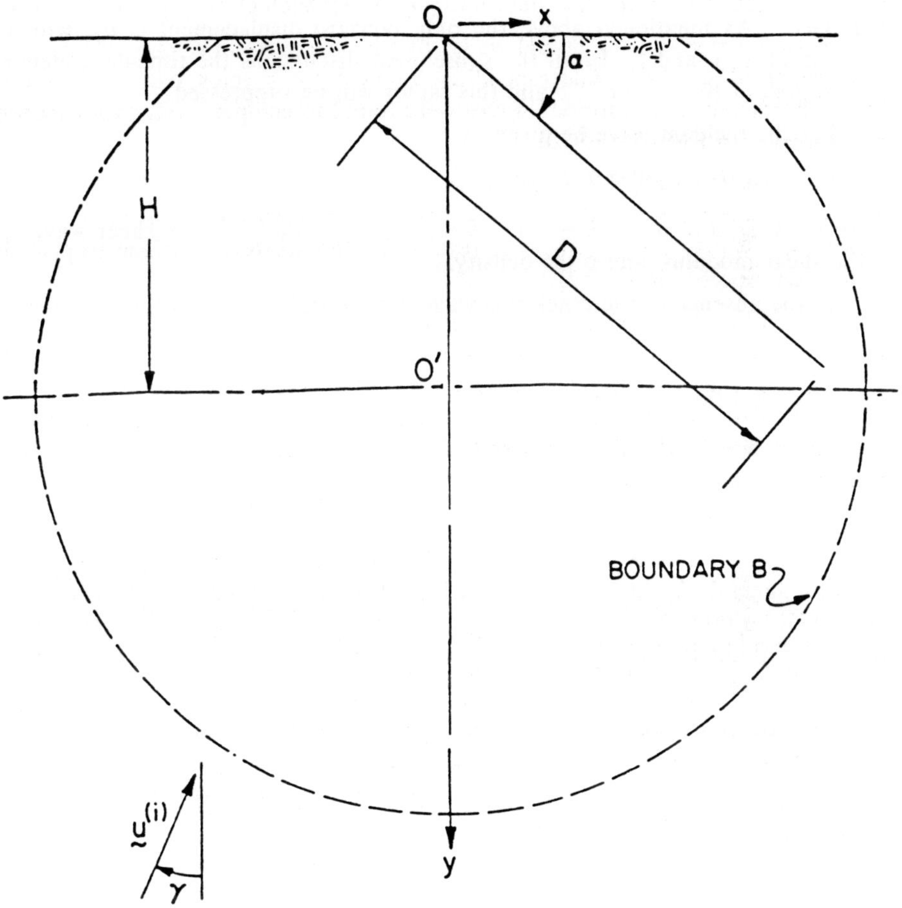

Figure 1: Geometry of surface-breaking planar crack

Now two problems can be considered: one is the *SH* wave in which the only non-zero displacement component is in the z-direction and the other is the in-plane wave in which the displacement is either parallel (P) or perpendicular (SV) to the direction of incidence. The first problem was analyzed in [2] using a combined finite element and eigenfunction expansion method. A low-frequency asymptotic solution was presented earlier in [25]. The second problem is studied in [12] using the numerical method discussed in [2]. In the following we first summarize our findings reported in [2] and then we present some results from [12] as well as from [11] for near-surface cracks.

SH-Wave. As mentioned above the only nonzero displacement is u_z, which is a function of x, y and t. In all the subsequent discussions the time-dependence will be assumed of the form $e^{-i\omega t}$ and this factor will be suppressed.

Let the incident wave be given by

$$u_z^{(i)}=w_0 e^{i\varepsilon(\bar{x}\sin\gamma-\bar{y}\cos\gamma)} \quad . \tag{1}$$

Here $\varepsilon=k_2 D$, $k_2=\omega/c_2$, $\bar{x}=x/D$, $\bar{y}=y/D$, $c_2(=\sqrt{\mu/\rho})$ is the shear wave speed, μ the shear modulus, and ρ the density.

In the absence of the crack this wave will be reflected from the free surface as

$$u_z^{(r)}=w_0 e^{i\varepsilon(\bar{x}\sin\gamma+\bar{y}\cos\gamma)} \quad . \tag{2}$$

Thus the total field seen by the crack is

$$\bar{u}_z=2w_0 e^{i\varepsilon\bar{x}\sin\gamma}\cos(\bar{y}\cos\gamma) \quad . \tag{3}$$

The hybrid method used to solve the scattering problem requires that the region near the crack (inside the fictitious boundary B) is divided into finite elements. Because this region is divided into finite elements the crack geometry can be quite general. In fact the case of a branched crack is reported in [12]. Also the crack can be taken to be lying in a material with properties different than those of the material outside B. In [2] calculations were reported for a case when the crack was in an anisotropic region.

Let the field inside B be denoted by $u_z^{(2)}$ and that outside by $u_z^{(1)}$. If the region bounded inside and on B is divided into finite elements having N_I+N_B nodes, where N_B is the number of nodes on B, then the field $u_z^{(2)}$ can be represented in terms of shape functions $L_j(x,y)$ and its nodal values W_j as

$$u_z^{(2)}(x,y)=\Sigma L_j W_j$$

Following the usual procedure it then follows that

$$[S_{II}]\{W_I\}=-[S_{IB}]\{W_B\} \tag{4}$$

where

$$S_{MN}=\int_{R_2}[\{\nabla L_M\}^T\{\nabla L_N\}-k'^2_2\{L_M\}^T\{L_N\}]\,dxdy \tag{5}$$

Note that $k'_2=\omega\sqrt{\rho/C_{55}}$, C_{55} being the appropriate elastic constant of the material in the region R_2 inside B.

To solve for W_I and W_B it is now necessary to match $u_z^{(2)}$ with $u_z^{(1)}$ on B. Now $u_z^{(1)}$ is composed of $\bar{u}_z$ and the scattered field $u_z^{(s)}$. Also $u_z^{(s)}$ can be written in terms of multipolar solutions, which satisfy the wave equation in R_1 (outside B), the boundary condition on $y=0$ and the radiation condition as $r(=\sqrt{x^2+y^2})$ tends to infinity. Such a solution is

$$u_z^{(s)}=\sum_{n=0}^{\infty}A_n\cos n\theta' H_n^{(1)}(k_2r) \tag{6}$$

where θ' is the polar angle measured from the x-axis and $H_n^{(1)}(k_2r)$ is the Hankel function of the first kind.

For the purpose of matching the two solutions $u_z^{(2)}$ and $\bar{u}_z+u_z^{(s)}$ only N_B number of terms are kept in (6). Enforcing the continuity of the displacement and traction at the nodes on B we get the required equations for the determination of A_n ($n=0,\ldots,N_B-1$ and the boundary values $\{W_B\}$. Details can be found in [2]. Figures 2 and 3 show the far-field amplitude of $u_z^{(s)}$ on $y=0$. Also shown in these figures are the results for the case of a crack lying in an anisotropic material with materials properties

$$\rho_2/\rho_1=1.01,\ C_{55}/\mu=1.63,\ C_{44}/\mu=1.04\ .$$

Figure 2 shows comparison with experiment and Figure 3 with the long wavelength matched asymptotic expansion (MAE) results. It is seen that the calculations agree well with experiments. An interesting feature is that the signal amplitude rises rapidly with increasing crack length up to about $\lambda/4$, λ being the wavelength, after which the signal reaches a plateau. Thus maximum sensitivity for crack length estimation is in the long wavelengths range.

In-plane motion. Similar technique is used also for scattering of P, SV and Rayleigh waves by near-surface and surface-breaking cracks. Near-surface cracks with depth of center, H, were analyzed in [11] and results for the surface-breaking cracks can be found in [12]. The results show that for both buried and surface-breaking cracks there are large deviations in the signals scattered in the forward direction from cracks of different orientations. It is interesting to note that for surface breaking cracks signals from planar and branched cracks are not significantly different. This is found also to be true for P and SV wave incidence. For buried cracks, however, large deviations were found in signals scattered from planar and branched cracks (with branches pointing towards the free surface).

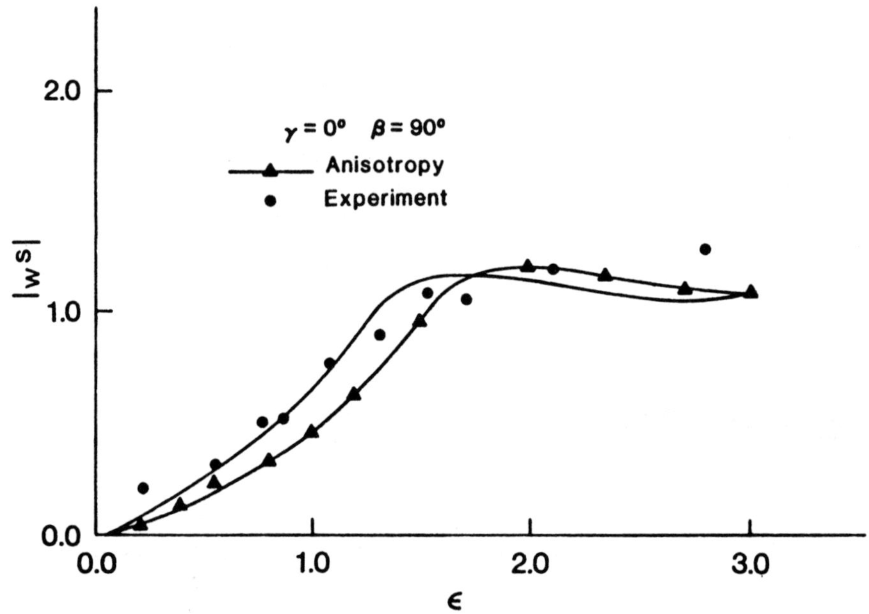

Figure 2: *Far-field scattered displacement amplitude on $y=0$. γ, β are angles made by the incident wave and the crack, respectively, with the x-axis.*

2.2. *Thick Plate*

Scattering of *SH*-waves by planar cracks of different orientations in a plate has been reported in [5-8]. Results for vertical surface-breaking cracks were presented in [5,7] and those for buried normal cracks can be found in [8] (see also [6]). Comparison of reflected amplitude measured at a point on the plate surface was found to agree with computations. This is shown in Figure 4. Buried cracks with different orientations were also studied in [6]. Figure 5 shows the amplitudes of the scattered surface displacements due to normal and canted buried cracks. It is seen from Figure 4 that the reflection amplitude increases rapidly first as the crack length increases, but then shows plateaus after the length of the crack reaches about a quarter of the plate thickness. Figure 5 shows the reflected displacement amplitude on $y=0$ when zeroth mode *SH*-wave is incident on buried normal cracks. For small normal cracks the signal amplitudes have maxima and minima at about the same locations. These locations shift towards the left when the crack is canted. In the far field the signals from a canted crack rise rapidly and monotonically as the crack length increases. For a normal crack, however, the rise is rapid initially but levels off. Also the distribution of amplitude on the surface due to a normal crack changes significantly as the crack gets very long.

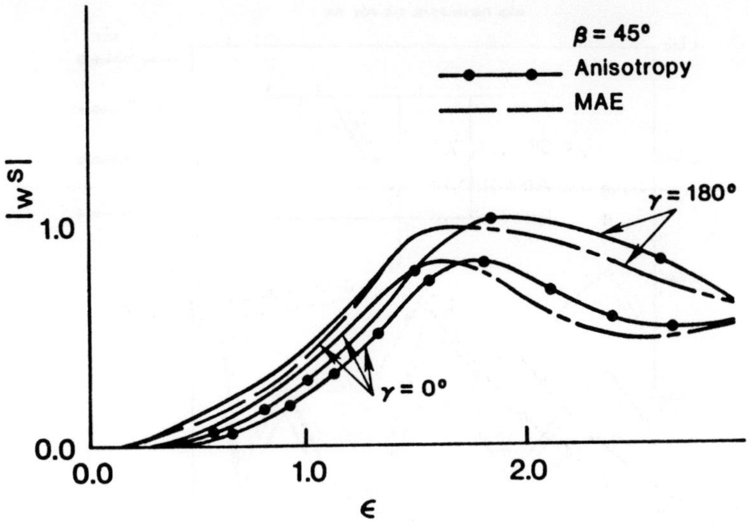

Figure 3: *Far-field back-scattered displacement amplitude on $y=0$ for a canted crack.*

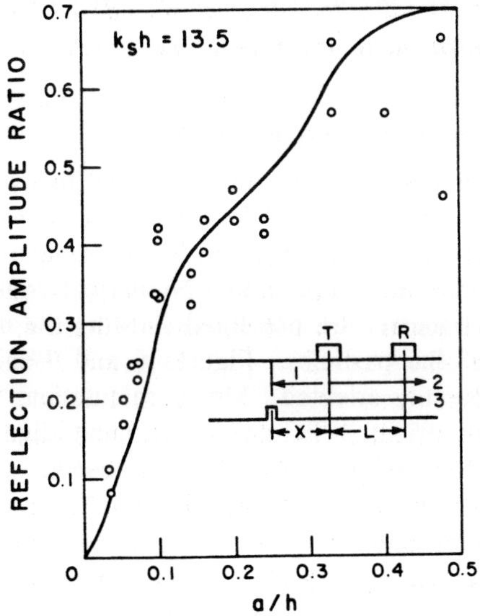

Figure 4: *Comparison of observed and calculated SH wave reflection amplitude ratios for a normal surface-breaking crack in a plate of thickness h. $h=15.9$ mm, $x=100$ mm, and $L=210$ mm.*

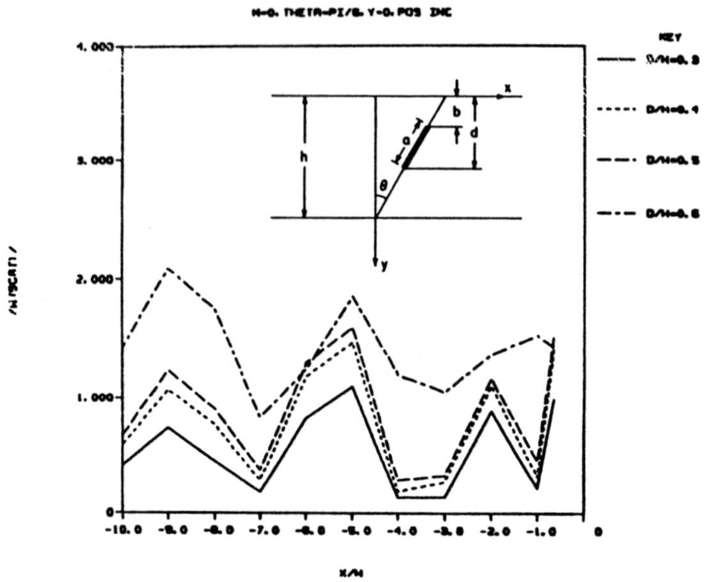

Figure 5: *Back-scattered displacement amplitudes for 0th mode SH-wave incident on buried normal cracks.* $k_2h = 13.5$ *and* $b/h = 0.1$.

3. Scattering by a Distribution of Inclusions and Voids

In this section we summarize our work on wave propagation in a composite material with a distribution of inclusions and voids.

For the details of the model used to calculate the effective phase velocities of plane wave the reader is referred to [21]. There expressions are derived for longitudinal and shear wave speeds when the inclusions are spheroidal in shape. Both random and aligned orientations are considered. Numerical results for a particular case of SiC particles in Al matrix are presented showing the dependence of phase velocities on the shapes of the particles. Figures 6 and 7 show this dependence when the particles are randomly oriented. Model calculations for a nonhomogeneous distribution of randomly oriented SiC particles around aligned oblate spheroidal islands are shown in Figures 8 and 9. Comparison of observed and predicted elastic stiffnesses for a specific nonhomogeneous distribution is made in Table 1. The agreement is found to be very good.

It is to be noted that the expressions for a distribution of voids in a matrix material are derivable as special cases by letting the elastic stiffnesses and densities of the inclusions go to zero. Such model calculations for a particular case of void distribution in porous Olivine are depicted in Figure 10. Phase velocities are found to be highly sensitive to void aspect ratio when the voids are oblate spheroidal in shape. In these calculations it was assumed that Olivine has 5 volume percent spherical pores.

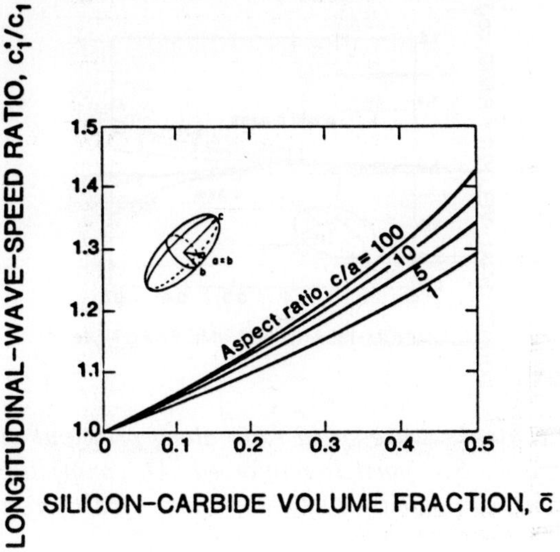

Figure 6: *Longitudinal - wave-speed ratio in a SiC-particle-reinforced Al composite (random orientation). For Al matrix, $C_1=2\,km/s$, $C_2=0.99\,km/s$, $\rho=2.706\,g/cm^3$; for SiC particles, $C_1=12.2\,km/s$, $C_2=7.7\,km/s$, $\rho=3.181\,g/cm^3$.*

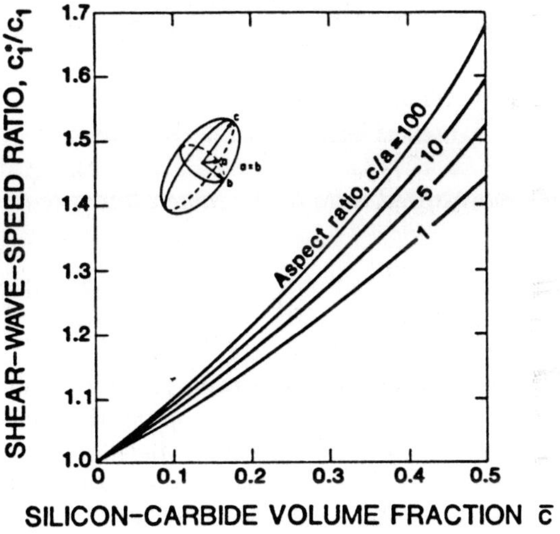

Figure 7: *Shear - wave-speed ratio in the same composite as described in Figure 6.*

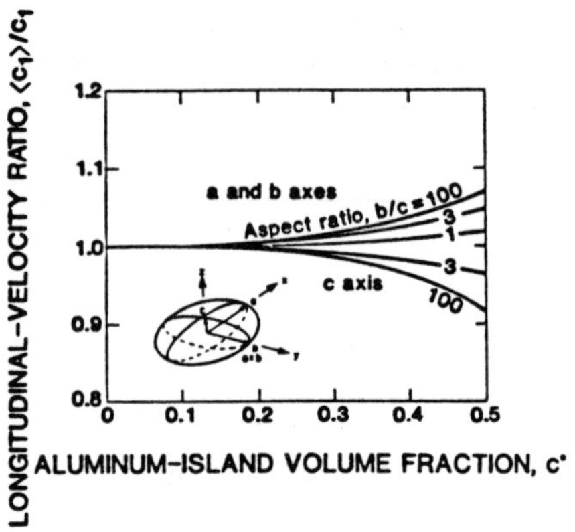

Figure 8: *Longitudinal - wave-speed ratio along principal directions in a nonhomogeneous SiC-particle-reinforced Al composite. SiC particles are randomly oriented at 30% volume fraction around oriented Al islands.*

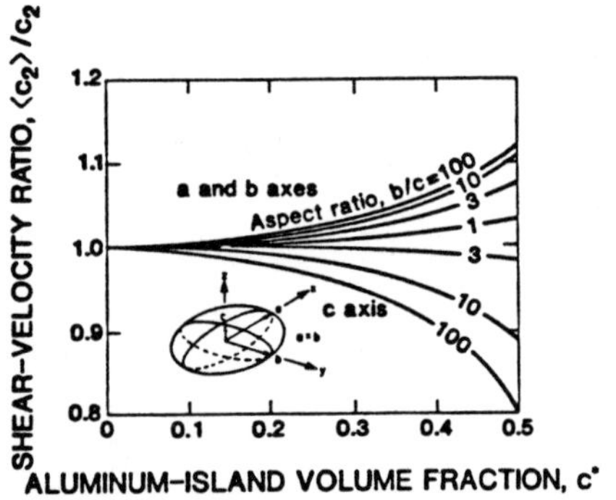

Figure 9: *Shear - wave-speed ratio in the same composite as described in Figure 8.*

Finally model calculations and observations of longitudinal wave velocity in a polycrystalline copper at high temperature and under tensile loading are shown in Figure 11. Details of the experiment can be found in [25]. Predicted values assuming oblate spheroidal voids of aspect ratio 1:10 seem to conform with experiments. It is seen that there is an initial anisotropy in the copper bars used for the experiments. This anisotropy is due to texture. Model calculations do not take into account this anisotropy.

Table 1: Observed and predicted elastic stiffnesses, C_{ij}, of 30-volume-percent SiC/6061 Aℓ at room temperature. Units are $10^{11}\ N/m^2$.

	6061	SiC	30-volume-percent SiC/6061-Al			
			Obs.	R-o-m	Homog.	Nonhomog.
Spec. Grav.	2.706	3.181	2.838	2.849	2.849	2.849
C_{11}	1.105	4.742	1.659	2.195	1.583	1.727
C_{22}	1.105	4.742	1.651	2.195	1.583	1.727
C_{33}	1.105	4.742	1.483	2.195	1.583	1.480
C_{44}	0.267	1.881	1.433	0.751	0.443	0.429
C_{55}	0.267	1.881	0.434	0.751	0.443	1.429
C_{66}	0.267	1.881	0.487	0.751	0.443	0.513
C_{12}	0.571	0.980	0.685	0.693	0.697	0.701
C_{13}	0.571	0.980	0.662	0.693	0.697	0.677
C_{23}	0.571	0.980	0.662	0.693	0.697	0.677

Because the analysis applied to waves propagating along principle directions, it did not provide values for C_{13} and C_{23}, where we assumed $C_{13}=C_{23}=0.966\,C_{12}$. We based the latter assumption on observation. We assume $C_{23}=C_{13}$.

Predictions in this table are based on $c = 0.30$, $c^* = 0.50$, SiC-particle aspect ratio = 3.0, Aℓ-island aspect ratio = 0.33.

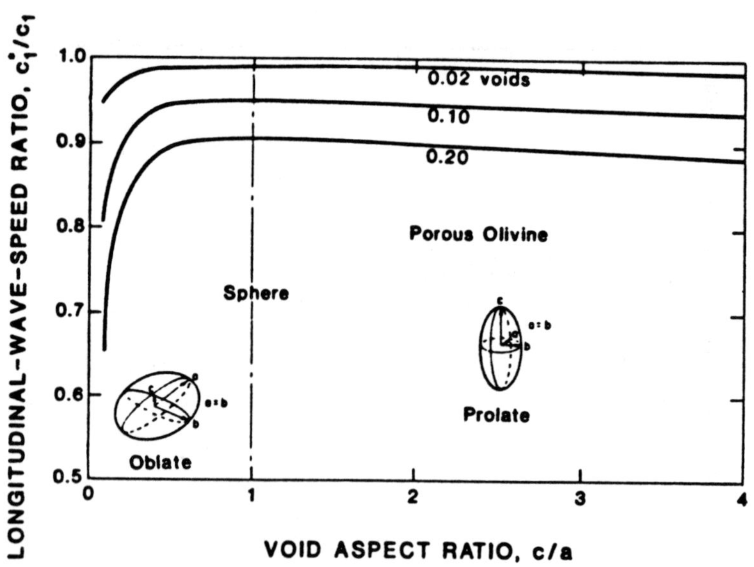

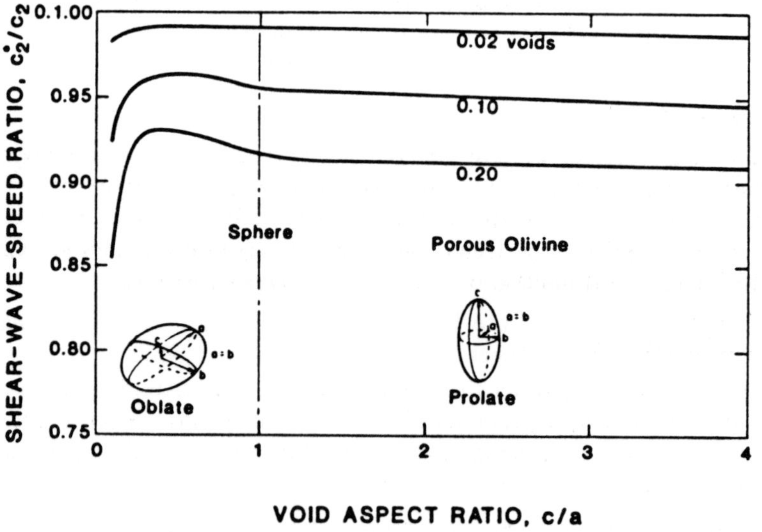

Figure 10: *Wave-speed ratios in a porous olivine containing randomly-oriented spheroidal cavities. For the porous olivine, Young's modulus = 1.781×10^{11} Pa and shear modulus = 0.72×10^{11} Pa.*

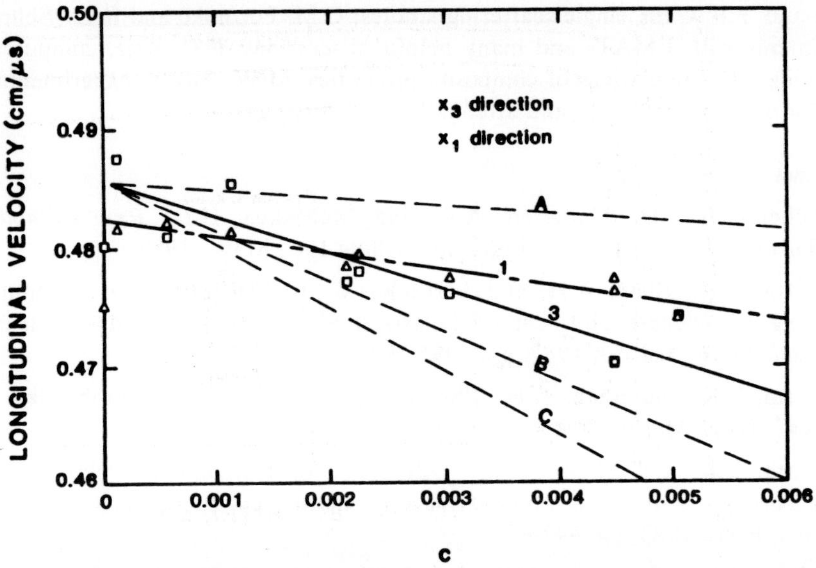

Figure 11: Measurement and modeling of longitudinal wave-speeds in a polycrystalline copper undergoing high-temperature creep. Lines 1 and 3 represent velocities in x_1 and x_3 directions, respectively. Lines A, B and C represent calculations: A for spheres, B and C for oblate spheroids with a 1:10 aspect ratio, B corresponding to random orientation and C to orientation perpendicular to x^3.

4. Conclusion

Results presented above show the versatility and success of the various modeling studies we have carried out in the past on single and multiple scattering of elastic waves in an initially isotropic homogeneous medium. Our current studies are directed to extending the techniques to model scattering in anisotropic and layered media. The hybrid method is well suited for such extensions.

Acknowledgements

The single scattering studies reported here were supported in part by the National Science Foundation, a grant (A-7988) from the Natural Science and Engineering Research Council of Canada, and by the Department of Transportation, Office of Pipeline Safety Regulation.

The reported work on wave propagation in composite media received partial support from the National Science Foundation, DARPA and the NBS office of Nondestructive Evaluation.

Many people contributed substantially to these studies: Z. Abduljabbar, K.C. Wong and Y.F. Chin, single scattering studies; C.M. Fortunko and R.E. Schramm, experiments with EMATs and many helpful discussions; R.D. Kriz, computer programming and calculations of composite properties; M.W. Austin, experimental systems for ultrasonic velocity and attenuation measurements.

References

[1] Achenbach, J.D., Gautesen, A.K. and McMaken, H.L., *Ray Methods for Waves in Elastic Solids*, Pitman Publishing Inc., Mass., 1982.

[2] Datta, S.K., Shah, A.H. and Fortunko, C.M., "Diffraction of Medium and Long Wavelength Horizontally Polarized Shear Waves by Edge Cracks", *J. Appl. Phys.*, Vol. 53, 1982, pp. 2895-2903.

[3] Datta, S.K. and Shah, A.H., "Scattering of SH Waves by Embedded Cavities", *Wave Motion*, Vol. 4, 1982, pp. 265-283.

[4] Datta, S.K. and Shah, A.H., "Diffraction of Horizontally Polarized Shear Waves", in *Numerical Methods in Engineering*, ed. P. Lascaux, Vol. 2, Pluralis, Paris, 1983, pp. 447-457.

[5] Abduljabbar, Z., Datta, S.K. and Shah, A.H., "Diffraction of Horizontally Polarized Shear Waves by Normal Edge Cracks in a Plate", *J. Appl. Phys.*, Vol. 54, 1983, pp. 461-472.

[6] Abduljabbar, Z., *Diffraction of Horizontally Polarized Shear Waves in a Plate*, Ph.D. Thesis, University of Colorado, Boulder, 1983.

[7] Abduljabbar, Z. and Datta, S.K., "Diffraction of Horizontally Polarized Shear Waves by Normal Edge Cracks in a Plate", in *Review of Progress in Quantitative Nondestructive Evaluation*, Vol. 2A, eds. D.O. Thompson and D.E. Chimenti, Plenum Press, New York, 1983, pp. 385-411.

[8] Abduljabbar, Z. and Datta, S.K., "Diffraction of Horizontally Polarized Shear Waves by Cracks in a Plate", in *Numerical Methods for Transient and Couple Problems*, eds. R.W. Lewis, E. Hinton and P. Bettess, Pineridge Press, Swansea, 1984.

[9] Fortunko, C.M. and Schramm, R.E., "Ultrasonic Nondestructive Evaluation of Butt Welds Using Electromagnetic Acoustic Transducers", *Welding Journal*, Vol. 61, 1982, pp. 39-46.

[10] Fortunko, C.M. and Schramm, R.E., "Evaluation of Pipeline Girth Welds Using Low-Frequency Horizontally Polarized Shear Waves", *J. Nondestructive Eval.*, Vol. 3, 1983, pp. 155-173.

[11] Shah, A.H., Wong, K.C. and Datta, S.K., "Surface Displacements due to Elastic Wave Scattering by Buried Planar and Non-Planar Cracks", *Wave Motion*, Vol. 7, 1985, pp. 319-333.

[12] Shah, A.H., Chin, Y.F. and Datta, S.K., "Elastic Wave Scattering by Surface-Breaking Planar and Non-planar Cracks", to be published.

[13] Datta, S.K., "Propagation of SH-Waves through a Fiber-reinforced composite - Elliptic Cylindrical Fibers", *J. Appl. Mech.*, Vol. 42, 1975, pp. 165-170.

[14] Datta, S.K. and Ledbetter, H.M., "Elastic Constants of a Fiber-reinforced Boron-Aluminum: Observation and Theory", *Int. J. Solids and Str.*, Vol. 19, 1983, pp. 885-894.

[15] Datta, S.K., Ledbetter, H.M. and Kriz, R.D., "Calculated Elastic Constants of Composites Containing Anisotropic Fibers", *Int. J. Solids and Str.*, Vol. 20, 1984, pp. 429-438.

[16] Datta, S.K., "Scattering by a Random Distribution of Inclusions and Effective Elastic Properties", in *Continuum Models of Discrete Systems*, ed. J. Provan, University of Waterloo Press, 1978, pp. 111-127.

[17] Datta, S.K., "Wave Propagation in the Presence of a Random Distribution of Inclusions", in *Continuum Models of Discrete Systems*, eds. E. Kröner and K.-H. Anthony, University of Waterloo Press, 1980, pp. 565-582.

[18] Ledbetter, H.M., Datta, S.K. and Kriz, R.D., "Elastic Constants of an Anisotropic, Nonhomogeneous Particle-reinforced Composite", *Acta. Metall.*, Vol. 321, 1984, pp. 2225-2231.

[19] Ledbetter, H.M. and Datta, S.K., "Young's Modulus and Internal Friction of an SiC-particle-reinforced Aluminum Composite", *Mat'ls. Sci. Engrg.*, Vol. 67, 1984, pp. 25-30.

[20] Datta, S.K. and Ledbetter, H.M., "Waves, Microstructures, and Effective-Medium Approximation", in *The Mechanics of Dislocations*, eds. E.C. Aifantis and J.P. Hirth, American Society for Metals, Metals Park, 1985, pp. 213-223.

[21] Ledbetter, H.M. and Datta, S.K., "Effective Wave Speeds in an SiC-particle-reinforced Al Composite", *J. Acoust. Soc. Am.*, to be published.

[22] Datta, S.K. and Ledbetter, H.M., "Multiple Scattering of Elastic Waves and Effective Properties in Materials Containing Inclusions", in *Wave Propagation in Inhomogeneous Media and Ultrasonic Nondestructive Evaluation*, ed. G.C. Johnson, AMD-Vol. 62, The American Society of Mechanical Engineering, New York, 1984, pp. 123-139.

[23] Ledbetter, H.M. and Datta, S.K., "Damping in Metal-Matrix Composites: Measurement and Modeling", in *Vibration Damping 1984 Workshop Proceedings*, Ed. L. Rogers, AFWAL-TR-84-3064, Air Force Wright Aeronautical Laboratories, Dayton, 1984, pp. W.2-W.18.

[24] Datta, S.K. and Ledbetter, H.M., "Effect of Interface Properties on Wave Propagation in a Medium with Inclusions", in *Mechanics of Materials Interfaces*, eds. A.P.S. Selvadurai and G. Voyiadjis, Elsevier Science Publishers, Amsterdam, 1986, to be published.

[25] Ledbetter, H.M., Fields, R.J. and Datta, S.K., "Creep Cavities in Copper: An Ultrasonic-Velocity and Composite-Modeling Study", to be published.

HYBRID RAY-MODE AND WAVEFRONT-RESONANCE TECHNIQUES FOR ACOUSTIC EMISSION AND SCATTERING IN MULTIWAVE LAYERED MEDIA

L.B. Felsen, I.T. Lu and H. Shirai
Department of Electrical Engineering & Computer Science
Polytechnic University
Farmingdale, New York

Abstract

Analytical modeling of acoustic detection of flaws in composite laminates poses a difficult problem in wave propagation and scattering. Wave excitation and propagation in the laminate itself is complicated by the layering and by the generally multiwave character (due to anisotropy, elasticity, etc.) of each layer. Scattering from a flaw, even if it is understood in an unbounded material, is complicated by the presence of the layered environment. Dealing with the overall problem requires a self-consistent combination of phenomena due to layering and scattering. In this context, attention is here being called to self-consistent hybrid ray-mode and wavefront-resonance techniques – originally introduced in electromagnetics and underwater acoustics, and more recently in seismology – which incorporate a new approach to the multiwave, multilayer problem and the scattering problem, respectively. For the layered environment, a hybrid ray-mode Green's function seeks to remove the disadvantages of separate ray propagation and mode propagation models. Ray fields alone may be inconvenient due to the multiwave character (caused by elasticity, anisotropy, etc.) of the layers, which results in wave species coupling at boundaries, and consequent ray proliferation after a few multiple reflections. Mode fields alone may be inconvenient at high frequencies because many normal modes are then required. Thus, for general conditions encompassing a broad range of observables, neither rays nor modes alone provide an efficient algorithm for numerical implementation. For scattering of wide-band signals by targets modeling flaws or inclusions in the unlayered environment, a self-contained wavefront-resonance description seeks to reconcile the spiked wavefront-like responses observed at early times with the more smoothly oscillatory resonance-like responses observed at later times. These hybrid formulations are reviewed and illustrated by examples which clearly reveal the phenomenology noted above. Emphasis is placed on the implications of these concepts for NDE.

1. Introduction

In this paper, we introduce methods of analysis that may provide new options for analytical modeling of in situ location of flaws or cracks in elastic plate or laminate structures. These methods have been developed for electromagnetic and underwater acoustic wave propagation, but more recent extensions have dealt with multilayer, multiwave (for example, elastic) media.

In an elastic plate, the waves excited by a periodic vibrational (harmonic) or impact (transient) source can be analyzed either in terms of ray fields and wavefronts or in terms of the normal plate modes. Of particular concern is the behavior in the high frequency portion of the spectrum, corresponding to early observation times in the transient regime. Direct and multiple reflected wavefronts form a suitable and phenomenologically meaningful discriminant here because they can be individually resolved and are few in number. At later times, multiple arrivals proliferate and are complicated further by the coupling between dilatational and shear waves at each encounter with a boundary or traction-free surface. The field at these later times is described better in terms of the now phenomenologically meaningful normal modes. Conversely, normal mode expansions are unsuitable for early times since very many of them are required to synthesize the high frequency wavefields there. Thus, it has been customary to switch from ray fields to mode fields at some intermediate observation time. These general remarks for an elastic plate apply to any multiwave environment (for example, composite laminates). There is also another option of brute force integration over the plane wave spectral constituents in a Fourier integral decomposition of the localized source field, but this procedure is numerically inefficient at high frequencies, and it provides no physical interpretation of the result of the computations.

A new method, developed originally in electromagnetics, combines rays and modes self-consistently within a single framework so as to optimize the advantages of each. The method has been referred to as a hybrid ray-mode scheme [1]. Applications in structural mechanics have been primarily to wave processes like thickness shear or axial compression, which do not undergo coupling at the boundaries, but an extension to dilatational-shear coupling has also been developed [2-4]. The results obtained so far reveal significant advantages obtained by use of this procedure. We now outline their possible implications for nondestructive evaluation (NDE).

When a flaw, such as a crack, is located in a plate or a more general multiwave multilayer structure, the incident field, whether expressed as wavefronts or as normal modes, is scattered. At high frequencies and short response times, the edges of the crack act as scattering centers which send diffracted fields to the receiver [5]. These diffracted fields undergo multiple reflection at the layer boundaries, and are therefore subject to the difficulties described above. By employing the hybrid concept for the scattered field, one may attempt to construct an entirely new theory of scattering, applicable for short as well as long observation times. At short times, the wavefronts, which then predominate in the hybrid scheme, would serve by their different arrival times and paths to locate the edges of the crack [5]. At later times, the multiple fields diffracted along the crack, between the two edges, can be considered collectively to generate resonances which contain global information about the crack structure [6]. The same considerations can be applied to other flaws like voids and inclusions. These concepts have been employed in electromagnetics to structure a new hybrid wavefront-resonance theory of target identification, using broadband incident fields [7]. Because of the promising results obtained there, it seems worthwhile to explore these ideas in reference to in situ location of cracks, inclusions, interlamina debonding or weakened regions, etc., within an elastic layered environment.

2. Hybrid Green's Function for Layered Structures

Analytical modeling of scattering from flaws in a multilayer material requires the availability of the green's function for the unflawed environment. Any perturbation in the medium can then be expressed in terms of equivalent sources, equivalent boundary conditions, etc., in the unperturbed background. It is desirable to explore alternative formulations of the multilayer dyadic Green's function since the wave constituents employed in a particular formulation may favor the modeling of phenomena associated with a certain class of parameters and observables, both as regards computational efficiency and the most cogent physical interpretation of the result. The conventional available options are:

a) plane wave spectral representation

b) guided mode representation (discrete plus continuous spectrum)

c) (generalized) ray representation

For plane layered media, these options are inherently exact. Option a) highlights the features associated with the various plane waves (propagating and evanescent) that synthesize the source distribution. Peaks in the plane wave spectra expressed as a function of horizontal slowness occur at modal resonances and thereby serve to identify the mode spectral pattern. By direct numerical integration (for example, via the Fast Fourier Transform), one can generate the field solution. This method of calculation is numerically time consuming at high frequencies; it can create computational difficulties because singularities of the spectral integrand are located on the (real) axis of integration: branch points at the spectral edges of the propagating wave spectrum and poles at the modal resonances (unless dissipation in the layers is included, which displaces the resonance poles into the complex domain).

The guided mode representation in b) results from the spectral integral in a) by contour deformation combined with residue and branch cut integral calculus in the complex wavenumber plane [2,4]. For multilayered multiwave structures, the complete spectrum of the propagating modes – and of the virtual or leaky modes, if any, that can represent dominant effects of the continuous spectrum – may be difficult to ascertain, especially in those parameter regimes where the modes are clustered. Moreover, even after the modes have been identified, modal summation may be inefficient, especially at high frequencies where many modes are required. The generalized ray representation in c) resolves the response into traveling wave events that are tracked along multiply reflected ray paths to the observer. Evidently, such a scheme becomes intractable and physically obscure if very many multiples must be included.

The hybrid ray-mode method seeks to overcome the disadvantages associated with the various individual options listed above. By combining rays and modes in a self-consistent framework, one may choose a mix that exploits the best features of each. Since ray fields and mode fields are, in a sense, each other's Fourier transforms, their convergence properties are complementary: spectral domains containing clusters of the one are sparsely filled by the other. The bilateral ray-mode equivalence is formalized by the Poisson summation which expresses a finite (or infinite) sum of elements as a sum of finite (or infinite) Fourier transforms of these elements, plus a remainder [1]. If the initial elements are rays, the transformed elements are modes, and vice versa. The computable remainder is often small and can then be neglected. The ray-mode equivalent is useful not only for expressing groups of poorly convergent ray fields in terms of a better behaved group of mode fields, and vice versa, but it also allows treatment of clusters of multiply reflected ray fields in terms of a single "collective" ray field which follows a ray trajectory but whose amplitude is "loaded" with the internal multiples within a layer (Figure 1). This establishes further options that make possible the tracking of otherwise intractable events.

Because of wave species coupling and the consequent ray splitting at each encounter with a boundary or interface in a multiwave medium, new spectral objects have had to be devised to implement the hybrid ray-mode method in such an environment. These are what we have called "eigenrays": unique combinations of plane waves representing each of the wave species in the multiwave bulk, which propagate through multiple reverberations without change, like a ray field in a

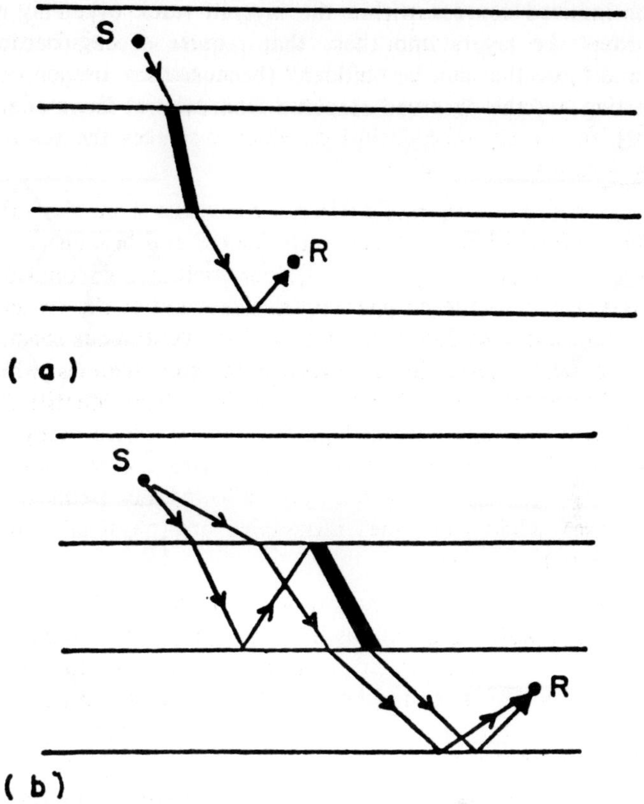

Figure 1: *Schematization of collective scheme for a particuar ray. The collective ray accounts in composite form for all omitted multiple reflections in the hidden layer. The heavy segments denote the collective loading. Collective treatment can be applied when all affected multiples in the hidden layer have approximately the same inclination angles. (a) Source and receiver with small lateral separation. Ray from source to receiver passing directly through the hidden layer can be weighted collectively to describe the composite effect of that layer. (b) Source and receiver with larger lateral separation. Retention of direct and first multiple ray in the hidden layer, with collective weighting of the latter for all higher order multiples.*

single-wave medium. We have also developed a new form of the multiwave multilayer dyadic Green's function which favors *simultaneous* treatment of *multiple* actual or induced sources within the layered stack [4]. This new representation, which orders the layers into those that require access (because they contain a source) and those that can be "hidden" (because they are source free), may be useful in sorting out the observed response from several flaws or sources of emissions (Figure 2).

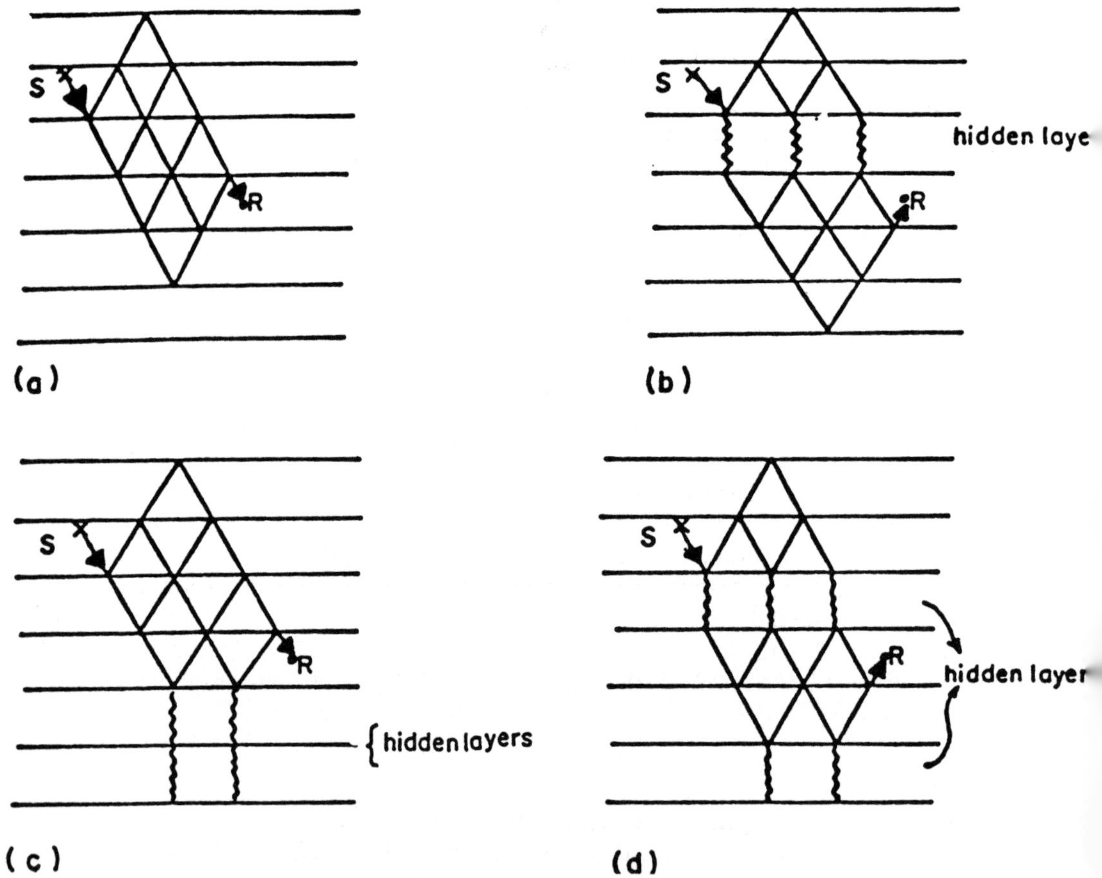

Figure 2: Typical ray traces with 3 reverberations (n=3) for various models with and without "hidden" layers. Ray splitting at boundaries into various wave species is omitted in order not to complicate the sketch. (a) No hidden layers. (b) Rays with reverberations in 3rd layer treated collectively. (c) Rays with reverberations in 5th and 6th layers treated collectively. (d) Rays with reverberations in 3rd and 6th layers treated collectively.

The new representations described above have been completed in exact form. They have also been reduced by asymptotic approximations, which simplify the numerical treatment substantially and impart physical content into the formal spectral integrals. Details may be found in the cited references. A frequency domain numerical test for a single elastic plate, performed to assess the utility of the hybrid scheme and the range of validity of asymptotic approximations, has been reported in reference [2]. For a sample calculation involving a high frequency pulsed source, results are shown in Figure 3 [3].

3. Hybrid Wave-Front-Resonance Formulation of Transient Scattering

Progressing waves (wavefronts) and oscillatory waves (resonances) may be employed as alternative building blocks to synthesize the transient fields scattered by an object [1]. In a progressing wave formulation, one tracks a causal wavefront from the source to the scatterer, and thereafter observes contributions corresponding to multiple wavefront passes around the object and (or) multiple diffractions from scattering centers located on the object. This locally sensitive description becomes cumbersome at late observation times when many wavefronts have had time to reach the observer. Moreover, analytic tractability, by asymptotic considerations, usually is limited to moderate times after passage of a wavefront, with consequent difficulties when early arrivals are monitored at late times.

The oscillatory representation, formalized by the Singularity Expansion Method (SEM) [8], expresses *global* phenomena due to the scatterer as a whole. These global responses, which take the form of (damped sinusoidal) resonances, may be shown to account for the cumulative effect of multiple wavefront arrivals [7,9]. Therefore, this formulation is most convenient for late observation times, whereas at early times, convergence problems may arise due to the many resonances needed to synthesize the abruptly changing local surface field near a wavefront or the causal zero field before the first wavefront has arrived. In the formal structure of the SEM, based on the solution of an integral equation formulation, the SEM resonance series for the scattered field has to be augmented by an entire function [10]. The entire function, in turn, influences the excitation coefficients of the resonances.

Our recently developed hybrid representation avoids the difficulties associated with either of the above descriptions by incorporating the well-behaved portions – wavefronts at early times and resonances at later times – within a single self-consistent framework that is applicable and convenient for all times. The hybrid approach – illustrated in references [6], [7] and [9] for two-dimensional objects – formalizes the intimate relation between wavefronts and resonances as cumulative phenomena of the one in terms of the other, and thereby furnishes a cogent physical picture of the scattering process. The bilateral equivalence is established directly from an analysis based on the concepts of the Geometrical Theory of Diffraction (GTD) and avoids the use of integral equations entirely. The connection between wavefronts and resonances in terms of the physically motivated constructs

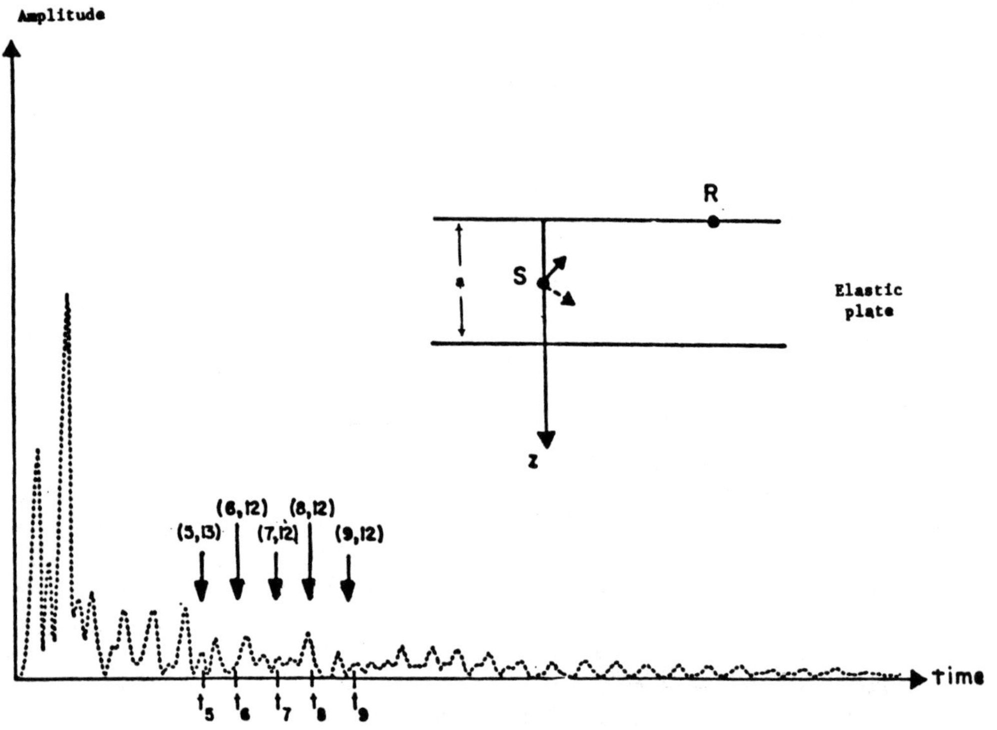

Figure 3: *Transient response due to a high frequency Gaussian pulsed compressional line forcing function at S observed at R in an elastic plate. Various hybrid combinations* (N,M_N) *containing N ray fields,* M_N *modes and remainder are calculated for* $N=5,6,...,9$*, where N is the number of reflections of the* $p-SV$ *coupled ray fields. The arrival times* t_N *of the ray fields are also shown. Differences between responses for each of the various hybrid combinations are too small to be depicted in the drawing. Each hybrid solution also agrees with an independent reference solution. It has been confirmed that the multiple arrivals at late times are well expressed collectively by 12 modes plus remainder (i.e., omitting rays entirely). 125 modes would be required if mode summation were used exclusively. The peaked ray arrivals at early times and the smoothed modal arrivals at later times, combined selfconsistently within the hybrid format, evidently provide a cogent interpretation of the observed response.*

of GTD furnishes an interpretation that has remained hidden within the integral equation format.

For a smooth convex object, resonances are established by multiple wavefront passes of individual traveling creeping waves [9], which arrange themselves by phase coherent closure into standing wave patterns that yield damped oscillations. Assuming a time dependence $exp(-i\omega t)$, the resonant frequencies ω_m are arranged in layers in the lower half of the complex ω-plane, each layer being generated by a particular creeping wave. These resonances are poles in the Laplace inversion integral. For a flat strip, or more generally on any surface contour with edges, edge diffraction provides the interaction mechanism (Figure 4a) [6],[11]. Self-consistent closure of these edge-diffracted GTD ray fields results for the dominant resonances in an equation of the form

$$1-D^2C^2(kd)=0 \tag{1}$$

where d is the strip width, D represents the known edge diffraction coefficient for a surface ray diffracted back along the surface, and $C(kd)$ is the cylindrical wave field generated at one edge by an equivalent line source of unit strength located at the other edge. When this equation is solved numerically for the acoustically hard and soft boundary conditions, one obtains the resonances in Figure 4b. These results, labeled GTD, are compared with independently computed results, using the method of moments on the strip integral equation [12]. Agreement is seen to be excellent, even for resonances with low frequencies, although GTD is a *high-frequency* method! It is to be noted that the method of moments is most reliable at low frequencies but less so at higher frequencies.

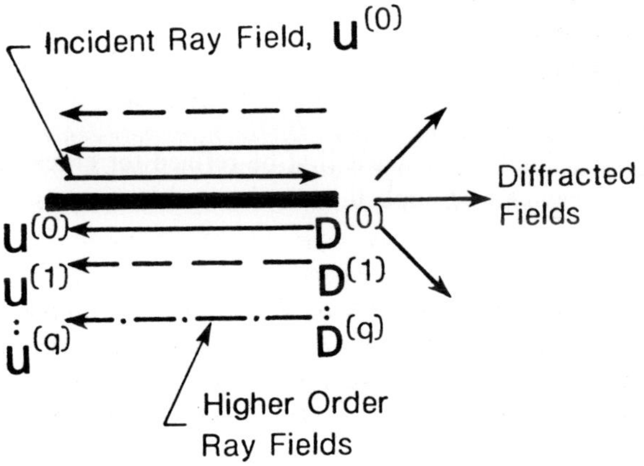

Figure 4(a): Multiple interaction mechanism, and complex resonance map, for an acoustically hard flat strip. n-th order diffracted ray field $U^{(n)}$ with edge diffraction coefficient $D^{(n)}$.

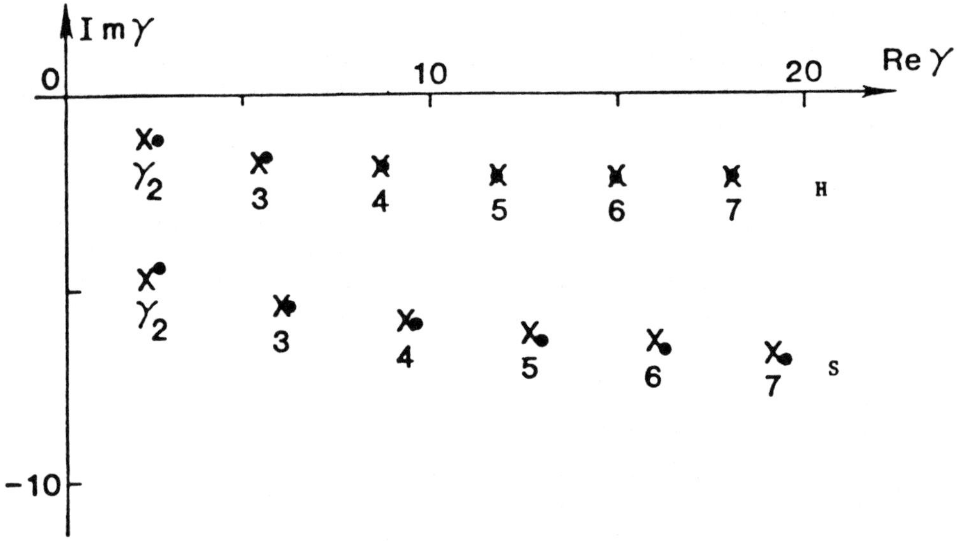

Figure 4(b): Multiple interaction mechanism, and complex resonance map, for an acoustically hard flat strip. Solutions of the resonance equation in (1) for the SEM poles γ_m in the $\gamma = kd$ plane. Only those in the fourth quadrant are shown. ××× *GTD method; -- moment method [12]. H = acoustically hard; S = acoustically soft.*

The conventional GTD format for the strip obstacle furnishes only a single layer of resonances, the dominant layer closest to the real frequency axis. Higher order layers, which are known to exist from the moment method solutions, are generated by newly introduced higher order diffracted surface rays [11], schematized by the dashed arrows in Figure 4a. These higher order resonances have been calculated in reference [11].

The concepts outlined above must first be refined for cracks in an infinite elastic medium before they can be applied to the layered laminate configuration.

4. Summary

In this paper, we have summarized new approaches to time-harmonic and time-dependent radiation and propagation in multiwave multilayer media, and to transient scattering by various target shapes. These separately developed techniques, when extended to elastic materials and layered structures simultaneously, appear to be promising for analytical modeling of NDE. It is hoped to explore these aspects in the future and also to generalize them, by localization, to configurations that depart weakly from the plane stratified prototype.

Acknowledgement

The research summarized here has been supported by the National Science Foundation under Grant No. EAR-8213147 and by the Office of Naval Research under Contracts No. N-00014-79-C-0013 and N-00014-83-K-00214.

References

[1] Felsen, L.B., "Progressing and Oscillatory Waves for Hybrid Synthesis of Source Excited Propagation and Diffraction", invited paper, AMD/ASME Symposium on Wave Propagation in Inhomogeneous Media and Ultrasonic NDE, San Antonio, Texas, June 17-21, 1984, published in *J. Appl. Mechanics*, 52, 1985, pp. 27-32.

[2] Lu. I.T. and Felsen, L.B., "Ray, Mode and Hybrid Options for Source Excited Propagation in an Elastic Layer", *J. Acoust. Soc. Am., 78*, 1985, pp. 701-714.

[3] Lu, I.T. and Felsen, L.B., "Ray, Mode and Hybrid Options for Transient Source Excited Propagation in an Elastic Layer", *Geophys. J. Roy. Astron. Soc., 86*, 1986, pp. 177-201.

[4] Lu, I.T. and Felsen, L.B., "Matrix Greens' Functions for Array-Type Sources and Receivers in Multiwave Layered Media", *Geophys. J. Roy. Astron. Soc., 84*, 1986, pp. 31-42.

[5] Achenbach, J.D. and Norris, A.M., "Ray Methods for Scattering of Ultrasonic Waves by Cracks", *Germany-U.S. Workshop on Research and Development Toward New Procedures in Non-destructive Testing*, Saabrucken, W. Germany, 1982, p. 369.

[6] Shirai, H. and Felsen, L.B., "Wavefront and Resonance Analysis of Transient Scattering by a Perfectly Conducting Flat Strip", *IEEE Trans. Antennas and Propag., AP-34*, 1986, p. 1196-1207.

[7] Heyman, E. and Felsen, L.B., "A Wavefront Interpretation of the Singularity Expansion Method", *IEEE Trans. Antennas and Propag.*, AP-33, 1985, pp. 706-718.

[8] Baum, C.E., "The Singularity Expansion Method", *Transient Electro-megnetic Fields*, L.B. Felsen (Ed.), Springer Verlag, 1976, New York.

[9] Heyman, E. and Felsen, L.B., "Creeping Waves and Resonances in Transient Scattering by Smooth Convex Objects", *IEEE Trans. on Antennas and Propag.*, AP-31, 1983, pp. 426-437.

[10] Pearson, L.W., "The Singularity Expansion Method Representation of Surface Currents on a Perfectly Conducting Scatterer", *Hybrid Formulation of Wave Propagation and Scattering*, L.B. Felsen, (Ed.), Nijhoff Publishing Co., The Hague, Netherlands, 1984.

[11] Shirai, H. and Felsen, L.B., "Modified GTD for Generating Complex Resonances for Flat Strips and Disks", *IEEE Trans. Antennas Propag., AP-34*, 1986, pp. 779-790.

[12] Hwu, S.U. and Pearson, L.W., "Numerically Determined Natural Resonances for an Infinitely Long Perfectly Conducting Strip", paper presented at *International IEEE/AP-S and URSI Symposium*, Boston, Mass., June, 1984.

MONITORING ELASTIC STIFFNESS DEGRADATION IN GRAPHITE/EPOXY COMPOSITES*

R.D. Kriz**
Fracture and Deformation Division
National Bureau of Standards
Boulder, Colorado

Abstract

Many stiffness-critical aerospace structures exploit the high specific stiffness of graphite/epoxy composites. Elastic-stiffness degradation of these materials is therefore important. Here a nondestructive technique is described that measures stiffness degradation of the graphite-fibers and epoxy-matrix. This technique monitors variations in the direction of stress-wave propagation (energy-flux) corresponding to a change in composite stiffness.

** NRC-NBS Postdoctoral Research Associate, 1981-1982.

1. Theoretical Background

1.1. *Energy-flux Vector*

Using Huygens' Principle, Musgrave graphically demonstrates how the energy associated with a stress wave can move along a path oblique to the wave-front normal [1]. This direction of stress-wave propagation is referred to as the energy-flux vector, which is defined as the amount of energy flowing per unit time per unit area out of a control volume.

$$E_j = -C_{ijk\ell}\, u_{k,\ell}\, \dot{u}_i$$

where

$C_{ijk\ell}$, elastic stiffness tensor

u_i, displacement vector

$\dot{u}_i$, time derivative $(\partial u_i/\partial t)$

$u_{k,\ell}$, differentiation with respect to coordinate (x_ℓ)

Love also derives this relationship by minimizing the sum of kinetic and potential energies [2].

Merkaulov and Yakoviev use the Schlieren technique to permit visualization of the direction of stress-wave propagation in quartz crystals [3]. Using this technique, simultaneous generation of quasi-longitudinal (QL) and quasi-transverse (QT) waves can be observed to propagate in directions that deviate from the wave normal.

Green provides a discussion that demonstrates how the texture of polycrystalline aggregates can influence the solid's mechanical properties, including ultrasonic wave propagation [4]. Unidirectional graphite/epoxy composites, which like textured polycrystalline aggregates are anisotropic and exhibit a deviation in the energy-flux vectors for both QL and QT waves [5].

1.2. *Stiffness Degradation*

Extreme environmental conditions (temperature, moisture, and radiation) can degrade graphite/epoxy elastic stiffnesses. This elastic stiffness degradation is caused by a degradation of the fiber-matrix components. The epoxy-matrix used in this study is a two-part diaminodiphenyl-sulphone (DDS-cured tetraglycidyl 4,4′ diaminodephenyl methane (TGDDM) that was saturated with moisture. When moisture is absorbed into TGDDM-DDS resin by diffusion ("wet"), the elastic properties degrade as shown in Table 1 [6]. The graphite-fiber elastic properties, obtained from Ref. 5, are assumed not to degrade. The composite elastic properties, listed in Table 1, are calculated from fiber and matrix elastic properties using Hashin's transversely isotropic relations [7]. Elastic-constant subscripts used in Table 1 are defined in Figure 1: for example, the plane-strain bulk modulus, K_{23}, occurs in the 2-3 plane.

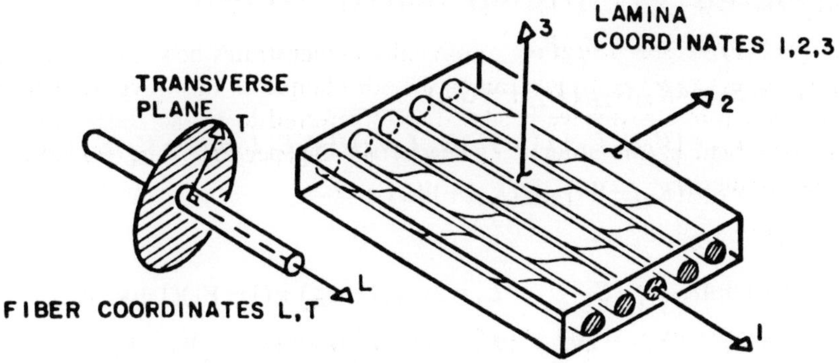

Figure 1: Coordinate axes definition

Table 1: Dry and degraded wet elastic properties

Material	Elastic Properties (GPa) Dry/Wet					
Matrix TGDDM-DDS	Young's Modulus, E		Shear Modulus, G		Poisson's Ratio, ν	
	5.35/3.96		1.95/1.43		0.353/0.388	
Graphite	E_L	E_T	G_{LT}	G_{TT}	ν_{LT}	ν_{TT}
Fiber Ref. 5	232/232	15.0/15.0	24.0/24.0	5.02/5.02	0.290/0.290	0.490/0.490
Composite Lamina V_f= 0.55	E_1	E_2	G_{12}	G_{23}	ν_{12}	ν_{23}
	130/129	9.72/8.69	5.39/4.16	3.25/2.83	0.308/0.323	0.492/0.323

V_f, Fiber Volume Fraction

$$K_{23}=K_m+V_f/[1/(K_{TT}-K_m)+(1-V_f)/(K_m+G_m)]$$

$$G_{23}=G_m\frac{(1+\alpha_1 V_f^3)(\alpha_2+\beta_1 V_f)-3V_f(1-V_f)^2\beta_1^2}{(1+\alpha_1 V_f^3)(\alpha_2-V_f)-3V_f(1-V_f)^2\beta_1^2}$$

where

$$\alpha_1=(\beta_1-\gamma\beta_2)/(1+\gamma\beta_2),\quad \alpha_2=(\gamma+\beta_1)/(\gamma-1)$$

$$\beta_1=1/(3-4\nu_m),\ \ \beta_2=K_{TT}/(K_{TT}+2G_{TT}),\ \ \alpha=G_{TT}/G_m$$
$$G_{12}=G_{13}=G_m+V_f/[1/(G_{LT}-G_m)+(1-V_f)/2G_m]$$
$$E_1=V_fE_L+(1-V_f)E_m$$
$$E_2=E_3=4K_{23}G_{23}/(K_{23}+\Psi G_{23})$$
$$\nu_{12}=\nu_{13}=[V_fE_LL_1+(1-V_f)E_mL_2\nu_m]/[V_fE_LL_3+(1-V_f)E_mL_2]$$
$$\nu_{23}=\nu_{32}=(K_{23}-\Psi G_{23})/(K_{23}+\Psi G_{23})$$

where

$$\Psi=1+4K_{23}\nu_{12}^2/E_1\ ,\qquad L_1=2\nu_{LT}(1-\nu_m^2)+(1-V_f)(1+\nu_m)\nu_m$$
$$L_2=V_f(1-\nu_{LT}-2\nu_{LT}^2)\ ,L_3=2(1-\nu_m^2)V_f+(1+\nu_m)(1-V_f)$$

2. Monitoring Stiffness Degradation

Moisture degradation of TGDDM-DDS resin can alter elastic anisotropy of the graphite/epoxy composite. This change in texture caused by absorbed moisture ("wet") will increase the direction of the QT stress-wave propagation (energy-flux) as illustrated in Figure 2. Energy-flux deviation angles, shown in Figure 3, are calculated over a range of unidirectional fiber orientations (0°−90°) using the method outlined in Ref. 1 for hexagonal materials. As shown in Figure 3, energy-flux vectors were also measured experimentally from the same batch of material used in Ref. 5. Stress waves were propagated at 5 MHz with 0.318-cm-dia. transducers. Specimens were all 0.318-cm thick and of proper length and width to receive both QL-deviated and QT-deviated wave component. The peaks of the wet and dry received wave forms illustrated in Figure 2 were assumed to be coincident with the centerline of the acoustic beam. All received wave forms were extremely narrow (0.40 cm or less), giving reproducibility in measuring the deviation angles to within ±1.0 degree. An array of transducers would provide a more accurate measurement of these deviation angles.

This technique does not measure a stiffness degradation directly but measures a change in anisotropy caused by preferential degradation of the fiber-matrix components of the composite. This preferential degradation of the fiber-matrix elastic properties can depend on environmental conditions. A decrease in the flux deviation angle of 5° is predicted, but not experimentally verified, when the set of fiber elastic properties listed in Table 1 are degraded 30 percent. It is therefore possible, depending on the environmental conditions, that both fiber and matrix elastic properties could degrade resulting in no net change in anisotropy, which could not be measured using this technique.

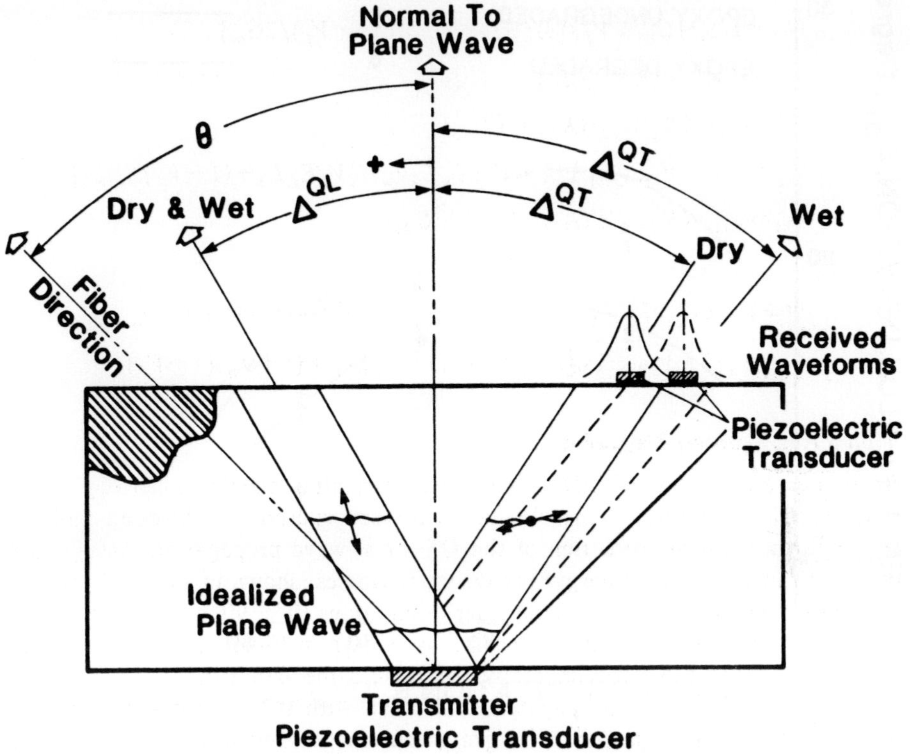

Figure 2: *Shift in quasi-transverse stress-wave propagation. Polarization of quasi-longitudinal (QL) and quasi-transverse (QT) waves is represented by double-sided arrows.*

3. Conclusions

It is possible to monitor epoxy degradation in graphite/epoxy composites by measuring an increase in the flux-deviation angle. Degradation owing to absorbed moisture has been experimentally verified to increase the energy-flux deviation angle by 3°.

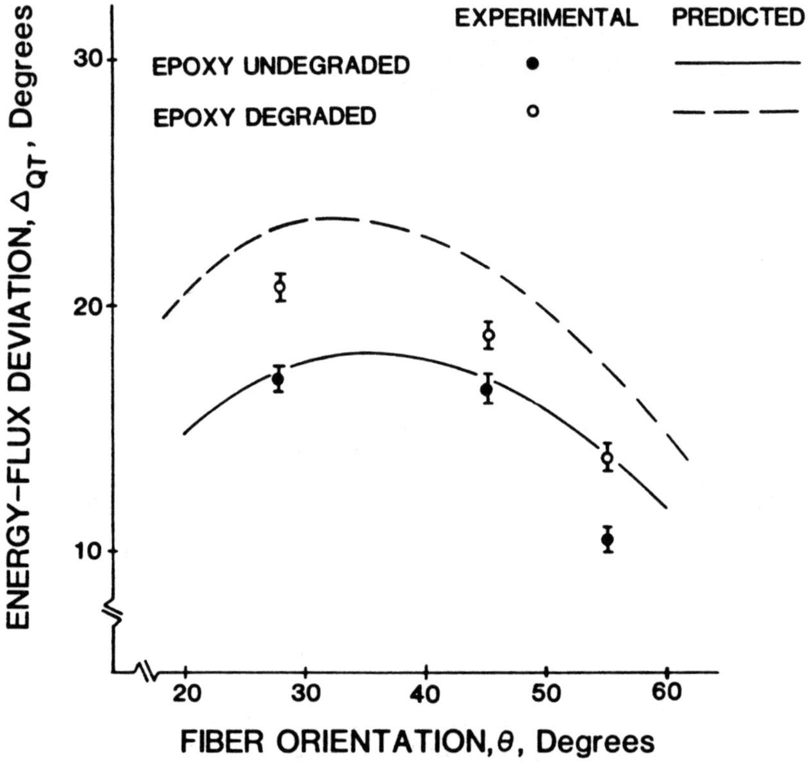

Figure 3: *Energy-flux deviations in graphite/epoxy, $V_f = 0.55$.*

References

[1] Musgrave, M.J.P., *Crystal Acoustics*, Holden-Day, 1970.

[2] Love, A.E.H., *A Treatise on the Mathematical Theory of Elasticity*, Dover 4-th edition, 1927, (Article 126), pp. 177-178.

[3] Merkaulov, L.G. and Yakoviev, L.A., Propagation and Reflection of Ultrasonic Beams in Crystals, *Sov. Phys. Acoust.*, 8, 1962, pp. 72-77.

[4] Green, R.E., Jr., *Treatise Materials Science and Technology*, Vol. 3, Academic Press, 1973.

[5] Kriz, R.D. and Stinchcomb, W.W., Elastic Moduli of Transversely Isotropic Fibers and Their Composites, *J. Exp. Mech.*, 19, 1979, pp. 41-49.

[6] Kriz, R.D. and Stinchcomb, W.W., Effects of Moisture, Residual Thermal Curing Stresses, and Mechanical Load on the Damage Development in Quasi-Isotropic Laminates, *Damage in Composite Materials*, *ASTM-STP*, 775, K.L. Feifsnider, (Ed.),, American Society for Testing and Materials, Philadelphia, Penn., 1982, pp. 63-80.

[7] Hashin, Z., Analysis of Properties of Fiber Composites with Anisotropic Constituents, *J. Applied Mech.*, 46, 1979, pp. 543-550.

MECHANICAL MODELING AND MEASUREMENTS ON FIBROUS COMPOSITES

Adnan H. Nayfeh
Department of Aerospace Engineering
and Engineering Mechanics
University of Cincinnati
Cincinnati, Ohio

D.E. Chimenti
Materials Laboratory
Wright-Patterson Air Force Base
Dayton, Ohio

Abstract

The mechanical behavior of fiber-reinforced composite has been investigated theoretically and experimentally. As an example, ultrasonic leaky Lamb wave propagation has been studied in detail to assess the validity of the modeling. From measurements of the reflected acoustic field from composites plates over a range of incident angles, experimental velocity dispersion curves for Lamb wave modes have been constructed. Because of the high degree of elastic anisotropy of composite laminates, these modes display behavior not typically found in isotropic plates. The theoretical model, based on a continuum-mixture approach, is found to account remarkably well for the data.

1. Introduction

Elastic wave propagation in fibrous composite materials has been the subject of numerous investigations in recent years. However, the morphology of fiber-reinforced composites can seriously complicate the calculation of their wave propagation properties. Typically, these materials are composed of brittle, high-stiffness fibers embedded in a matrix of more ductile material which bonds the fibers together and acts as a load-transfer medium. Since it is clearly not practical to attempt a solution of the completely general elastic-wave problem, most prior work [1-4] has employed various approximations to render the calculations tractable. Our own approach [5,6] to interacting continua offers an alternative procedure for modeling the response of composites, where in particular, a rational construction of the mixture momentum and constitutive-relation interaction terms is given. This theory leads to simple wave propagation equations which potentially contain the full influence of the microstructure, that is, the distribution of displacements and stresses within individual constituents of the composite.

Interest in the acoustic properties of these inhomogeneous and generally anisotropic materials is heightened by their increasing use in applications requiring high stiffness-to-weight ratios. Ultrasonic nondestructive evaluation is one useful means to insure the structural integrity of fibrous composites. In order to take full advantage of the information contained in an ultrasonic test [7], it is helpful to have a clear understanding of wave propagation characteristics in the material under examination. In a recent paper [8] we presented results of experimental measurements and theoretical calculations on ultrasonic leaky Lamb wave propagation in fiber-reinforced, unidirectional composite laminates. With the Lamb wave vector oriented parallel to the fiber direction, dispersion curves of phase velocity versus frequency and plate thickness have been constructed from measurements of ultrasonic reflection from fluid-loaded glass-epoxy composite plates. The experimental results are supported by a theoretical model of Lamb wave propagation in the composite plate. The model begins with an approximate calculation of the effective homogeneous transversely isotropic elastic behavior of a unidirectional composite laminate in the long-wavelength limit, using a two-step procedure based on alternating layered media. This intermediate continuum result is then incorporated into a calculation of the ultrasonic reflection coefficient of a fluid-loaded anisotropic plate, which is assumed to approximate the fibrous composite laminate. Good quantitative agreement with the model is found if the fiber volume fraction is taken to be an adjustable parameter.

In the following we shall extend our analysis to the case in which the fibers are assumed to be anisotropic. Specifically, we treat the case which exhibits transverse isotropy in the direction normal to fiber direction. An important composite example which exhibits this property is the unidirectional graphite-epoxy composite. In order to derive the effective composite properties we shall extend the analytical procedure outlined in [8].

2. Theoretical Summary

We consider the case of an acoustic wave incident on a fiber-reinforced composite plate which is immersed in a fluid. In practice, the plate consists of small, nearly parallel transversely isotropic fibers of circular cross section laid out on an approximately hexagonal array. A coordinate system is chosen such that the x-axis coincides with the average fiber direction and that the z-axis is normal to the plate-fluid boundaries. We further assume that the plate has infinite extent in the y-direction. The plan is to approximate the highly structured composite by a continuum, retaining the appropriate elastic anisotropy, and to analyze the ultrasonic reflection from such a plate to study the behavior of Lamb waves in these materials.

We now follow the procedure of [8] and employ a building-block approach to arrive at the two-dimensional results we seek. First, a layered structure composed of fiber and matrix is analyzed, deriving from this parallel-stress model the properties of "compound" layer 1. In the second step, the final model for the fibrous composite is established by treating the composite as consisting of the two-dimensional "compound" layer 1. In the second step, the final model for the fibrous composite is established by treating the composite as consisting of the two-dimensional "compound" layer 1 stacked in series with the (matrix) layer 2 (see [8], for details).

2.1. *Layered Model*

Consider the bilayered composite of Figure 1, composed of two types of laminates in welded contact and stacked normal to the y-direction. The translational invariance of the problem permits us to isolate a repeating unit, as indicated in Figure 1. On this cell all continuity and symmetry relations are displayed. The relevant equations of motion are given by

$$\frac{\partial \sigma_{ij}}{\partial r_i} = \rho \ddot{v}_j \quad , \quad i,j=1.2,3 \tag{1}$$

which apply for each constituent. The appropriate constitutive relation for the isotropic matrix is given by

$$\sigma_{jk} = \lambda \delta_{jk} \frac{\partial v_\ell}{\partial r_\ell} + \mu \left(\frac{\partial v_j}{\partial r_k} + \frac{\partial v_k}{\partial r_j} \right) , \tag{2}$$

and for the transversely isotropic fibers by

$$\sigma_{ij} = F_{ijk\ell} e_{k\ell} \quad , \quad i,j,k,\ell=1,2,3 \tag{3}$$

where appropriate restrictions on the elastic properties $F_{ijk\ell}$ to reflect transverse isotropy are assumed; these will be discussed later on in the analysis. In Equations (1), (2) and (3) summation over repeated variable indices in the same term is assumed. v_i are the components of the displacement u,v,w; r_i are the spatial coordinates x,y,x; and σ_{jk} are the elements of the elastic stress tensor; the material density is ρ, and λ and μ are the Lame constants for the isotropic material.

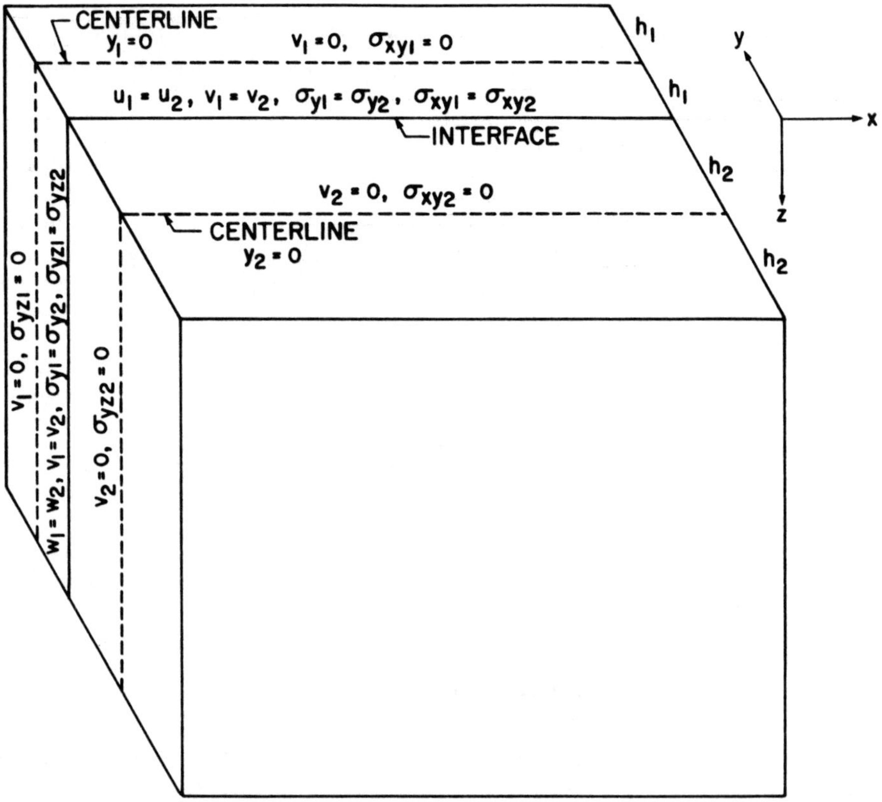

Figure 1: Schematic representation of the layered model generating unit cell with all symmetry and interface conditions

Following [8], applying the boundary conditions of Figure 1 and averaging Equation (1) across the cell thickness according to

$$(\bar{\ })_\alpha = h_\alpha^{-1}\int_0^{h_\alpha}(\ )_\alpha dy \quad (\alpha=1,2) \ ,$$

followed by invoking the low frequency approximation (i.e., $\bar{u}_1 \rightarrow \bar{u}_2 = \bar{u}$ and $\bar{w}_1 \rightarrow \bar{w}_2 = \bar{w}$ [8]) and summing the resulting two equations yields the momentum mixture equations in the x- and z-directions

$$\rho_c \frac{\partial^2 \bar{u}}{\partial t^2} = \frac{\partial \bar{\sigma}_{xx}}{\partial x} + \frac{\partial \bar{\sigma}_{xz}}{\partial z} \tag{4}$$

$$\rho_c \frac{\partial^2 \bar{w}}{\partial t^2} = \frac{\partial \bar{\sigma}_{zz}}{\partial z} + \frac{\partial \bar{\sigma}_{xz}}{\partial x} \tag{5}$$

where ρ_c and $\bar{\sigma}_{ij}$, $i,j = x,z$ are the weighted sums of the respective quantities or materials 1 and 2, e.g.,

$$\rho_c = \rho_1 n_1 + \rho_2 n_2 \tag{5a}$$

$$\bar{\sigma}_{ij} = n_1(\bar{\sigma}_{ij})_1 + n_2(\bar{\sigma}_{ij})_2 \tag{5b}$$

with $n_\alpha = h_\alpha/(h_1+h_2)$ being the volume fraction of materials 1 and 2 in the laminated plates model.

It now remains to drive expressions for the weighted stresses $\bar{\sigma}_{ij}$. This will be obtained from consideration of the constitutive relations (2) and (3). To this end, we specialize the relation (3) to that of the transversely isotropic case, and for facilitating the following analysis, write it in the expanded form

$$\sigma_{xx} = F_{11}\frac{\partial u}{\partial x} + F_{12}\frac{\partial v}{\partial y} + F_{12}\frac{\partial w}{\partial z} \tag{6a}$$

$$\sigma_{yy} = F_{12}\frac{\partial u}{\partial x} + F_{22}\frac{\partial v}{\partial y} + F_{23}\frac{\partial w}{\partial z} \tag{6b}$$

$$\sigma_{zz} = F_{12}\frac{\partial u}{\partial x} + F_{23}\frac{\partial v}{\partial y} + F_{22}\frac{\partial w}{\partial z} \tag{6c}$$

$$\sigma_{yz} = F_{44}\left(\frac{\partial w}{\partial y} + \frac{\partial v}{\partial z}\right), \quad \sigma_{xz} = F_{55}\left(\frac{\partial w}{\partial z} + \frac{\partial u}{\partial z}\right) \tag{6d,e}$$

$$\sigma_{xy} = F_{55}\left(\frac{\partial v}{\partial x} + \frac{\partial u}{\partial y}\right) \tag{6f}$$

where F_{11}, F_{12}, F_{22}, F_{23}, F_{44} and F_{55} stand for F_{1111}, F_{1122}, F_{2222}, F_{2233}, F_{2323} and F_{1313}, respectively.

If the two sets of Equations (2) and (6a-f) are averaged across laminates thickness for $(\sigma_{ij})_\alpha$, $i,j=x,z$ and $\alpha=1,2$ followed by satisfying the appropriate continuity conditions of Figure 1 and finally constructing the sums (5b) one gets

$$\bar{\sigma}_{xx} = (F_{11}n_1 + E_2 n_2)\frac{\partial \bar{u}}{\partial x} + (F_{12}n_1 + \lambda_2 n_2)\frac{\partial w}{\partial z} + (F_{12} - \lambda_2)v^* \tag{7}$$

$$\bar{\sigma}_{zz} = (F_{12}n_1 + \lambda_2 n_2)\frac{\partial \bar{u}}{\partial x} + (F_{22}n_1 + E_2 n_2)\frac{\partial \bar{w}}{\partial z} + (F_{23} - \lambda_2)v^* \tag{8}$$

$$\bar{\sigma}_{xz} = \mu_c\left(\frac{\partial \bar{w}}{\partial x} + \frac{\partial \bar{u}}{\partial x}\right) \tag{9}$$

where $E_2 = \lambda_2 + 2\mu_2$, $\mu_c = (F_{55}n_1 + \mu_2 n_2)$ is the mixture shear modulus and v^* is the constitutive relations interaction term given by

$$v^* = v_1(h_1)/(h_1+h_2) \tag{10}$$

with $v_1(h_1)$ being the interfacial transverse displacement component of material 1. By equating the average stress components $(\bar{\sigma}_{yy})_1$ and $(\bar{\sigma}_{yy})_2$ we can solve for v^* as

$$v^* = \frac{1}{G}\left[(\lambda_2 - F_{12})\frac{\partial \bar{u}}{\partial x} + (\lambda_2 - F_{23})\frac{\partial \bar{w}}{\partial z}\right] \tag{11}$$

where

$$G = \frac{F_{22}}{n_1} + \frac{E_2}{n_2}\,. \tag{12}$$

Substituting from (11) into (7) and (8) one finally gets

$$\bar{\sigma}_{xx} = E_c\frac{\partial \bar{u}}{\partial x} + \lambda_c\frac{\partial \bar{w}}{\partial z} \tag{13}$$

$$\bar{\sigma}_{zz} = \bar{E}_c\frac{\partial \bar{w}}{\partial z} + \lambda_c\frac{\partial \bar{u}}{\partial x} \tag{14}$$

where

$$E_c = F_{11}n_1 + E_2 n_2 - (\lambda_2 - F_{12})^2/G \tag{15a}$$

$$\lambda_c = F_{12}n_1 + \lambda_2 n_2 - (\lambda_2 - F_{12})(\lambda_2 - F_{23})/G \tag{15b}$$

$$\bar{E}_c = F_{22}n_1 + E_2 n_2 - (\lambda_2 - F_{23})^2/G\,. \tag{15c}$$

Equations (4), (5), (9), (13) and (14) constitute the two-dimensional mixture equations for the laminated plates model.

2.2. *Fibrous Model*

Building on the above calculation we now construct the effective elastic properties for a fibrous composite by considering the "compound" layer 1 stacked in series with the matrix material of layer 2 normal to the z-axis. Similar to the procedure of [8] by following the development that led to Equations (9), (13) and (14), we define new elastic constants appropriate to mixtures in parallel and series of the properties of compound layer 1 and matrix layer 2. These relations take the series combination form

$$\rho = \rho_c n_c + \rho_2 n_2^* \tag{16a}$$

$$E = E_c n_c + E_2 n_2^* \tag{16b}$$

and the parallel combination form

$$E^* = \bar{E}_c E_2/(\bar{E}_c n_2^* + \mu_2 n_c) \tag{17a}$$

$$\mu^* = \mu_c \mu_2/(\mu_c n_2^* + \mu_2 n_c) \tag{17b}$$

whereas

$$\lambda^* = (\lambda_c n_c/E_c + \lambda_2 n_2^*/E_2)(n_c/\bar{E}_c + n_2^*/E_2)^{-1}\,. \tag{18}$$

Here n_c is the volume fraction of the compound layer and $n^*{}_2$ is that of the matrix layer (see [8]). Now a new set of two-dimensional field equations can be constructed which are formally identical to Equations (1), (or Equations (4) and (5)), but where instead of Equations (2) and (3) we have the mixture constitutive relations

$$\sigma_x = E\frac{\partial \bar{u}}{\partial x} + \lambda^* \frac{\partial \bar{w}}{\partial z} \tag{19a}$$

$$\sigma_z = E^* \frac{\partial \bar{w}}{\partial z} + \lambda^* \frac{\partial \bar{u}}{\partial x} \tag{19b}$$

$$\sigma_{xz} = \mu^* \left(\frac{\partial \bar{u}}{\partial z} + \frac{\partial \bar{w}}{\partial x} \right). \tag{19c}$$

3. Derivation of Reflection Coefficient

Consider the fibrous composite modeled above to be in the form of a plate of thickness d and immersed in an infinite fluid medium. The upper and lower surfaces of the plate are situated at $z=0$ and $z=d$, and an incident harmonic sound wave propagating in the $x-z$ plate insonifies the plate. The incident wave vector k_f makes an angle Θ with the z-axis, and all particle motion is assumed to be confined to the $x-z$ plane. The calculation of the plane-wave reflection coefficient from a fluid-loaded, transverseey isotropic plate proceeds in a straightforward manner. Substitution of the constitutive relations of Equations (19) into the equations of motion (4) and (5) yields coupled, second-order partial differential equations for the displacements. Making the usual ansatz for the formal solutions leads to a system of linear simultaneous equations for the displacement amplitudes. These equations are

$$D_{ij}A_j = Y_i \quad i,j=1,\ldots,6 \tag{20}$$

with

$$D_{1k} = \lambda^* + E^* \alpha_k W_k\,, \quad D_{15} = \rho_f c^2\,, \quad D_{16} = 0$$

$$D_{2k} = W_k + \alpha_k\,, \quad D_{25} = D_{26} = 0$$

$$D_{3k} = D_{1k} P_k\,, \quad D_{35} = 0\,, \quad D_{36} = D_{15}$$

$$D_{4k} = D_{2k} P_k\,, \quad D_{45} = D_{46} = 0$$

$$D_{5k} = W_k\,, \quad D_{55} = \alpha_f\,, \quad D_{56} = 0$$

$$D_{6k} = D_{5k} P_k\,, \quad D_{65} = 0\,, \quad D_{66} = -D_{55}$$

$$P_k = \exp[i\xi\alpha_k d]$$

$$W_k = (\rho c^2 - E - \alpha_k^2 \mu^*)/[(\lambda^* + \mu^*)\alpha_k]\,, \quad k=1,2,3,4$$

$$\alpha_f^2 = \frac{c^2}{c_f^2} - 1$$

$$Y_1 = -D_{15}\,, \quad Y_5 = D_{55}\,, \quad Y_2 = Y_3 = Y_4 = Y_6 = 0 \tag{21}$$

and α_k are the solutions of the quartic equation

$$(E-\rho c^2)(\mu^*-\rho c^2)+[(E-\rho c^2)E^*+(\mu^*-\rho c^2)\mu^*-(\lambda^*+\mu^*)^2]\alpha^2 + E^*\mu^*\alpha^4=0\,. \tag{22}$$

In the above relations the index k runs between 1 and 4, ξ is the wave number, c is the phase velocity, c_f and ρ_f are the fluids wave speed and density, respectively, and d is the plate thickness.

Using Cramer's Rule the reflection coefficient R can be obtained from the relation (20). This leads to the determinantal ratio

$$R=\frac{|D^*_{ij}|}{|D_{ij}|} \tag{23}$$

where all the entries of D^*_{ij} are the same as those of D_{ij} with the exception $D^*_{15}=-D_{15}$.

4. Experimental Technique

An experimental study has been performed in conjunction with our theoretical investigations to test the range of validity of the model calculation and its dependence on significant parameters. We have varied the plate thickness, composite fiber content, ultrasound frequency, and incidence angle in the course of many measurements on several different samples of unidirectional glass-epoxy composite plates. Elastic constants for two material constituents are collectd in [9].

A schematic of the experimental geometry is given in Figure 2. An ultrasonic beam from a damped piston transducer insonifies the plate at the selected incident angle. At frequencies corresponding to generation of Lamb waves in the plate, the reflected ultrasonic field will show the displacement and distortion characteristics of the presence of propagating waves in the plate, as illustrated schematically in Figure 2. This phenomenon, which is expected on the basis of the rapid phase variation in the reflection coefficient, is analogous to our earlier observations on leaky Rayleigh waves [10]. Not illustrated in Figure 2 is the transmitted field below the plate. Away from the resonance condition for efficient mode conversion from a compressional wave in the fluid to a Lamb wave in the plate, ultrasonic reflection should be nearly specular. By positioning our receiver at the strong minimum in the reflected beam profile of Figure 2, the finite-beam distortion effect permits detection of Lamb waves in the plate.

Data are acquired in these experiments by exciting a broadband ultrasonic transducer with low-level rf tone bursts from a function generator. The 20-μsec bursts propagate through a water path long enough to place the sample in the transducer's far field at all frequencies. The reflected acoustic beam is detected by a second, matched transducer positioned at the same angle and distance to the sample. Data are accumulated by discretely stepping frequency and recording the reflected signal with a broadband receiver and video detector, sampled by a gated integrating amplifier.

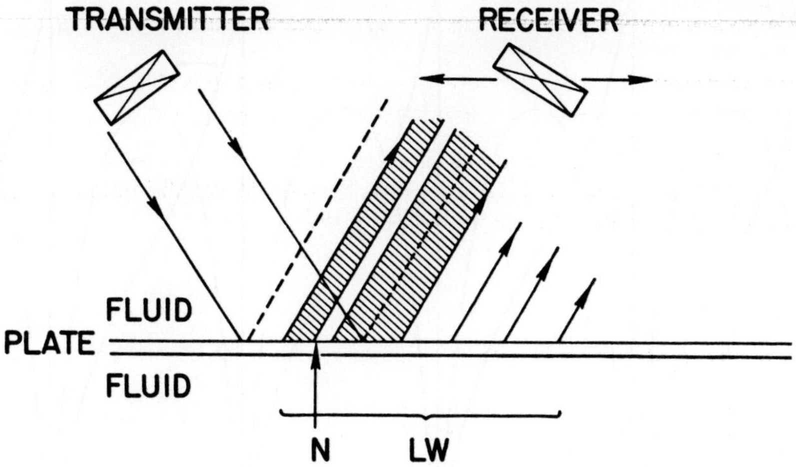

Figure 2: Experimental geometry of Lamb-wave measurements in fibrous composite plates. Transmitter is fixed with respect to sample, while receiver scans parallel to interface. Dashed lines show specular reflection. When Lamb wave is present, shaded region contains most of reflected energy. Null zone is denoted by N, and leaky wave reflected field by LW. A similar radiation pattern, not shown, exists below this plate.

To verify our experimental procedure and data analysis, we have undertaken some preliminary measurements on aluminum plates. Lamb waves in aluminum plates have been studied extensively and so offer a good opportunity for comparson. Very good detailed agreement at incident angles from 14 to 26 degrees gives us confidence in the validity of our approach.

5. Results and Discussion

For our illustrations we choose graphite-epoxy and glass-epoxy composites. Here, the graphite fibers are transversely isotropic whereas the glass fibers are isotropic. In Figure 3(a) we demonstrate the results of the calculations discussed above, where the magnitude and phase of the reflection coefficient for a 1-mm thick graphite-epoxy plate are presented as a function of frequency. Here, the incident angle is 22 degrees, corresponding to a mode phase velocity of 4 km/sec. In deriving the composite properties listed in [9] it is assumed that the fiber volume fracture is 0.63. The sharp minima and rapid phase reversals in the calculations of Figure 3(a) indicated the presence of propagating modes at those values of the frequency. In the lower curve, Figure 3(b), the phase velocity is 14 km/sec, and here the behavior is complicated by the pairwise occurrence of Lamb wave modes. In the dispersion curve at this incident angle the symmetric and antisymmetric modes approach each other closely.

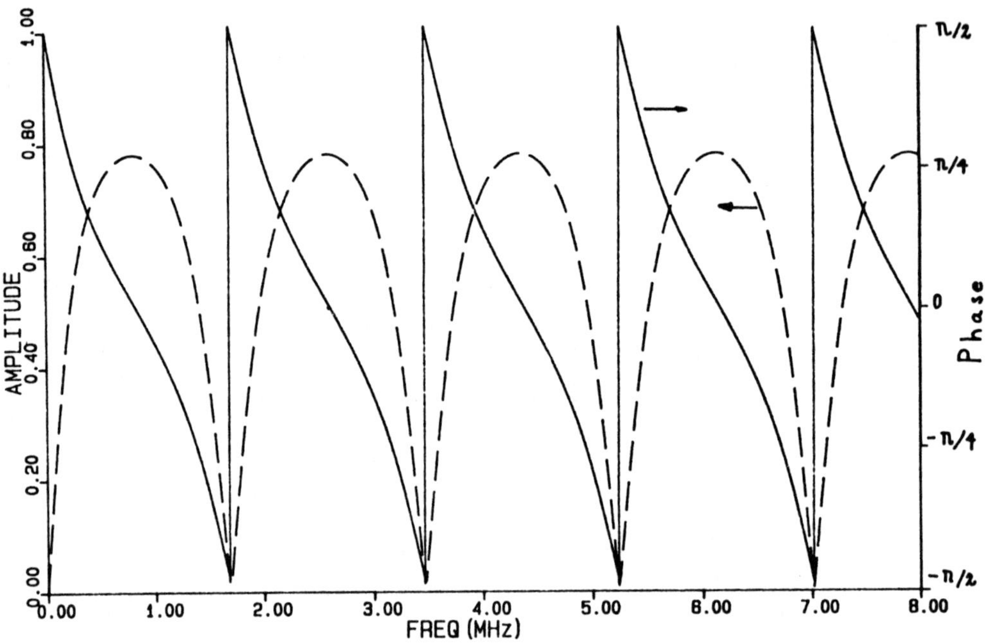

Figure 3(a): Ultrasonic plane-wave reflection coefficient from a graphite-epoxy composite plate as calculated according to Equation (23). Magnitude is solid curve; phase is plotted as dashed curve. Incident angle is 22 degrees.

In Figure 4(a) a similar example of the reflection coefficient for glass-epoxy with elastic properties collected in [9] is shown at an incident angle of 18 degrees, or phase velocity of 4.9 km/sec. Here also, the Lamb wave modes pairs are closely grouped. Referring to Figure 4(b) we see an example of the dispersion of Lamb waves in a composite plate, compared to the results of experimental measurements. At 4.9 km/sec the calculated modes are indeed in close proximity, and it has been established that each member of these pairs belongs to a different mode type, flexural and longitudinal. Above 5.2 km/sec there is an apparent convergence of the two curves, which persists until about 5.6 km/sec, when both are once again re-established as independent branches. Although the data are insufficiently detailed to permit a close comparison, they do not contradict this hypothesis.

In summary, we have calculated the effective homogeneous anisotropic elastic behavior of a unidirectional fibrous composite in which the fibers are assumed to be transversely isotropic. This was done in the long-wavelength limit using a two-step procedure based on alternating layered media. This intermediate result has then been incorporated into a calculation of the ultrasonic reflection coefficient of a fluid-loaded anisotropic plate, which has been taken as an approximation to the inhomogeneous composite laminate. Numerical illustrations in the form of the variation of the reflection coefficient with frequency are presented for graphite-epoxy

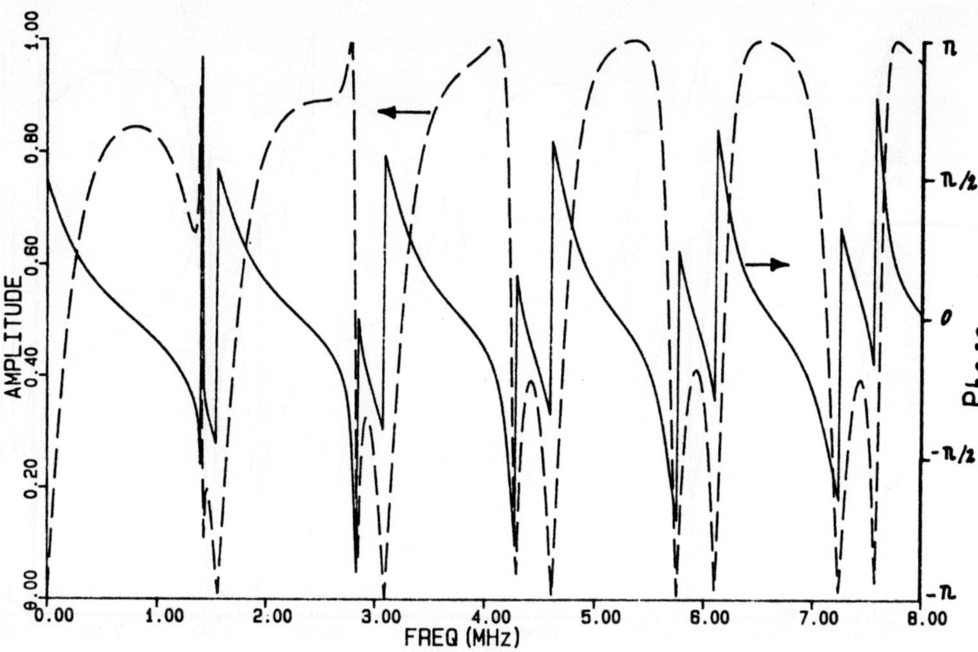

Figure 3(b): Ultrasonic plane-wave reflection coefficient from graphite-epoxy composite plate. Incident angle is 6 degrees.

and glass-epoxy composites. Also the experimental results on the Lamb wave dispersion in the glass-epoxy plate are compared with corresponding numerical results. Considering the complexity and potential nonideality of the real composite and the relative simplicity of our theortical model, the comparison between our calculations and experimental measurements has yielded remarkable agreement. This agreement gives further impetus to our intended future comparisons for the transversely isotropic graphite-epoxy composite.

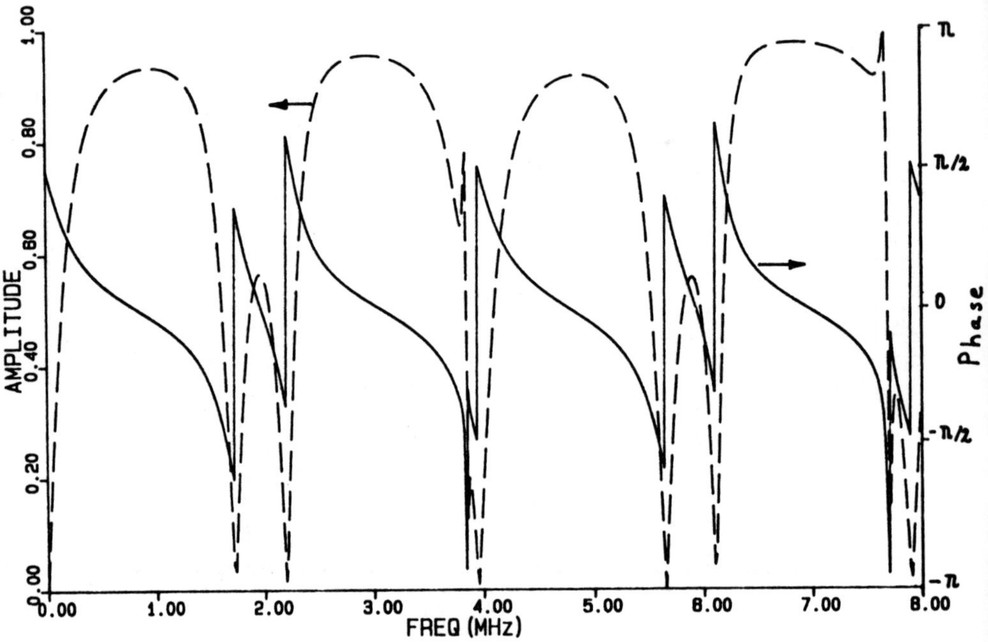

Figure 4(a): Ultrasonic plane-wave reflection coefficient from glass-epoxy composite plate. Incident angle is 18 degrees.

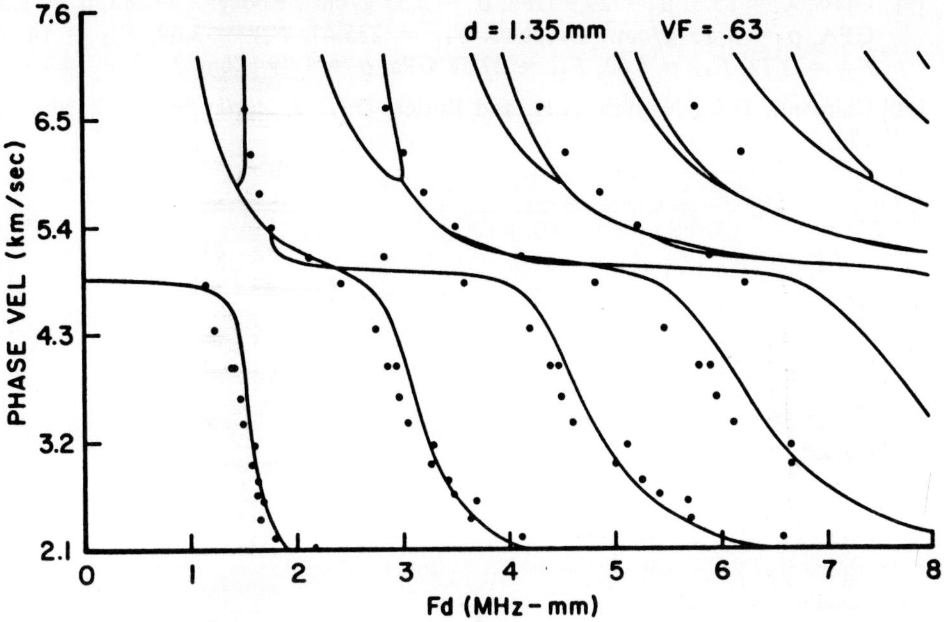

Figure 4(b): Velocity dispersion for Lamb waves in glass-epoxy composite plate 1.35 mm thick. Model calculation assumes fiber volume fraction of 0.63. Theory is solid curve; experimental data are discrete points.

References

[1] Postma, G.W., *Geophys., 20*, 1955, p. 780.

[2] Sun, C.T., Achenbach, J.D. and Herrmann, G., *J. Appl. Mech., 35*, 1986, p. 476.

[3] Bedford, A. and Stern, M., *J. Appl. Mech., 38*, 1971, p. 8.

[4] Minagawa, S., Nemat-Nasser, S. and Yamada, M., *Int. J. Num. Meth. Eng., 17*, 1981, p. 1335.

[5] Hegemier, G.A., Gurtman, G.A. and Nayfeh, A.H., *Int. J. Solids Struct., 9*, 1973, p. 395.

[6] Nayfeh, A.H. and Nasser, E.A., *J. Appl. Mech., 45*, 1977, p. 822.

[7] Bar-Cohen, Y. and Chimenti, D.E., *Review of Progress in Quantitative NDE*, Vol. 5, D.O. Thompson and D.E. Chimenti (Eds.), Plenum Press, New York, 1986, p. 1199.

[8] Chimenti, D.E. and Nayfeh, A.H., *J. Appl. Phys., 58*, 1985, p. 4531.

[9] Glass: $\lambda = 13.3$, $\mu = 29.9$ GPa, $\rho = 2.55$ g/cm^3; Epoxy: $\lambda = .89$, $\mu = 1.28$ GPA, $\rho = 1.25$ g/cm^3. Graphite: $F_{11} = 235.47$, $F_{12} = 3.69$, $F_{22} = 14.76$, $F_{23} = 3.72$, $F_{44} = 5.52$, $F_{55} = 27.58$ GPa, $\rho = 1.79$ g/cm^3.

[10] Chimenti, D.E., Nayfeh, A.H. and Butler, D.L., *J. Appl. Phys.*, *53*, 1982, p. 170.

ACOUSTIC EMISSION SOURCE CHARACTERIZATION OF MICROCRACKING IN A533B STEEL

T. Ohira and **Y-H Pao**
Department of Theoretical and Applied Mechanics
Cornell University
Ithaca, New York

1. Introduction

Acoustic Emission (AE) has been used fairly extensively as a laboratory tool to investigate the micromechanisms of deformation and fracture as well as a nondestructive evaluation technique [1,2]. The main advantages of AE techniques are high sensitivity and real time detectability. In fact, few other techniques are able to instantaneously detect a microscopic fracture inside a material such as the cracking of inclusions or intergranular fracture.

The main objectives of applying AE techniques to the study of microfracture are summarized as follows:

(1) source location,

(2) source identification (inclusion cracking, cleavage, intergranular fracture, etc.),

(3) criterion of cracking,

(4) source characterization in terms of crack size, orientation, fracture mode, formation velocity, etc.

Until recently, however, AE data have been interpreted in a statistical or empirical manner without an exact understanding of the physical processes involved during the generation, propagation and detection of AE. Although some of this research furnished useful information for the first three questions mentioned above, the fourth question has not been adequately addressed.

Recently, a method of AE source characterization has been developed [2-16] which is based on the following four steps:

(a) mathematical modeling of AE sources,

(b) understanding of wave propagation in a material,

(c) quantitative characterization of transducers,

(d) development of deconvolution processes.

Up to present, several sources such as intergranular or cleavage cracking in steels [7], macroscopic crack jump (pop-in cracking) in aluminum alloy [12] and the fiber fracture in composite materials [13] have been characterized by this method. It is recognized that, in general, multi-receiver detection system is necessary to characterize the source unambiguously, whereas some assumptions about the source (e.g., location, orientation and/or fracture mode) were required in the above-mentioned studies mainly because of experimental restrictions. Accordingly, the general scheme of source characterization still remains to be established.

In this study, ASTM A533B steel was chosen as a typical medium-strength steel exhibiting ductile fracture, and the initiation of microcracks at MnS inclusions and their coalescence during fracture toughness tests were detected by an 8-channel AE recording system. These cracks were characterized in terms of location, fracture criterion, size, orientation, fracture mode and time duration.

2. Theory

It is well known that AE sources such as deformation or cracking inside materials can be modeled as the generation of a displacement discontinuity across a surface A imbedded in an elastic body [3]. If the dimensions of A are small compared to the distance between the source and receiver and the shortest wavelength of interest, then the point source approximation is applicable. That is, the source is represented in terms of a moment tensor M_{jk},

$$M_{jk}(\mathbf{x},t)=C_{jklm}[u_1(\mathbf{x},t)]n_m\Delta A \tag{1}$$

where C_{jklm} are the elastic constants, $[u_1(\mathbf{x},t)]$ is the displacement discontinuity, n_m is the unit normal to A and ΔA is the area of A. The displacement response to the source is given by,

$$u_i(\mathbf{x},t)=\int_o^t G_{ij,k'}(\mathbf{x},t-t';\mathbf{x}')M_{jk}(\mathbf{x}',t')dt' \tag{2}$$

where $G_{ij}(\mathbf{x},t-t';\mathbf{x}')$ is the displacement in the i-th direction at $(\mathbf{x},t)$ due to an instantaneous point force of unit impulse in the j-th direction at $(\mathbf{x}',t')$ for a given boundary condition (the dynamic Green's function). $G_{ij,k'}=\partial G_{ij}/\partial x_k'$. If each component of M_{jk} has the same time dependence $S(t)$, $M_{jk}(\mathbf{x},t)$ can be rewritten as,

$$M_{jk}(\mathbf{x},t)=M_{jk}(\mathbf{x})S(t) \tag{3}$$

Next, the observed waveform $V^r(t)$ at the r-th receiver is generally different from the displacement because of the frequency response of the recording system (including transducer, amplifier, filter, etc.). Assuming that the transducer is sensitive to the i-th component of displacement, the observed waveform can be expressed as,

$$V^r(t)=\int_o^t T^r(t-t')u_i(\mathbf{x},t')dt' \tag{4}$$

where $T^r(t)$ is the transfer function of the receiver. Combining eqs. (2) to (4), we have,

$$V^r(t)=T^r(t)*G_{ij,k'}(\mathbf{x},t;\mathbf{x}')*M_{jk}(\mathbf{x}')S(t) \tag{5}$$

where the asterisk (*) denotes the convolution integral with respect to t. The ultimate objective of source characterization is then to recover M_{jk} and $S(t)$ from one or several waveform(s) arising from the same source which are measured at different points on the surface.

3. Experiment

Figure 1 shows the experimental setup for the AE measurements. Standard compact tension specimens of 25 mm thickness were machined from a hot rolled plate of 200 mm thick. The specimens are oriented in the ST direction so that the flattened MnS inclusions are subjected to the dominant normal tensile stress and can easily be separated from the matrix, thus resulting in detectable AE signals [17]. The transducers have an element of 1.35 mm in diameter and their outputs were amplified 60 dB and then digitized and stored via transient recorders with a sampling frequency of 30 MHz, resolution of 10 bits and data length of 2048 words per signal. In order to simultaneously capture the signals from the same event with the four digitizers, an additional transducer was used to provide a trigger as shown in Figure 1.

Prior to the test, the transfer function of each transducer were determined by measuring the response to the fracture of a glass capillary on a glass plate.

Fracture toughness tests were carried out at a crosshead speed of 0.5 mm/min. The elastic-plastic fracture toughness J_{Ic} was previously measured [18] and the reported value was 300kJ/m^2 according to the ASTM standards.

The location of each event was determined by measuring the difference in P-wave arrival time between transducers. Since only four transducers are needed to locate a source in three-dimensional space, there is a redundancy in this measurement, and the error in locating the source is reduced by means of the least squares procedure [19].

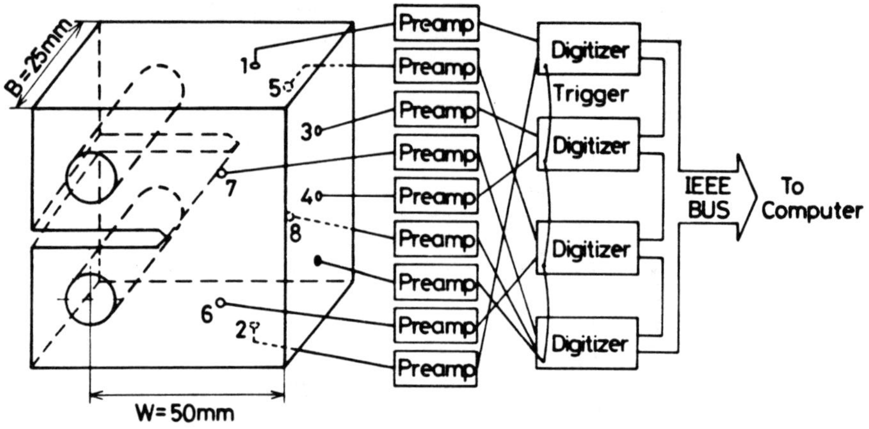

Figure 1: *Experimental setup for AE measurements.*

For the calculation of the Green's functions, the specimen was approximated as a half-space or infinite plate. That is, only the surface on which the transducer was mounted was considered for transducers 1 to 4, and only two side walls were taken into account for transducers 5 to 8 (see Figure 1). For each set of source and transducer coordinates, the Green's functions were calculated using a program developed by Ceranoglu and Pao [6].

4. Results

4.1. *Source Location*

Figure 2 shows the top and front views of the source locations together with the fatigue precrack front, which was observed after the test. In this experiment, the waveforms were clearly classified into two groups, called type I and type II in this paper. As shown in Figure 3(a), the type I signals are characterized by a distinctive sharp pulse (width of about $0.2\mu s$) at the first P-wave arrival. These events are attributed to the formation of microcracks (presumably the decohesion of MnS inclusions) because they occurred well before the onset of main crack growth J_{Ic}, and were located as far as 7 mm ahead of the precrack front and 5 mm away from the main crack plane. On the other hand, the type II signals shown in Figure 3(b) have a longer rise time ($\sim 1\mu s$), and they are probably due to the coalescence of microcracks (main crack growth) because they appeared later than the type I signals and were located close to the main crack plane. In view of the transducer size, the resolution of arrival time difference and the least squares error of the location algorithm, the error of source location was estimated to be within ± 0.5 mm [19].

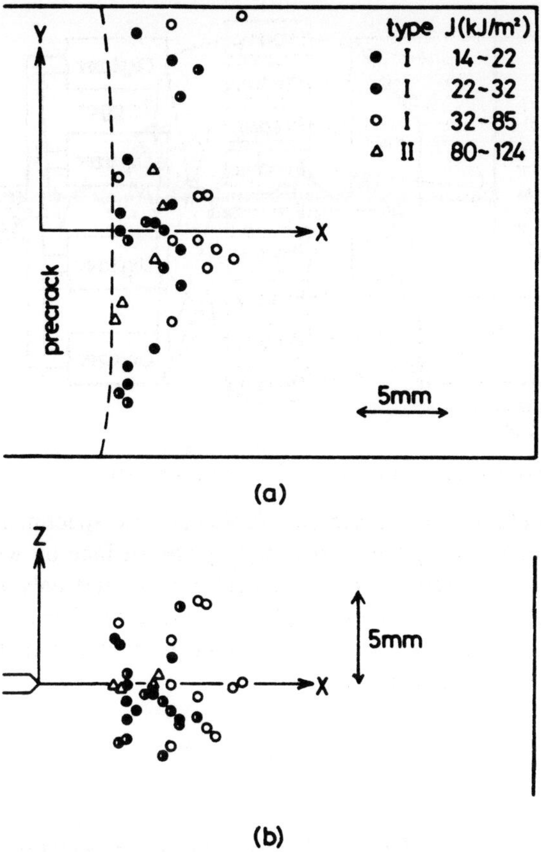

Figure 2: *Results of AE source location, (a) top view, (b) front view.*

4.2. *Observed Crack Profile*

In one test, the specimen was loaded up to $J = 600\mathrm{kJ/m^2}$ and then unloaded and cut along the midsection in order to observe the crack profile. As can be seen in Figure 4, the crack path consists of microcracks normal to the tensile axis (denoted as A) and coalescence of these microcracks by shear linkages (B). Since the orientation of the microcracks (A) coincides with that of flattened inclusions, it was concluded that these microcracks were nucleated at MnS inclusions and detected as type I AE, while it is likely that the coalescence (B) was detected as the type II AE [19].

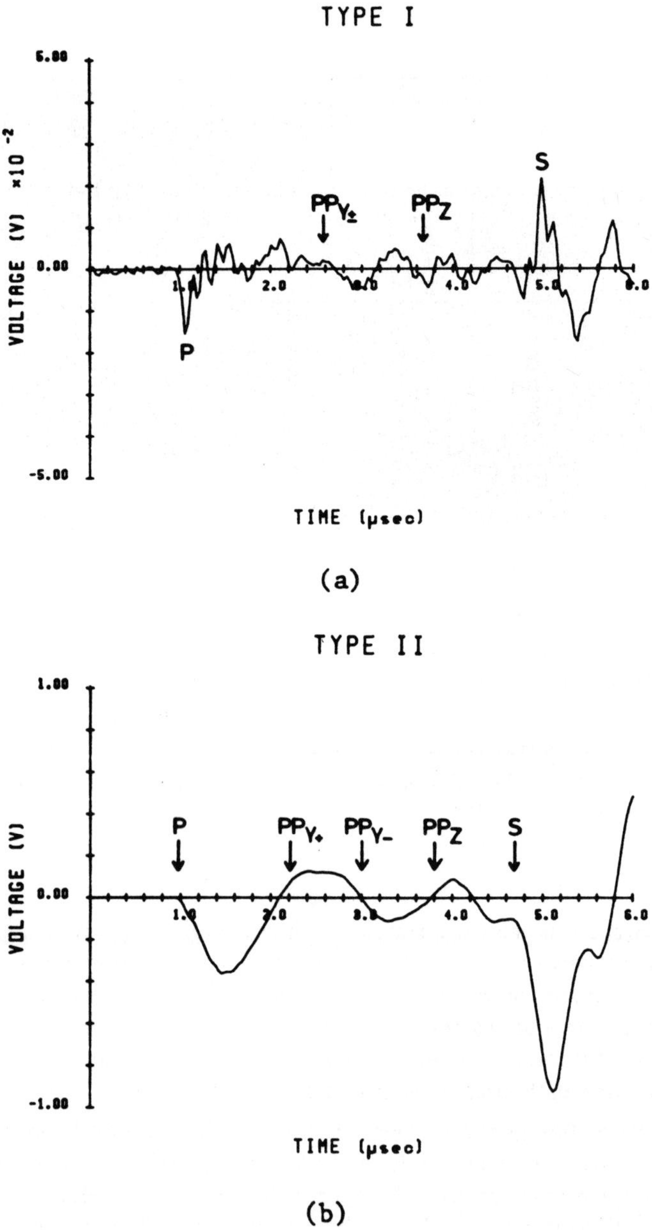

Figure 3: *Typical waveforms, (a) Type I waveform, (b) Type II waveform.*

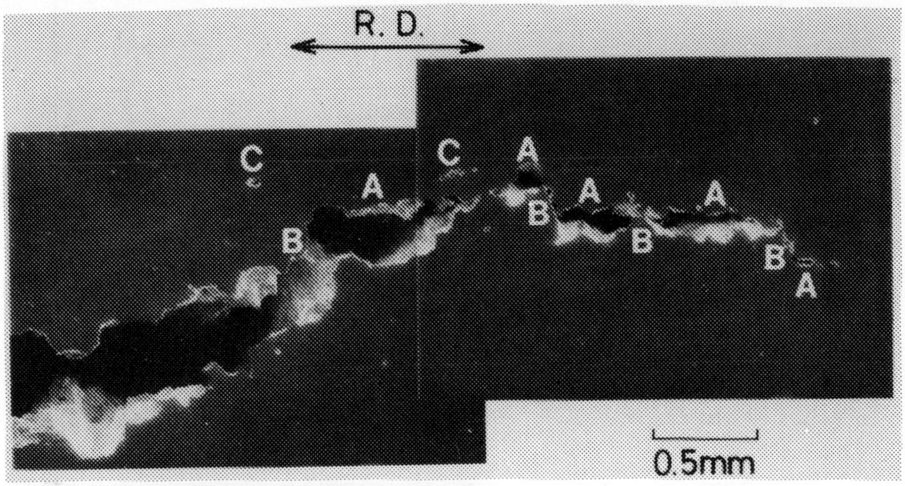

Figure 4: *Crack profile at midsection.*

4.3. *Criterion of Microcracking*

Applying the finite element analysis of the plastic zone by Larsson and Carlsson [20], it was found that most of the microcracks were generated before they were engulfed by the plastic zone. Therefore, the elastic stress field may be applicable to calculate the local stress condition at each microcrack. Since the microcracks lie on the plane parallel to the main crack as shown in Figure 4, the dominant normal tensile stress σ_z was calculated by eq. (6).

$$\sigma_z = \frac{K}{\sqrt{2\pi r}} \cos\frac{\theta}{2}(1+\sin\frac{\theta}{2}\sin\frac{3\theta}{2}) \tag{6}$$

where r and θ are the polar coordinates with the origin taken from the main crack tip. The relation between σ_z and θ is shown in Figure 5. In spite of some scattering, σ_z is independent of θ, and $\sigma_z = 400-800$MPa is obtained as a criterion of microcracking at MnS inclusions.

4.4. *Determination of Source Time Function*

The source time function $S(t)$ was calculated by deconvolving the observed waveform $V^r(t)$ with the transfer function $T^r(t)$ and the Green's function $G_{ij,k'}(t)$. Figure 6 shows examples of $S(t)$ for type I and type II waveforms. As shown in Figure 6(a), $S(t)$ for type I AE is a step-like function with a rise time of about 0.3 μs. On the other hand, $S(t)$ for type II AE has a longer rise time of about 1μs. Since the rise time of $S(t)$ is expected to correspond approximately to the time taken for each crack to attain its final shape, this result implies that the microcrack formation time is shorter than that of coalescence.

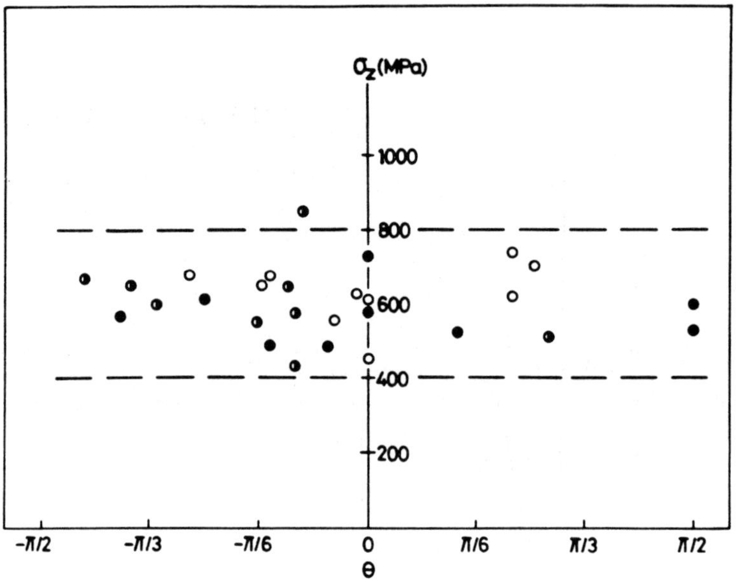

Figure 5: *Relation between normal stress σ_z and angular coordinate of micro-crack θ.*

4.5. *Determination of Moment Tensor*

The moment tensor M_{jk} has six independent components because of the symmetry of the elastic constants, $C_{jklm} = C_{kjlm}$. Therefore, if the radiation of elastic waves from the same source is detected by six or more transducers with identical transfer functions, the peak amplitude at the arrival of the P-wave can be used to determine M_{jk} by solving the simultaneous equations with respect to M_{jk}. In this experiment, however, each transducer has a different transfer function $T^r(t)$ and the sensitivity is quite dependent upon the coupling condition, possibly because of the small contact area. Also, it was found that slight errors in the $T^r(t)$ could result in significant errors in M_{jk}. Therefore an alternative scheme was developed in which the ratio of peak amplitude at P-wave and S-wave arrivals $R^r = V^r(t_p)/V^r(t_s)$ was measured and R^r was used to determine the ratio between the components of M_{jk} [21]. This scheme is based on the property that P-wave and S-wave have a different radiation pattern for a fixed M_{jk}, and it has an advantage that the result is not sensitive to the errors in the transducer sensitivity. As a check on the inversion procedure, the estimated M_{jk} were used to synthesize the waveforms, which were then checked against the observed ones.

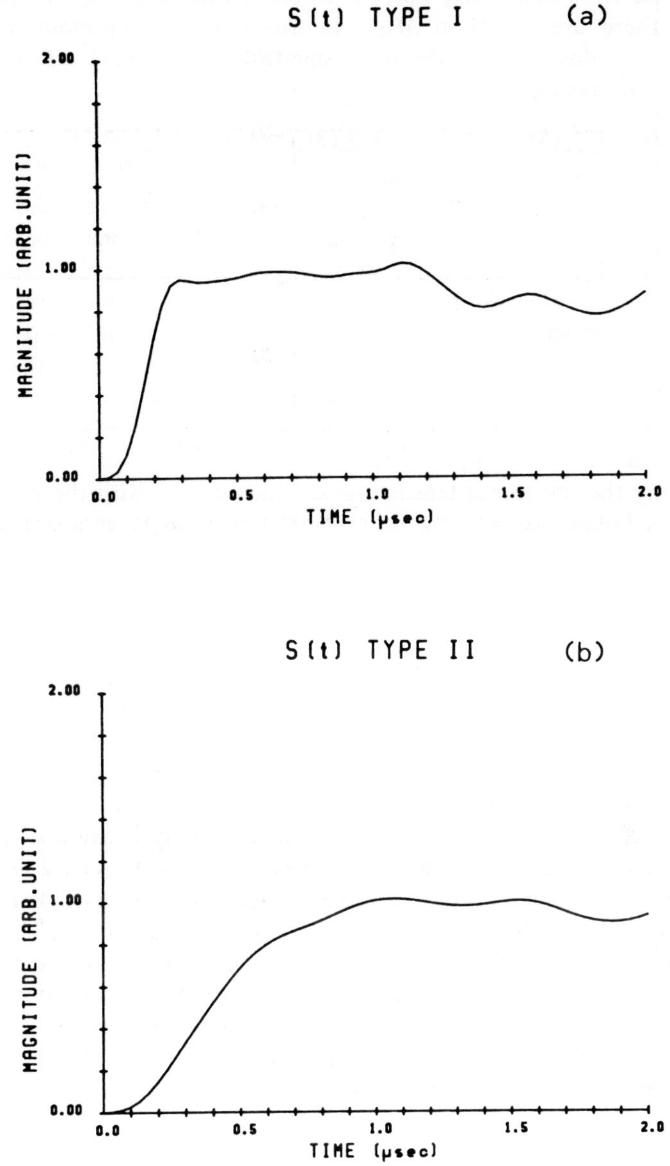

Figure 6: Source time function, (a) Type I, (b) Type II.

4.6. *Orientation of Crack and Fracture Mode*

In principle, the crack orientation **n** and the direction of displacement jump [**u**] can be determined from M_{jk} using eq. (1). In reality, however, eq. (1) may not have a solution if there are errors in M_{jk}. In this case, the optimum **n** and **u** were estimated by a double iterative least squares method [21]. That is, the least squares error E was defined as,

$$E=\sum_j\sum_k(M_{jk}-\hat{M}_{jk})^2=\sum_j\sum_k(M_{jk}-C_{jklm}[\hat{u}_1]\hat{n}_m\Delta A)^2 \tag{7}$$

The [$\hat{\mathbf{u}}$] and $\hat{\mathbf{n}}$ associated with the smallest value of E are then the optimum estimates of [**u**] and **n**. Table 1 gives examples for type I and type II AE. Here, M_{jk}, $\hat{M}_{jk}$ and [$\hat{\mathbf{u}}$] are normalized with M_{33}, $\hat{M}_{33}$ and $|[\hat{\mathbf{u}}]|$, respectively. The direction cosine between [$\hat{\mathbf{u}}$] and $\hat{\mathbf{n}}$ is calculated, which is a measure of the fracture mode. That is, $\cos\phi=1$ corresponds to a tensile crack and $\cos\phi=0$ to a shear crack. The average $\cos\phi$ of 20 type I AE signals was 0.95 and that of 3 type II AE was 0.32. The projection of [$\hat{\mathbf{u}}$] and the crack plane onto the $x-z$ plane is shown in Figure 7. In this figure, the lines show the crack plane and the arrows indicate the direction of [$\hat{\mathbf{u}}$]. It can be seen that most of the type I data are approximately the horizontal tensile crack, which agrees with the expected orientation of MnS inclusions. On the other hand, the type II data are inclined shear crack. This result is in accordance with the expected model of ductile fracture and AE which predicts that microcracking will occur in tensile mode while coalescence will proceed predominantly in shear mode [8,23].

Table 1: Examples of M_{jk}, $\hat{M}_{jk}$, $\hat{u}_1$, $\hat{n}_m$ and $\cos\phi$.

	M_{jk}	$\hat{M}_{jk}$	$\hat{u}_1$	$\hat{n}_m$	$\cos\phi$
type I	$\begin{bmatrix}0.59 & 0.15 & 0.01\\ 0.15 & 0.66 & 0.07\\ 0.01 & 0.07 & 1.00\end{bmatrix}$	$\begin{bmatrix}0.42 & 0.02 & 0.06\\ 0.02 & 0.45 & 0.14\\ 0.06 & 0.14 & 1.00\end{bmatrix}$	$\begin{Bmatrix}0.10\\ 0.23\\ 0.97\end{Bmatrix}$	$\begin{Bmatrix}0.10\\ 0.23\\ 0.97\end{Bmatrix}$	1.00
type II	$\begin{bmatrix}1.9 & 0.0 & 1.1\\ 0.0 & 1.1 & 0.61\\ 1.1 & 0.61 & 1.00\end{bmatrix}$	$\begin{bmatrix}1.9 & 0.19 & 1.2\\ 0.19 & 0.79 & 0.50\\ 1.2 & 0.50 & 1.00\end{bmatrix}$	$\begin{Bmatrix}0.43\\ -0.04\\ 0.90\end{Bmatrix}$	$\begin{Bmatrix}0.91\\ 0.41\\ 0.09\end{Bmatrix}$	0.46

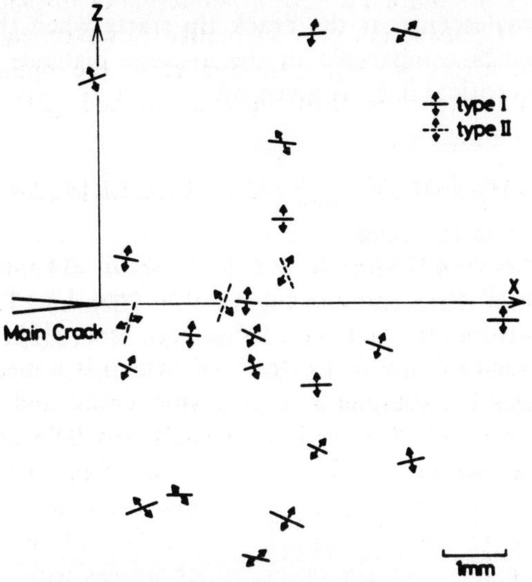

Figure 7: *Crack orientation and opening direction.*

4.7. *Crack Size*

Knowing the absolute magnitude of $T^r(t)$, $G_{ij,k'}$ $(x,t;x')$ and the ratio between the components of M_{jk}, we can estimate the absolute magnitude of M_{jk} and the crack size. From eq. (1), we have,

$$M_{jk} = C_{jklm} \frac{[\hat{u}_1]}{|\,[\hat{\mathbf{u}}]\,|} n_m \,|\,[\hat{\mathbf{u}}]\,|\, \Delta A \tag{8}$$

The crack volume v is,

$$v = |\,[\hat{\mathbf{u}}]\,|\, \Delta A \cos\phi \tag{9}$$

For type I AE, v estimated from the source functions was 730 to 4400μm^3. Since type I AE is attributed to the decohesion of flattened MnS inclusions, we may relate v to the crack radius Δa by applying a model of penny-shaped crack subjected to normal tensile stress σ_z and shear stresses τ_{xz}, τ_{yz}. As τ_{xz} and τ_{yz} do not contribute to the crack volume, the relation between v and Δa is given by,

$$v = \frac{8\sigma_z(1-\nu)}{3\mu} \Delta a^3 \tag{10}$$

where ν is Poisson's ratio. As previously mentioned, $\sigma_z = 400-800$MPa, Δa is estimated as 40 to 75μm. This value of Δa agrees well with the order of microcrack size shown in Figure 4, and also with the inclusion size ($\sim$100μm in diameter).

On the other hand, type II AE, which is probably from the coalescence of microcracks, requires another model. According to the analysis by Rice and Johnson [22], the void coalescence at the crack tip starts when the crack tip opening displacement δ becomes comparable to the average distance between the second phase particles. The critical δ, δ_c, is given by,

$$\delta_c = \frac{J_{Ic}}{\alpha \sigma_{ys}} \tag{11}$$

where α is a constant in the range of 2 to 4, and σ_{ys} is the yield strength. Using $J_{Ic} = 300 kJ/m^2$ and $\sigma_{ys} = 477$MPa, δ_c is 150 - 300μm. This is also in good agreement with the length of shear crack in Figure 4 ($\sim$200μm). The product of $|[\hat{\mathbf{u}}]|$ and ΔA calculated from M_{jk} was $4 \times 10^4 \mu m^3$ to $2 \times 10^5 \mu m^3$. Therefore, if we assume the area of shear crack to be $200^2 = 4 \times 10^4 \mu m^2$, the displacement discontinuity across the crack is estimated as 1 to 5μm.

5. Summary and Conclusions

In this study, the initiation of microcracks at MnS inclusions and their coalescence during fracture toughness tests of A533B steel were detected by an 8-channel AE recording system, and these cracks were characterized in terms of location, fracture criterion, time duration, size, orientation and fracture mode. It should be noted that there are several factors which may affect the accuracy of the results. For instance, the specimen was approximated as a half-space or infinite plate for the calculation of the Green's functions and the reflections of waves from other surfaces were not considered. In particular, the effect of precrack can be significant as it intervenes immediately after the direct waves if the source is close to the crack tip [10,11]. Detailed discussions regarding these and other problems are reported elsewhere [19,21]. In spite of these problems, however, this approach has been successfully applied to quantitatively characterize the cracks. It is expected that these source characterization methods will in future be applied more extensively to enhance our understanding on the microfracture processes of materials.

Acknowledgement

This research is supported by the Materials Science Center of Cornell University under a grant from the National Science Foundation. The first author (T.O.) also acknowledges a fellowship for research abroad provided by the Japan Society for the Promotion of Science.

References

[1] Sano, K. and Fujimoto, K., *Fundamentals of Acoustic Emission*, K. Ono, ed., (UCLA School of Engineering and Applied Science), 1973, pp. 131.

[2] Wadley, H.N.G. and Mehrabian, R., *Mat. Sci. Eng., 65*, 1984, pp. 245.

[3] Burridge, R. and Knopoff, L., *Bul. Seis. Soc. Am., 54*, 1964, pp. 1875.

[4] Breckenridge, F.R., Tschiegg, C.E. and Greenspan, M., *J. Acoust. Soc. Am., 57*, 1975, pp. 626.

[5] Hsu, N.N., Simmons, J.A. and Hardy, S.C., *Mat. Eval., 35*, 1977, pp. 100.

[6] Ceranoglu, A. and Pao, Y.-H., *J. Appl. Mech., 48*, 1981, pp. 125.

[7] Wadley, H.N.G., Scruby, C.B. and Shrimpton, G., *Acta Met., 29*, 1981, pp. 399.

[8] Kuribayashi, K. and Kishi, T., *Progress in Acoustic Emission*, (Japanese Society for N.D.I.), 1982, pp. 295.

[9] Scruby, C.B., *Review of Progress in Quantitative Nondestructive Evaluation, 4*, D.O. Thompson and D.E. Chimenti, eds., (Plenum Press, New York), 1985, pp. 643.

[10] Achenbach, J.D., Hirashima, K. and Ohno, K., *J. Sound. Vib., 89*, 1983, pp. 523.

[11] Wadley, H.N.G. and Scruby, C.B., *Int. J. Fract., 17*, 1983, pp. 111.

[12] Kishi, T. and Ohira, T., *Trans. Japan Inst. Metals, 24*, 1983, pp. 255.

[13] Ohira, T., Kishi, T., Miyashita, H., Kagawa, Y. and Nakata, E., *J. Japan Inst. Metals, 48*, 1984, pp. 248.

[14] Ohtsu, M. and Ono, K., *J. Acoustic Emission, 3*, 1984, pp. 27.

[15] Kim, K.Y. and Sachse, W., *Progress in Acoustic Emission II*, (Japanese Society for N.D.I.), 1984, pp. 163.

[16] Michaels, J.E. and Pao, Y.-H., *J. Acoust. Soc. Am., 77*, 1985, pp. 2005.

[17] Ono, K. and Yamamoto, M., *Mat. Sci. Eng., 47*, 1981, pp. 247.

[18] Yuuki, R., et al., *J. Japanese Society for N. D. I., 30*, 1981, pp. 858.

[19] Ohira, T., and Pao, Y.-H., *Met. Trans.*, in press.

[20] Larsson, S.G. and Carlsson, A.J., *J. Mech. Phys. Sol., 21*, 1973, pp. 263.

[21] Ohira, T. and Pao, Y.-H., *Met. Trans.*, to be published.

[22] Rice, J.R. and Johnson, M.A., *Inelastic Behavior of Solids*, M.F. Kanninen, et al., eds., McGraw-Hill, New York, 1970, pp. 641.

[23] Clark, G. and Knott, J.F., *Metal Sci., 11*, 1977, pp. 531.

DIFFUSE FIELD DECAY RATES FOR MATERIAL CHARACTERIZATION

Richard L. Weaver
Department of Theoretical & Applied Mechanics
University of Illinois
at Urbana-Champaign

Abstract

While measurements of internal friction and ultrasonic attenuation have long been known to be sensitive to material damage, applications in materials characterization NDE have been disappointingly few. The usual limitation to testing on specially designed specimens is certainly partly responsible for the lack of field application. Low accuracy in ultrasonic pulse-echo attenuation measurements may be another reason.

Ultrasonic diffuse field decay rate measurements offer a new method for internal friction assessment. The method applies in the frequency range from about 0.1 to 2.0 MHz – depending on the size and absorptivity of the sample – and imposes few restrictions on specimen shape. The diffuse waves generate a stochastic signal and the processing therefore entails statistical analysis; nevertheless the resulting accuracies far exceed those of conventional techniques. The technique is illustrated by application to measurements of damage incurred by machining in 6061 aluminum.

1. Introduction

Ultrasonic attenuation measurements are commonly proposed for nondestructive materials evaluation. However measurements of attenuation α (usually quoted in units of *db* per centimeter) are plagued by many uncertainties. The usual measuring scheme involves assessing ultrasonic pulse heights as functions of distance travelled [1-4]. The diminishment in amplitude, after correction for geometrical spreading dispersion and imperfect reflections and transmissions as the pulse meets interfaces, is presumably a function of the material alone. Because of the short path lengths, usually of the order of centimeters, and because of the need for correction factors, attenuations are rarely quoted with accuracies better than about $\pm 0.1 db/cm$ [4].

Interpretation of the measured attenuations is customarily conducted by writing the total attenuation as the sum of a scattering loss attenuation α_s and an attenuation due to true losses (viscoelasticity for example), or absorption, α_A. Apparent attenuation due to scattering of the pulse off irregular inhomogeneities such as grain boundaries, inclusions, voids and fibers is not a true loss in that mechanical energy is not converted to heat but merely lost from the main beam or pulse. It is worth noting here though that the usual theoretical calculations of α_s in randomly heterogeneous media [5-7] are calculations of the attenuation in an ensemble averaged ultrasonic field. Comparisons to attenuations measured in spatially averaging experiments implies an assumption of equivalence between the averages which has generally gone unaddressed. Extraction of microstructural parameters from attenuation measurements is complicated further by uncertainties about the absorption contribution α_A. Nevertheless, because the attenuation is sensitive to microstructure, and therefore also to important mechanical properties such as fracture toughness [2] it continues to be a candidate as a nondestructive testing tool.

Ultrasonic attenuation measurements in single crystals and other homogeneous materials are often used for α_A measurements, as in such media α_s can be assumed zero [3].

Internal friction is the term usually used to describe the absorption contribution to attenuation. An extensive body of literature on the measurement of internal friction is provided by the proceedings of a series of conferences [8,9] largely oriented towards using internal friction measurement as a probe of nanostructure, dislocation and impurity dynamics, in nonengineering materials, as a contribution to metallurgy and solid state physics. Techniques reported include cycling of thin wires in torsional pendula at frequencies of order one Hz, the decay of the lowest modes of resonant bars at frquencies of tens of kiloHertz, and ultrasonic attenuation measurements in single crystals in the tens of MHz range as discussed above [10]. There currently seem to be no techniques applicable in an intermediate frequency regime.

Here a new method for internal friction measurement is proposed. The new method consists of the introduction, into an isolated specimen, of a broadband ultrasonic pulse. The ultrasonic field is allowed to distribute itself diffusely across the specimen. The decay of the resulting incoherent field is then monitored as a function of frequency.

The concept of a diffuse field is well established in applications to architectural acoustics but it has received little attention in ultrasonics. Recent work [12-14] in elastodynamic diffuse fields has suggested applications in acoustic emission source characterization. The experimental work [14] has pointed out another possible application. Inasmuch as a diffuse ultrasonic field suffers losses only due to absorption mechanisms, scattering processes merely enhancing the rate at which a field becomes diffuse, the free decay of the energy of the apparently random motions must be a measure of internal friction.

Barring ambiguities regarding losses into supports, transducers, the air, and into surface contaminants, the measured volume averaged decay rate, a function of frequency, is presumably a measure of the specimen's material condition. Absorption is known in at least some cases to be a function of dislocation density, chemical composition, moisture content and many other parameters of engineering and mechanical significance.

It is thus clear that the new method for internal friction measurement may, if convenient, substantially aid attempts to make nondestructive evaluations of materials. It is furthermore clear that inasmuch as the method can provide absolute absorption data over a wide frequency range it should also be useful in metallurgical and condensed matter studies of internal friction.

2. Application

The technique is illustrated here by application to measurement of material property changes induced by machining in aluminum. The results are preliminary and not intended as a contribution to the metallurgy of aluminum; they are merely intended as an illustration of the ultrasonic technique.

A schematic of the experimental configuration is shown in Figure 1. A single shot broad band pulse was generated with a piezoelectric source. Typical strain amplitudes generated were below 10^{-6}. The present application employed a Valpey-Fisher pin transducer for a source but data has also been taken using the simulated acoustic emission broken pencil lead source. The small diameter pinducers are weakly coupled to the specimen by using a minimal amount of oil couplant, and by letting the contact take place at an angle slightly removed from the normal. The weak coupling attempts to assure that the observed decay rates are functions of the specimen rather than the transducer. The contribution to the decay rate due to the transducers is discussed further in another communication [15]. The specimen is supported by three sharp thumbtack points and the loss thereinto found to be insignificant. Losses into the air can be estimated theoretically; they are also found to be insignificant.

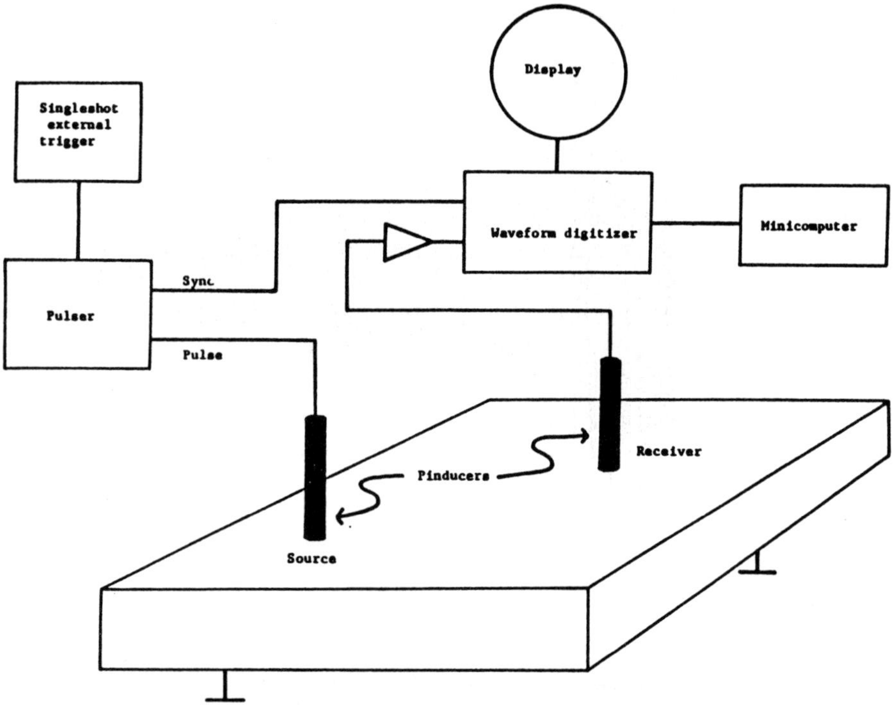

Figure 1: Schematic of testing apparatus. The specimen is supported by three dry thumbtack points. Wide-band pin transducers of small cross section are weakly coupled to the surface. The received signals are amplified, filtered and digitized at 2 megasamples/seconds.

The surface response is measured by another pinducer, also weakly coupled. The same pinducer could in principle be used for both source and receiver. The signal is amplified and digitized at 2 megasamples/second for 8192 μsec, a total of 16384 words. Capture is delayed by one or two milliseconds so as to assure that the sampled wave corresponds to a fully developed diffuse field. A typical captured waveform is shown in Figure 2. Note that the response has the appearance of noise; detailed views on expanded time scales show that this apparently stochastic process has no obvious nonstationarity beyond the simple slow decay.

Signal processing begins by artificially breaking the waveform into sixteen time windows of 512 μsec each, within any one of which the decay is negligible. Each time window is then independently Fourier transformed and the total power (mean square spectral amplitude) in each of 64 frequency bins each of width about 16 kHz ascertained. The result of this processing is, for each of 64 frequency bins, a picture of a power density evolving in time. The power in a typical bin is shown in Figure 3 as a function of time. Note that there are 16 data points corresponding to the 16 time windows. The evolution fits well to a decaying exponential. The slope

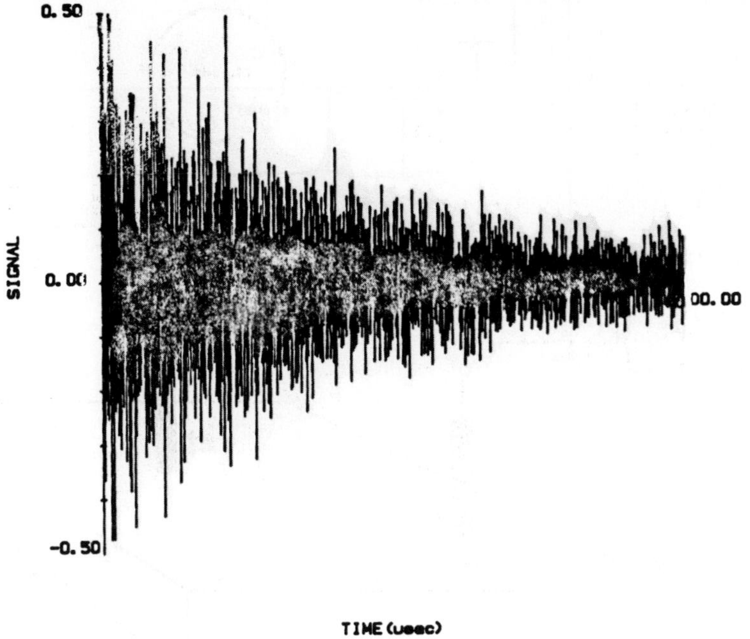

Figure 2: A typical captured waveform-8192 microseconds of data are captured starting at a time shortly after, not at, the time of the main bang, in order to assure the presence of a fully developed diffuse field. The signal is artificially broken into successive time windows of width Δt = 512 microseconds and then Fourier transformed window by window.

of the decay is set to 2σ where σ is the decay rate in nepers per millisecond. The quality of the fit to exponential decay is tested with the chi-square test. The zero time intercept corresponds to the original energy deposition, and is a function of source and receiver character and thus not of interest in the present application. It has however been used for source identification in other work [14].

It is important to recognize that the signal is stochastic; thus spectral power densities can be estimated with only limited accuracy. Standard arguments [16] for stationary random processes indicate that each data point is accurate only to a part in $(\Delta f \Delta t)^{1/2} = (16kHz \cdot 512\mu sec)^{1/2}$ or 35%, hence the error bars in Figure 3. The fit of a line of slope 2σ to the data of Figure 3 is thus somewhat uncertain; σ is determined only to within $\pm$ .02 nepers/millisecond.

The determined decay rates σ are shown as functions of frequency (that is, for each of the 64 bins) in Figures 4 and 5. A few points, especially at the lower frequencies, are missing. The computer automatically rejects the data from any bin for which the least square fit gave an unacceptably high or low chi-square. The reasons for the occasional bad fits are not yet clear. Poor results in small samples at

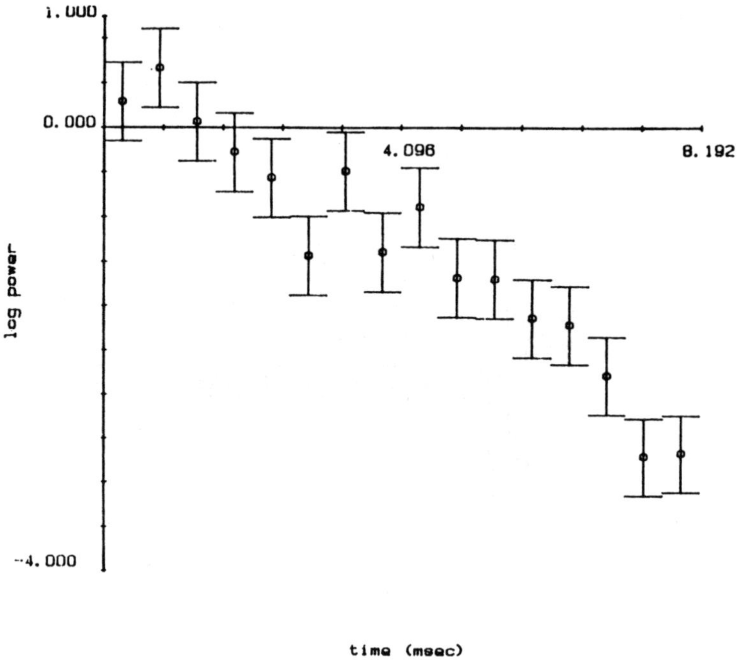

Figure 3: The spectral power densities in each time window and in frequency bands of width $\Delta f = 16$ kHz are then studied as a function of time. Due to the stochastic nature of the signal the power in each band and window is known only to an rms accuracy of a part in $(\Delta f \Delta t)^{1/2}$. The evolving power density is fit to a decaying exponential of slope 2σ, where σ is the absorption rate.

low frequencies are not unexpected however, the diffuse field concept having problematic validity at low fequencies. It should be noted that there seems to be no compelling reason for supposing that a diffuse field decays exponentially. Room acoustics is known to show deviations from simple exponential decay. Thus existence of a few poor fits is not too disturbing.

It should be noted that the accuracies obtained by this method, ± 0.02 neper/millisecond, are greater than those of conventional techniques. Typical attenuation accuracies [4] of ± 0.1 db/cm correspond to ± 3.5 neper/millisecond if one uses a typical wave speed of 0.3 cm/microsecond for a conversion factor. The enormous increase in accuracy is largely due to the greatly increased path lengths employed in the present method.

Greater accuracies can be obtained by averaging the measured decay rates of neighboring bins or by taking wider bins at the start. Such a procedure of course assumes that σ varies only slowly with frequency.

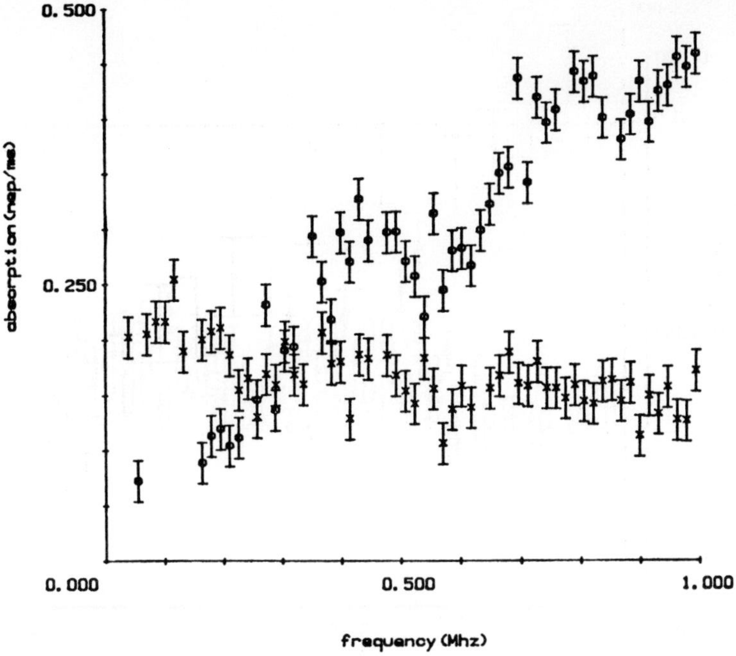

Figure 4: The absorption rates σ, in a specimen of as received 6061-T651 aluminum (x's) and after milling (o's).

A 1.2×1.2×7.5 cm piece cut on a high speed lubricated band saw from a half inch plate of 6061T651 aluminum is used as a specimen. The original absorption is displayed along the nearly horizontal locus shown in Figure 4 with the data points shown with x's. The absorption at low frequencies lies at about 0.19 nepers/ms and drops to 0.15 nepers/ms at 1 MHz. A similar piece cut on one side by use of an in-miller displays an absorption which rises to 0.5 nep/msec at 1 MHz. The increase in absorption is clear. The decrease in σ at low frequencies is not readily explained, but as discussed above, the low frequency data is suspect anyway in small specimens.

The specimen was then vigorously cleaned with soap, water and finally with acetone. The resulting absorption is shown in the upper curve in Figure 5. The peice was then polished and the absorption measured once more. Light polishing produced the lower curve in Figure 5; further polishing had no additional effect. Note that the cleaned and polished piece has not been returned to its original state. Solution and precipitation heat treatments have been found to have no measureable effect.

These results are preliminary and conclusions regarding damage in machined aluminum are premature, but there is room for conjecture. One might suppose that the machining process leaves either an oily residue or a fragile rough surface easily removed with cleaning, and that this highly superficial region contributes strongly

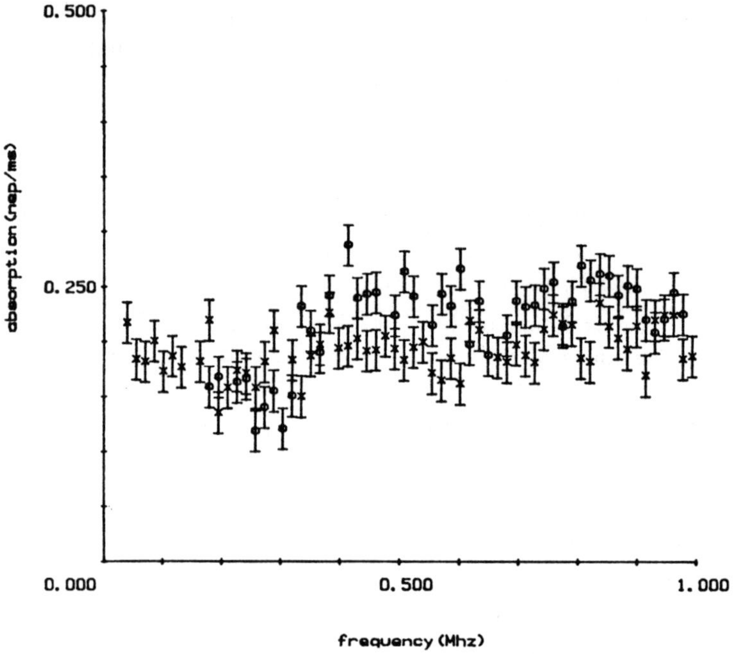

Figure 5: The absorption rates after vigorously cleaning the machined specimen (o's) and after polishing the machined surface (x's). Compare with Figure 4 and note that the cleaning and polishing seems to remove most, but not all, of the machining induced changes.

to the volume averaged internal friction. One conjectures further that the machining induced a damage confined to near the surface which is removable by polishing. One supposes that there is also a deeper damage lying beyond the range of even vigorous polishing. The hypothesis that this increased absorption is due to dislocation drag is not consistent with the results from solution and precipitation treatment. A proper discussion of these conjectures must, though, await further laboratory work.

3. Summary

The diffuse field decay technique for measuring internal friction is by no means a substitute for any current techniques. It has many limitations. It measures a spatially averaged absorption and it requires that the specimen be isolated. There are severe ambiguities in regard to the concept of a diffuse field, especially for small samples at low frequencies. Losses into supports and transducers remain to be quantified and it is clear that a comparison in homogeneous specimens between absorption as measured by this technique and by attenuation measurements should be carried out.

Nevertheless the present technique is promising in many ways. It offers measurements in a new, and wide, frequency domain. It makes few demands on specimen shape. It applies to homogeneous and heterogeneous materials; it is highly accurate. Applications are envisioned in nondestructive evaluation of engineering materials and in metallurgical and condensed matter studies of defect dynamics.

References

[1] Papadakis, E.P., "Ultrasonic Velocity and Attenuation Measurment Methods with Scientific and Industrial Applications", in *Physical Acoustics*, Vol. XII, W.P. Mason and R.N. Thurston (Eds.), Academic Press, 1979, 277-374 pp.

[2] Vary, A., "Interrelation of Material Microstructure, Ultrasonic Factors and Fracture Toughness of a Two Phase Titanium Alloy", *Mats. Eval.*, *41*, 1983, 309-314 pp.

[3] Truell, R., Elbaum, C. and Chick, B.B., *Ultrasonic Methods in Solid State Physics*, Academic Press, 1969.

[4] Generazio, E.R., "The Role of the Reflection Coefficient in Precision Measurement of Ultrasonic Attenuation", *Mats. Eval.*, *43*, 1985, 995-1004 pp.

[5] Weaver, R.L., "Waves in Random Media", *Int. J. Eng. Sci.*, *22*, 1984, 1149-1157 pp.

[6] Weaver, R.L., "A Variational Principle for Waves in Discrete Random Media", *Wave Motion*, *7*, 1985, pp. 105-121.

[7] Sobczyk, K., *Stochastic Wave Propagation*, Elsevier, 1985.

[8] *Proceedings of the Seventh International Conference on Internal Friction and Ultrasonic Attenuation in Solids*, J. de Physique, *42*, Colloq. C-5, 1981.

[9] *Proceedings of the Fifth International Conference on Internal Friction and Ultrasonic Attenuation in Solids*, Springer-Verlag, 1973.

[10] DeBatist, R., *Internal Friction of Structural Defects in Crystalline Solids*, North-Holland, 1972.

[11] Morse, P.M. and Bolt, R.H., "Sound Waves in Rooms", *Rev. Mod. Phys.*, *16*, 1955, 69-150 pp.

[12] Egle, D.M., "Diffuse Waves in Solid Media", *J. Acoust. Soc. Am.*, *70*, 1981, p. 476.

[13] Weaver, R.L., "Diffuse Waves in Finite Plates" *J. Sound Vibr.*, *94*, 1984, p. 319.

[14] Weaver, R.L., "Laboratory Studies of Diffuse Waves in Plates", submitted to *J. Acoust. Soc. Am.*

[15] Weaver, R.L., "On Diffuse Field Decay as a Measure of Internal Friction", in preparation.

[16] Newland, D.E., *Random Vibrations and Spectral Analysis*, 2nd Edition, Longman Inc., 1984.

EPILOGUE

THEY MET TO ARGUE NDE†

On Evanston they did converge,
They met so that there might emerge
A picture, clear for all to see,
Why we should work on NDE.

The meeting started out just fine.
The night before, with snacks and wine,
We sharpened wits from drink to drink
And learned what other fellows think.
Some stressed the role of NDT.
Some others argued NDE.
Some questioned what we ought to do
When we prefix the letter "Q".
The *Quantity* in Q has clout.
The *Quantitative* Q leaves doubt.

Thus well prepared did we depart
Next morning for the meetings start.
The setting was designed to teach.
The forum, where one gave his speech,
Was in a classroom fully packed.
But there emerged one small defect:
Although the speakers spoke with poise
Some did not generate such noise \m
As local profs are known to do -
To sonify a distant pew.

The lesson learned was crystal clear:
No processing, if you can't hear.
Eventually, with threshold reached
The silence barrier was breached
And one arrived at the condition
Of strong acoustical emission.

The first few speakers made it plain
That DOD needs to sustain
Activities in NDE,
On land, in air, and on the sea.
The audience thus being briefed
Liked what it heard, and felt relieved.

The meeting did then plunge with vigor
Into AE, which tries, with rigor,
To analyze the sound dispatched
In solid samples being stretched,
And to find methods that relate
The sound to breaks that radiate.
The problem is far from unique.
Does what we find match what we seek?
One has to use a bag of tricks
But, woe, there is no easy fix.
With forward problems well in hand,
One may begin to understand
How models can be put in place
That leads to the inversion trace.

Inversion yielded to the punch
Delivered by the buffet lunch.
Emission was the game to play
After ingesting what was on your tray.
There was some help to make you thinner:
Discussion scheduled after dinner.

To generate strong repercussion,
"Discussors" organized discussion.
They did this with elan and flair,

†Presented at the banquet of the ONR Symposium on Solid Mechanics Research for QNDE, held at Northwestern University, Evanston, Illinois, September 18-20, 1985.

Much like Agents Provocateurs.
Some attendees rose up and spoke,
But others, one could not provoke.
Yet much discussion did ensue,
Suggesting what we ought to do.

From what emerged, one fact was clear;
Whether your game is compression or shear,
Whether you analyze, measure, apply,
Coming together has helped each to try
Better to know what the others achieved.
It has been useful to so have been briefed.

We thank the Navy for lending support.
Northwestern's efforts, we also exhort.
They had much work, they've been a good host.
This to acknowledge, I offer a toast.

L. Felsen

LIST OF ATTENDEES

Achenbach, J.D., Northwestern University
Adler, L., Ohio State University
Angel, Y.C., Rice University
Awerbuch, J., Drexel University
Bar-Cohen, Y., Douglas Aircraft Company
Batra, N.K., Naval Research Laboratory
Bechtel, S.E., Ohio State University
Briggs, G.A.D., University of Oxford
Brock, L.A., University of Kentucky
Buck, O., Ames Laboratory
Budreck, D.E., Northwestern University
Chang, C.I., Naval Research Laboratory
Chiang, F.P., State University of New York
Chimenti, D.E., Wright-Patterson Air Force Base
Clark, A.V., Jr., National Bureau of Standards
Cohen, J.K., Colorado School of Mines
Daniel, I.M., Northwestern University
Datta, S.K., University of Colorado, Boulder
Dayal, V., Texas A & M University
Dempsey, J.P., Clarkson College
deNale, R., Naval Sea Systems Command
Djordjevic, B., Martin Marietta
Dunayevsky, V., Sohio Petroleum Company
Dundurs, J., Northwestern University
Eitzen, D.G., National Bureau of Standards
Ellingson, W.A., Argonne National Laboratory
Felsen, L.B., Polytechnic Institute of New York
Fukuoka, H., Osaka University
Garroway, A.N., Naval Research Laboratory
Gautesen, A.K., Iowa State University
Gericke, O.R., Army Materials & Mechanics Research Center
Goff, J.F., Naval Surface Weapons Center
Green, R.E., Johns Hopkins University
Gubernatis, J.E., Los Alamos National Laboratory
Hahn, T.H., Washington University
Hamstad, M., University of Denver
Hargrove, L.E., Office of Naval Research
Harris, J.G., University of Illinois-Urbana
Henneke, E.G., Virginia Polytechnic Institute
Heymann, J.S., NASA Langley
Johnson, G.C., University of California, Berkeley
Kamath, S.M., University of Illinois, Urbana
Katz, E., Drexel University
Kechter, G.E., Northwestern University

Keller, M., General Electric Corporation
Khuri-Yakub, B.T., Stanford University
Kinra, V.K., Texas A & M University
Kitahara, M., Tokai University
Kuo, M.K., National Taiwan University
Kupperman, D., Argonne National Laboratory
Kushner, A., Office of Naval Research
Matzkanin, G.A., Southwest Research Institute
Mendelsohn, D.A., Ohio State University
Mikata, Y., Northwestern University
Mura, T., Northwestern University
Nayfeh, A., University of Cincinnati
Norris, A.N., Rutgers University
Ohira, T., Cornell University
Pao, Y.H., Cornell University
Prine, D.W., GARD Division, Chamberlain National
Qu, J., Northwestern University
Rajapakse, Y., Office of Naval Research
Roberts, R., Argonne National Laboratory
Rokhlin, S.I., Ohio State University
Rudnicki, J.W., Northwestern University
Sabina, F.J., National University of Mexico
Sachse, W., Cornell University
Salama, K., University of Houston
Sayers, C.M., University of Bath
Scott, W.R., Naval Air Development Center
Segal, E., Argonne National Laboratory
Sendeckjy, G.P., Wright-Patterson Air Force Base
Simmons, J.A., National Bureau of Standards
Sotiropoulos, D.A., Northwestern University
Thomas, R.L., Wayne State University
Thompson, D.O., Ames Laboratories
Thompson, R.B., Ames Laboratories
Ting, T.C.T., University of Illinois, Chicago
Vary, A., NASA Lewis Research Center
Wadley, H.N.G., National Bureau of Standards
Weaver, R.L., University of Illinois, Urbana
Weglein, R.D., Hughes Aircraft Company
Weitsman, Y., Texas A & M University
Welsh, L.B., Signal Research Center
Yang, J.N., George Washington University
Yuhas, D.E., Magnaflux Corporation